AF618395

PET STUDIES ON AMINO ACID METABOLISM AND PROTEIN SYNTHESIS

Developments in Nuclear Medicine

VOLUME 23

Series Editor: Peter H. Cox

The titles published in this series are listed at the end of this volume.

PET Studies on Amino Acid Metabolism and Protein Synthesis

Proceedings of a Workshop held in Lyon, France within the framework of the European Community Medical and Public Health Research

edited by

B. M. MAZOYER

Groupe d'Imagerie Neuro-fonctionelle,
Service Hospitalier Frédéric Joliot,
CEA-DRIPP Orsay,
et Hôpital R. Debré,
Paris, France

W. D. HEISS

Max-Planck-Institute for Neurological Research,
Cologne, Germany

and

D. COMAR

E.E.C. Concerted Action on PET Investigations
of Cellular Regeneration and Degeneration,
Service Hospitalier Frédéric Joliot,
Hôpital d'Orsay,
Orsay, France

Springer Science+Business Media, B.V.

Library of Congress Cataloging-in-Publication Data

PET studies of amino acid metabolism and protein synthesis : proceedings of workshop held in Lyon, france within the framework of the European Community medical and public health research / edited by B.M. Mazoyer, W.-D. Heiss, and D. Comar.
p. cm. -- (Developments in nuclear medicine ; 23)
Includes bibliographical references and index.
ISBN 978-0-7923-2076-0 ISBN 978-94-011-1620-6 (eBook)
DOI 10.1007/978-94-011-1620-6
1. Amino acids--Metabolism--Congresses. 2. Tomography, Emission--Congresses. 3. Proteins--Synthesis--Congresses. 4. Brain--Tumors--Diagnosis--Congresses. I. Mazoyer, B. M. II. Heiss, W.-D. (Wolf-Dieter), 1939- III. Comar, D. IV. European Economic Community. V. Series.
[DNLM: 1. Amino Acids--metabolism--congresses. 2. Proteins--biosynthesis--congresses. 3. Tomography, Emission-Computed--congresses. W1 DE998KF v.23]
QP561.P48 1993
612.3'98--dc20
DNLM/DLC
for Library of Congress 92-48951

ISBN 978-0-7923-2076-0

Printed on acid-free paper

MEDICAL AND HEALTH RESEARCH PROGRAMME OF THE EC

BIOMEDICAL ENGINEERING IN THE EUROPEAN COMMUNITY

The involvement of the European Community (EC) in the field of Medical and Health Research started in 1978 with the first Programme which contained three projects. Since then, it has steadily expanded and it will include around 120 projects by the end of the fourth Programme (1987–1991).

The *general goal* of the programme is clearly to contribute to a better quality of life by improving health, and its *distinctive feature* is to strengthen European collaboration in order to achieve this goal.

The main objectives of this collaboration are:

- increase the *scientific efficiency* of the relevant research and development efforts in the Member States through their gradual coordination at Community level following the mobilization of the available research potential of national programmes, and also their *economic efficiency* through sharing of tasks and strengthening the joint use of available health research resources,
- *improve scientific and technological knowledge* in the research and development areas selected for their importance to all Member States, and promote its *efficient transfer into practical applications*, taking particular account of potential industrial and economic developments in the areas concerned,
- *optimize the capacity and economic efficiency of health care efforts* throughout the countries and regions of the Community.

The current programme consists of *six research targets*. Four are related to major health problems: CANCER, AIDS, AGE-RELATED PROBLEMS, and PERSONAL ENVIRONMENT AND LIFE-STYLE RELATED PROBLEMS; two are related to health resources: *MEDICAL TECHNOLOGY DEVELOPMENT* and HEALTH SERVICES RESEARCH.

Funds are provided by the Community for relevant "concerted action" activities which consist of research *COLLABORATION and COORDINATION* in EC Member States and/or in other European participant countries. *NETWORKS* of research institutes can be set up and supported by means of meetings, workshops, short-term staff exchanges/visits to other countries, information dissemination and so on; centralized facilities such as data banks, computing, and preparation and distribution of reference materials can also be funded. The funds are *not* direct research grants; the institutes concerned must fund the research activities carried out within their own countries – it is the international coordination activities which are eligible for Community support. Each such research network is placed under the responsibility of a *PROJECT LEADER* chosen from among the leading scientists in the network, with the assistance of a *PROJECT MANAGEMENT GROUP* representing the teams participating in the network.

The Commission of the European Communities is assisted in the execution of this programme by a Management and Coordination Advisory Committee (CGC – Medical and Health Research), and by Concerted Action Committees (COMACs) and Working Parties, composed of representatives and of scientific experts respectively, designated by the competent authorities of the Member States.

Other European countries, not belonging to the EC but participating in COST (Cooperation on Science and Technology) may take part in the Programme.

The present work was conducted according to the advice of *COMAC-BME* which supervises the coordination of research in *biomedical engineering* (BME) within the Medical Technology Development target.

More information may be obtained from: Commission of the European Communities
Directorate General XII-F-6
200 Rue de la Loi
B - 1049 Brussels

CONTENTS

PREFACE

Parameters such as membrane transport, metabolism and protein incorporation govern the fate of amino acids in living tissue. Is it possible to use positron tomography to measure some of them, and what is their meaning in normal and pathological situations?
These questions have been addressed for a long time and no satisfactory answer has yet been given.

This book, which derives from an EEC workshop organized in the frame of the Concerted Action on 'PET Investigation of Cellular Regeneration and Degeneration', held in Lyon in February 1992, gives the present state of knowledge in this field based on the most recent studies. Contributions from 24 leading European and American scientists are presented and discussed in the following four parts:

- biochemistry and animal studies;
- amino acids labelling with positron emittors, quality control and metabolites measurement;
- kinetic modelling of amino acids transport, metabolism, and protein incorporation;
- clinical use of amino acids.

The book will aid and interest biochemists, radiochemists, pharmacologists, neurologists, oncologists and medical imaging scientists.

The editors wish to express their gratitude to Y. Guillet (Hôpital R. Debré, Paris) and D. Wietrich (E.C. Concerted Action, Orsay) for their invaluable assistance in preparing this book.

The Editors

LIST OF CONTRIBUTORS

G. Blomqvist, Department of Clinical Neurophysiology, Karolinska Hospital, S-104 01 STOCKHOLM, Sweden

P. Bobillier, Laboratory of Pathological Anatomy, CNRS URA 1195, Rue G. Paradin, F-69372 LYON Cedex 08, France
Co-authors: E. Grange, A. Gharib, M. Leclerc, N. Sarda and P. Lepetit

H.H. Coenen, Clinic for Nuclear Medicine, University of Essen, Hufelandstrasse 55, DW-4300 ESSEN, Germany

D. Comar, EEC Concerted Action on PET, Service Hospitalier Frederic Joliot, Hopital D'Orsay, F-91406 ORSAY, France

J.-M. Derlon, Cyceron, Boulevard Henri Becquerel, P.O. Box 5027, F-14021 CAEN Cedex, France

K. Ericson, Department of Neuroradiology, Karolinska Hospital, S-104 01 STOCKHOLM, Sweden,

W.-D. Heiss, Max-Planck-Institute for Neucological Research, Gleueler Str 50, DW-5000 COLOGNE, Germany

K. Herholz, Max-Planck-Institute, for Neurological Research, University Clinic, Gleueler Str 50, DW-5000 COLOGNE, Germany

A. Lajtha, N.S. Kline Institute for Psychiatric Research, Center for Neurochemistry, ORANGEBURG, NY 10962, U.S.A.
Co-authors: D. Dunlop and M. Banay-Schwartz

C. Lemaire, Liege University, B-30, Cyclotron Research Center, B-4000 LIEGE, Belgium

S. Leskinen-Kallio, Department of Oncology and Radiotherapy, Turku University Central Hospital, SF-20520 TURKU, Finland

H. Lundqvist, Department of Radiation Sciences, P.O. Box 535, S-751 21 UPPSALA, Sweden

B.M. Mazoyer, Groupe d'Imagerie Neuro-fonctionelle, Service Hopitalier Frédéric Joliot, CEA-DRIPP , Hôpital R. Debré, 45, Boulevard Serrurier, F-75019 PARIS, France

G.J. Meyer, Department of Nuclear Medicine and Special Biophysics, DW-3000 HANNOVER 61, Germany
Co-authors Chapter 13: J. van den Hoff, W. Burchert and H. Hundeshagen
Co-authors Chapter 18: W. Burchert, K.-F. Gratz and H. Hundeshagen

K. Någren, Turku Medical Cyclotron-PET Center, Radiochemistry Laboratory, Turku University, Porthaninkatu 3, SF-20500 TURKU, Finland

A.M.J. Paans, PET Center, University Hospital Groningen, P.O. Box 30.001, 9700 RB GRONINGEN, The Netherlands
Co-authors: P.H. Elsinga and W. Vaalburg

M.-C. Petit-Taboué, Cyceron, Boulevard Henri Becquerel, P.O. Box 5027, F-14021 CAEN Cedex, France
Co-authors: F. Dauphin and J.-M. Derlon

A.M. Planas, INSERM U 334, SHFJ 4, Pl. du Général Leclerc, F-91406 ORSAY, France
Co-authors Chapter 4: C. Prenant, B.M. Mazoyer, S. Chadan, D. Comar, and L. DiGiamberardino
Co-authors Chapter 22: B. Kaschten, B. Sadzot, A. Stevenaert, L. DiGiamberardino and D. Comar

U. Roelcke, PET Program, Paul Scherrer Institute, CH-5232 VILLIGEN, Switzerland
Co-authors: E.W. Radü and K.L. Leenders

B. Sadzot, Department of Neurology, Cyclotron Research Center, University of Liege, Bat. B30, B-4000 LIEGE, Belgium
Co-authors: B. Kaschten, G. Delfiore, J.M. Peters, A. Stevenaert, G. Franck and D. Comar

C. Beebe Smith, Department of Health and Human Services, National Institute of Mental Health, BETHESDA, MD 20892, U.S.A.

W. Vaalburg, PET Center, University Hospital Groningen, P.O. Box 30.001, 9700 RB GRONINGEN, The Netherlands
Co-authors: P.H. Elsinga and A.M.J. Paans

K. Wienhard, Max-Planck-Institute for Neurological Research, Gleueler Str 50, DW-5000 COLOGNE, Germany
Co-authors: K. Herholz, H.H. Coenen, J. Rudolf, P. Kling, G. Stöcklin and W.-D. Heiss

Cerebral Protein Turnover: Aspects and Problems

A. Lajtha, D. Dunlop, and M. Banay-Schwartz

Introduction

Our ideas and our knowledge about the mechanisms of cerebral protein synthesis and degradation, the metabolic rates of these processes, and the factors which alter these rates have undergone major changes in the past few decades. Because of the complexity of the task, brain has not been examined in this respect until comparatively recently. Due to the technical difficulties, such as the limited rates of access of labeled amino acids across the blood-brain barrier, it was thought that protein turnover is very slow or absent in the nervous system, a not unreasonable conclusion, considering that the brain lacks significant regenerative capacity and is the site of permanent information (memory). Now we know that protein metabolism is highly active in the brain and is involved in many crucial functions such as neuropeptide formation and enzyme activation. Both synthesis and degradation rates undergo numerous changes in development and aging, and they can be influenced by pathological and environmental factors. It is hoped that further knowledge of the mechanisms involved will open up a variety of therapeutic and diagnostic possibilities. With PET it may now be possible to study protein metabolism in the living human brain in vivo, a possibility only dreamed of till recently.

We have examined various aspects of brain protein metabolism in animals for many years, and in this chapter we will briefly review some of the findings, discuss some of the problems, and consider what implications these have for PET studies.

Among the important questions are whether the metabolism of protein in the brain is different in significant aspects from that of other organs, and whether studies focused on brain protein metabolism are justified. Starvation is one of a number of examples demonstrating differences in metabolic response. Most body tissues in starving animals suffer severe losses of protein, but not the brain. Also, the presence of the blood-brain barrier introduces a number of technical problems, and perhaps the disease phenylketonuria, which are specific to the brain and to measurements of metabolic rates therein. Finally protein synthesis rates, like glucose consumption, must be general indicators of the activity and

B. M. Mazoyer et al. (eds.), PET Studies on Amino Acid Metabolism and Protein Synthesis, 1–17.

health of nervous tissue. Neurons are reported to synthesize protein at several fold the rates of glial cells, so that neuronal loss, followed by gliosis, should be easily detectable by a determination of the local rate of protein synthesis. Thus, aside from the specificity of protein metabolism in the brain, rate measurements should have a very practical application.

We will discuss here synthesis and breakdown in separate sections. Although the two must be in balance to maintain overall equilibrium, they obviously result from two entirely different mechanisms, with different controls and modulating factors. Proteolysis, which has been thought to play only a minor regulatory role, is now recognized as having importance in physiology and pathology. One particularly interesting finding is that brain calpain content and life span are inversely proportional [1] across a number of species. It is possible then that cerebral proteolysis may even have a role in aging.

Problems in measuring protein turnover

To measure rates of synthesis, specific activities must be determined over the entire period of incorporation for the labeled precursor (transfer RNA-bound amino acid) and at the end point for the protein. Many of the earliest attempts at measuring brain protein synthesis rates were based on pulse labeling. It was generally assumed that the specific activity of the tissue free amino acid pool was equal to or at least close to the actual precursor specific activity, i.e., the specific activity of the amino acid bound to the tRNA. This assumption seemed entirely reasonable since the size of the amino acyl tRNA pool was known to be very small compared to the rate at which amino acids are incorporated into protein. Thus, the entire pool would have to turnover many times a minute, and must reflect the specific activity of its source. Since the tRNA is enzymatically acylated within the cell, the intracellular specific activity also seemed to be the logical choice. It was also well known that degradation rate measurements generally give much longer half-lives than are obtained from studies based on incorporation rate, and this difference could be attributed to recycling [e.g., 2], which was also assumed to be intracellular. Later the question was raised as to whether amino acids might not be imported directly from the plasma and thus the specific activity of the amino acyl tRNA might not more nearly equal that of plasma rather than intracellular pools, for both brain and other tissues. The question is of consequence, since infusion studies had demonstrated that even when infusions are continued to equilibrium, the tissue specific activity may well be only 50 to 70% of the plasma value, and rate calculations based on the two pools could differ by a factor of two. During this period a number of techniques were developed to stabilize precursor specific activities to allow more accurate measurements of synthesis rates. These included multiple injections of precursor, various types of infusion, administration of large or flooding amounts of precursor, and tyrosine pellet implantations. From an analysis of the rates calculated from both plasma and tissue pools, it seems clear that consistent rates could only be obtained from using the tissue pool as precursor [e.g., 3, 41], not the plasma pool. Recent measurements of the specific activity of tRNA bound precursor [5] support the assumption that the precursor, in those cases, has a specific activity equal to, or very close to, that of the brain free amino acid pool. Nevertheless, the compartmentation of amino acids could be a problem with specific amino acid precursors. Protein breakdown within the lysosomes, for example, constitutes a continuing source of amino acids which might not be in immediate equilibrium with the cytoplasmic amino acid pools.

The heterogeneous regional distribution of amino acids may be relevant to these measurements. In a set of papers we recently measured the distribution of amino acids in 53 distinct areas of the rat brain [6-9]. Many significant differences were found. The highest level of a given amino acid was often over 10-fold the lowest regional concentration of that amino acid. The ratios of the highest level to the lowest level for the large neutral amino acids varied from 9:1 to 12:1. For leucine and valine, which are often used for measurement of protein synthesis rates, the ratios were 12:1 and 10:1 respectively. We are now studying the distribution of amino acids in human brain areas, where heterogeneity is somewhat less than in the rat brain. Amino acid levels are not constant. They generally decline with aging. Whether this heterogeneous and variable distribution influences the specific activity of the amino acid after a pulse or an infusion of label should be determined. Where the concentrations of precursor are especially low, the significance of any other pools, e.g., lysosomes, might be exaggerated.

Some of the problems encountered in measuring rates in vivo could be avoided by studying isolated systems. When we examined the optimal conditions for incorporation in brain slices, and compared rates in slices to those in vivo [10-11], we found (Table 1) that protein synthesis rates in slices of young brain are fairly similar in vitro and in vivo. But in slices from adult rat, rates are 10% or less those in vivo. Hence, it is very difficult to interpret the findings of synthesis rate studies based on slices of adult brain.

We have also compared rates of breakdown in slices to those in vivo [12]. In slices from young and adult brain the rates were 130%, and about 80%, respectively, of the in vivo rates. But the degradation rate in adult slices changes with time and slows greatly in the second hour, which does not correspond to the characteristics expected on in vivo proteolysis.

Table 1. Comparison of cerebral protein synthesis rates at different ages in vivo and in slices.

Young rats		Percent per hour cerebrum	cerebellum
Age, days			
3	in vivo	2.02	2.87
	slice	1.60	2.10
10	in vivo	1.61	2.45
	slice	1.21	2.34
23	in vivo	0.93	1.08
	slice	0.14	0.32
Adult rats	in vivo	0.62	0.70
	slice	0.07	0.04
Adult mice	in vivo	0.69	0.70
	slice	0.05	0.08

Protein turnover in the adult brain

Given the assumption that the brain intracellular pool is the precursor of the amino acyl tRNA pool, the brain protein synthesis rate measurements obtained by a number of laboratories using different techniques and different labeled precursors, can be compared. There is quite good agreement that the synthesis rate in young adult rodent brain was about 0.6 - 0. 7% per hour [13-17]. That the rate of synthesis is a biochemical parameter is more clear when it is expressed in terms such as "μmol of amino acid per gram brain per h" or "nmol per ribosome per second". When expressed in the traditional terms of product of the reaction, "%/h", it is obvious that the rate can be viewed as the average of all the different metabolic rates of every brain protein, some of which must be metabolized faster, others slower, than this average value. The synthesis rate in brain, 0.6 - 0.7 percent per h, is lower than that of some other tissues (liver 2.2, kidney 1.8, spleen 1.6, and lung 1.0% per h), but faster than others (muscle 0.5% per h) [18].

It is clear that most proteins are in a dynamic state, and are being broken down and resynthesized a number of times during the lifetime of the organism. In a study years ago [19] we fed lysine of constant specific activity to mice one week before and during pregnancy and lactation, and to their offspring until 60 days after birth, and then replaced the diet with a nonlabeled one. With this method all proteins were labeled, and if they had been permanent, would have retained their label. As shown in Table 2, very little stable protein can be detected.

Table 2. The permanence of brain protein.

	Protein Bound Lysine	
Experimental time	Percent of label in lysine	Lysine specific activity
(days)	(%)	(cpm/umol)
0	88	30
30	86	8.8
60	71	2.1
150	42	0.7

The specific activity of the dietary lysine was 30 cpm/µmol.
All brain proteins were initially labeled, but with time, most of the label was lost, demonstrating that the vast majority of brain protein in the mouse does turnover [19].

In another study, utilizing the low solubility of tyrosine, we administered a tyrosine suspension an then subcutaneously implanted a tyrosine pellet in young adult mice, and followed the label in free and protein-bound tyrosine in various organs for 5 days [20]. By studying the kinetics of incorporation over extended periods it is possible to derive some information about the different fractions of protein whose half-lives compose the overall rate. The best fit for the incorporation curves for these experiments was found by assuming that the brain protein was present as two fractions. The small fast factors (5.7% of the total) had a 15-h half-life, while the large slow one (94.3% of the total) had a 10-day half-life. Of the other tissues, liver could be treated as if it were a single protein fraction with à 26-h half-life, while kidney appeared to contain two fractions (41% with 18-h and 59% with 63-h half-lifes).

These values are obviously averages of large sets of proteins with different metabolic rates. When incorporation in myelin proteins was measured [21], various metabolic rates were found, with the Wolfgram protein labeled $2^{1/2}$ times more rapidly than myelin basic protein or proteolipid protein. The half-life for the Wolfgram protein was 10 days. Some indications were found of heterogeneous metabolism due to location in the myelin sheath. It is not clear exactly which characteristics of a protein determine its metabolic rate or how its location or function effect the rates. We found no direct relation to molecular weight and turnover rate in brain protein, altough there was some relationship to subunit size in SDS gel electrophoresis where subunits of 100 - 200 kDa were significantly more highly labeled than those under 60kDa [22, 23].

That there are different fractions, i.e., that the incorporation of labeled amino acid into adult brain protein over time is treated mathematically as if there were different groups of proteins with different half lives, rather than as a continuous spectrum, may suggest functional classes as well. The small fast fraction presumably contains highly regulated critical enzymes as opposed to the slower fraction which may contain structural elements. Each of the protein fractions, turning over at a different rate, will be labeled to a different degree. Since the synthesis rate of the fast fraction of protein in brain [20] is 15 to 20 times higher, while its protein content is only 1/20 that of the large slow protein pool, as much label appears, initially, in the fast pool (less than 5% of total proteins) as in the much larger slower pool. It follows that even a relatively large change in the synthesis rate of an abundant protein, such as myelin, may be difficult to detect in the overall synthesis rate if that protein has a long half-life because the measurement is heavily weighted to the more rapid proteins.

Table 3. Rates of protein metabolism in the developing brain.

		Percent per hour		
Species	Age	Synthesis	Degradation	Net growth
Mouse	Newborn	1.47	0.84	0.64
Rat	Newborn	1.37	0.74	0.64
Rat	5 days	1.71	1.02	0.64

From measurements of the synthesis rate and the growth rate, degradation rates can be calculated [24, 25].

Protein metabolism during development

The rate of protein synthesis, and also of breakdown, is considerably higher in the immature brain as compared to adult. During the first 4-8 days after birth incorporation rates are more than 2-fold higher than in adults. As can be seen in Table 3 [24, 25], net growth, as expected, results from a lower rate of degradation than of synthesis, while in the adult brain the rates of synthesis equal those of breakdown. Calculation of half-life in the immature brain gave a single pool with a half-life under 3 days. There are also some regional differences in metabolic rates in the immature brain. For example, proteins in the cerebellum have a higher average metabolic rate than those in the cerebrum [26]. The explanation for higher degradation rates in immature brain has not been elucidated. It may be that plasticity demands much reorganization, and there is a general increase in degradation, or much of the protein degraded may be concentrated in cells which die off during development.

We did a detailed study to find which fractions of proteins have a higher metabolic rate in the immature brain, comparing incorporation in 5-day-old rats to that in young adults (35-day-old). There were significant regional variations. In general, the relative incorporation rate in the young compared to the adult in most brain regions was between 2:1 and 3:1, except in hypothalamus, where rates were equal [23]. In regard to subcellular fractions, relative rates of synthesis in the young versus adult were also between 2:1 and 3:1 in the nuclear, mitochondrial, synaptosomal, and cytosol fractions. However, in the microsomal fraction, incorporation rates were equal [27]. When the incorporation ratio in young versus adults was compared in several organs, it was highest in brain, followed by muscle (2.2:1), heart, spleen, and lung (1.6:1), with incorporation in young and adult being similar in liver and kidney [13].

The developmental changes in rates of brain protein synthesis are determined by the level of RNA which, since most RNA is ribosomal, is a measure of the concentration of ribosomes. The concentration of RNA was proportional to the synthesis rate in the various brain regions analyzed at several ages [28]. Similar results are obtained when free and membrane bound ribosomes are prepared from young and adult brain. That is, the ratio of free to membrane bound ribosomes remains constant, and the total is proportional to the synthesis rate (Table 4) at each age [29].

Table 4. Ribosomal specific activity in the rat brain during development.

	Age		
	3 days	8 days	Adult (29-30 days)
Protein synthesis rate (%/h)	1.96	1.77	0.78
Protein concentration (mg protein/g wet wt)	62.8	64.8	104.5
Protein synthetic activity (mg protein synthesized/h/g wet wt)	1.23	1.15	0.815
Ribosomal content (mg free + bound RNA/g wet wt)	2.01	1.98	1.35
Ribosomal specific activity (mg protein synthesized/h/mg rRNA)	0.61	0.58	0.60

Changement of turnover in aging

The changes in rates of protein metabolism during aging, though less dramatic than those occuring during development, are nevertheless significant. In general, small decreases in metabolic rates in aging brain have been reported. The age-related changes found in in vitro incorporation systems, which are much larger [30-31] must therefore be viewed with caution. In vivo, a 9% decline was found (between 16.5 and 22.5 months of age) in one study [32] while others reported a 15-20% change (from 15-23 months of age), but only in about half the areas examined in rat brain [33]. In our studies on rats 4 to 24 months old [34], and mice between 28 and 39 weeks old [10], only a small (less than 5%) decrease was found, with some small regional differences.

Alterations of brain protein synthesis rates

Synthesis rates can also be influenced by external environmental factors. We examined the effects of a) temperature, b) malnutrition, and c) some frequently used compounds such as alcohol and nicotine on rates of synthesis in the brain.

The greatest effect we found was that of temperature. In goldfish brain the amino acid incorporation rate at 34°5C was 0.52%/h while at 10°C it was 0.026%/h. That is a 20-fold difference [35], and though these values are for extreme temperatures, it is likely that under normal conditions brain protein turnover rates in fish vary 8-to 10-fold with

Summer to Winter temperature changes. The temperature effect, about 7% decrease in incorporation for each decrease of 1°C in body temperature, appeared to be similar in an number of species (e.g., goldfish, lizard, and rat) [36-37]. Protein metabolic rates in lizard are below those in rats even at 38°C. These surprisingly large changes in cerebral metabolic rates are of more than theoretical significance, since they presumably occur under physiological conditions, and must result in lower protein metabolism in cold-blooded species (Table 5).

Table 5. Protein turnover rates in various species at different temperatures.

Species	Body temperature (°C)	Rate of incorporation (% per h) Brain	Liver	Muscle
Goldfish	22	0.23	0.57	0.07
Bullfrog	20	0.18	0.58	0.06
Lizard	26	0.13	0.43	0.04
	38	0.27	0.81	0.07
Chicken	39	0.70	2.17	0.24
Mouse	38	0.65	2.01	0.22

It is important to consider possible temperature effects on brain protein synthesis, since drug effects, at least in animal studies, are often not direct but due to a change in the body temperature.

While protein synthesis rates in the brain and other tissues are highly sensitive to temperature changes, other processes may be less temperature dependent. We found that amino acid transport across the blood-brain barrier changes very little at lowered body temperatures [37,38]. These differences in sensitivity raise the interesting possibility that pathological processes, such as those initiated by ischemia or stoke, which are dependent upon the formation of new proteins or the degradation of essential proteins, might be manipulated by temperature alterations, without inhibiting to the same extent the supply and removal of metabolites.

The effect of malnutrition is also of practical significance, since even today many people are still exposed to nutritional deficiency, often prolonged, both during and after development. The vulnerability of the young brain, and the relative resistance of the adult brain, to malnutrition are well established [39-41]. In exploring the mechanisms that protect the adult brain, we found that protein deficiency results in a decreased rate of brain protein synthesis, but also in a decreased rate of breakdown, so that the processes are in balance, except under long-lasting, extreme conditions [42], where there is some cell loss [43]. In a number of other tissues, protein deficiency causes decreased synthesis and increased breakdown, resulting in considerable tissue loss.

Protein-deficient diets do alter the free amino acid pool in the brain [44], but this in itself appears to have no effect on brain protein synthesis rates.

Of the drugs we tried, nicotine gave a slightly stimulatory effect on protein synthesis in the adult, but induced an inhibition in young. Smoking was slightly inhibitory, possibly via its hypothermic effects [45-46]. Alcohol caused a small inhibition of both synthesis [47] and breakdown [48].

Alterations of breakdown

Because of the technical difficulties involved in making the measurements, much less is known about the factors which alter degradation rates of brain protein. As noted, changes in degradation rates during development, though varying by brain region (Table 6), are large compared to the decline in turnover in aging. With in vitro systems we found that levels of cerebral proteinases do increase significantly with age. The changes are greatest for cathepsin D. Cathepsin D activity in rat increased 100% or more in each area of the aged brain studied [49]. Similar conclusions were reached from a study of in vitro microtubule protein degradation [50] in relation to age. An enzymatically inactive, but immunologically detectable, form of cathepsin D also increases [51]. Calpains, the calcium-dependent proteinases, and their endogenous inhibitor, calpastatin, as determined by enzyme assays, also increase with age in rat brain [52], but less so than cathepsin D. There is also an increase of cathepsin D with age in human brain [53] though less striking than that in the rat brain. These changes are of some interest as it has been proposed that these enzymes participate in aging [54], but they do appear to be disconnected from the actual in vivo rates of brain breakdown.

Table 6. Regional rates of protein degradation in vivo in immature and adult rats.

Area	Percent breakdown per hour Young (2 days old)	Adult (37 days old)
Hemisphere	1.3	0.8
Midbrain	0.9	0.7
Pons-Medulla	1.2	0.7
Cerebellum	1.9	0.8
Spina Cord	1.5	0.6

See reference 25.

Conclusions

Despite many false steps we are confident that protein metabolic rate measurements in animals have come of age, and a variety of techniques are available by which accurate and consistent synthesis rates can be determined. Rates of degradation are still difficult to obtain.

From these studies we might predict in human brain relatively high rates of both synthesis and degradation during development and small declines with age. Any changes in brain temperature, up or down, will have a dramatic effect. From the animal studies it would seem likely that there is not a sufficiently large, fast fraction of protein to introduce errors into synthesis rate measurements over the very short measurement periods involved in pulse labeling. Also, it is probably not possible to extract any information about the size and rates of various protein fractions in human brain for the same reasons.

Protein metabolic rate measurements, both synthesis and degradation, are dependent on an intact system. Only in some highly purified ribosomal systems do rates in vitro begin to approach those in vivo. It seems then that studies of rates in human brain will also have to be done largely in vivo. Measurements by PET offer great promise in that respect, but also face great obstacles. In the animal studies the procedures used to measure synthesis rates involve removing the tissue and determining the specific activity of the protein after all free labeled precursor has been removed. Since this is not possible in human studies, PET measurements generally rely on pulse labeling so that the free amino acid precursor is naturally removed from the brain over the experimental period. Thus the later measurements can be used to measure the specific activity of the brain protein.

Historically the first measurements of brain protein synthesis rates in animals also were done by pulse labeling, but is was soon clear that there were a host of technical problems with the procedure. The specific activity fluctuates very rapidly after the labeled amino acid is administered, increasing for a few minutes in the plasma, reaching a sharp peak, and then declining. The specific activity in the brain lags behind that of the plasma. By using many subjects the specific activity in the tissue can be compared to that in the plasma over time and curves constructed so that one can predict the specific activity in the brain from the plasma value. There are significant individual variations after pulse labeling in rats so that large numbers of animals are required, and the rapidity of the changes make numerous measurements in time necessary. These difficulties, though not insurmountable, led to the development of the newer techniques designed to give more reproductible and more stable precursor specific activities, but unfortunately none of these appears applicable to PET measurements. The major problem with pulse labeling in PET is that the specific activity in any area can be influenced by a large number of factors, such as blood flow to the area, transport into and out of the tissue, rates of metabolism of precursor, both inside and out, and, of course, rates of synthesis and degradation of protein. In brief there are significant sources of unlabeled precursor within the tissue itself, particularly from protein degradation, which we cannot see. In animal studies we can remove the tissue and determine the specific activity in the area so that this value is defined regardless of which factors may have changed in a given experimental situation. With PET we cannot see the endogenous pools of amino acid, or, more importantly, any changes in these pools resulting from individual variations or from the condition under study. We can determine whether more or less incorporation of label into protein has occurred, but the question then becomes whether a difference is due to a change in protein synthesis rate or to a change in precursor specific activity. This difficulty appears to be the major unresolved obstacle to obtaining precise rate measurements with PET in human brain.

References

1. Baudry M, DuBrin R, Beasley L, Leon M, Lynch G. Low levels of calpain activity in chiroptera brain: Implications for mechanisms of aging. Neurobiol Aging 1986; 7: 255-258.

2. Poole B. The kinetics of disappearance of labeled leucine from the free leucine pool of rat liver and its effect on the apparent turnover of catalase and other hepatic proteins. J Biol Chem 1971; 246: 6587-6591.

3. Fern EB, Garlick PJ. The specific radioactivity of the precursor pool for estimates of the rate of protein synthesis. Biochem J 1973; 134: 1127-1130.

4. Dunlop F, Lajtha A, Toth J. Measuring brain protein metabolism in young and adult rats. In: Roberts S, Lajtha A, Gispen WH, editors. Mechanisms, regulation, and special functions of protein synthesis in the brain. Amsterdam: Elsevier, 1977: 79-96;

5. Smith CB, Sun Y, Deibler GE, Sokoloff L. Effect of loading doses of L-valine on relative contributions of valine derived from protein degradation and plasma to the precursor pool for protein synthesis in rat brain. J Neurochem 1991; 57: 1540-1547.

6. Banay-Schwartz M, Lajtha A, Palkovits M. Changes with aging in the levels of amino acids in rat CNS structural elements: I. Glutamate and related amino acids. Neurochem Res 1989; 14: 555-562.

7. Banay-Schwartz M, Lajtha A, Palkovits M. Changes with aging in the levels of amino acids in rat CNS structural elements: II. Taurine and small neutral amino acids. Neurochem Res 1989; 14: 563-570.

8. Banay-Schwartz M, Lajtha A, Palkovits M. Changes with aging in the levels of amino acids in rat CNS structural elements: III. Large neutral amino acids. J Neurosci Res 1990; 26: 209-216.

9. Banay-Schwartz M, Lajtha A, Palkovits M. Changes with aging in the levels of amino acids in rat CNS structural elements: IV. Methionine and basic amino acids. J Neurosci Res 1990; 26: 217-223.

10. Dunlop DS, van Elden W, Lajtha A. Developmental effects on protein synthesis rates in regions of the CNS in vivo and in vitro. J Neurochem 1977; 29: 939-945.

11. Dunlop DS, van Elden W, Lajtha A. Optimal conditions for protein synthesis in incubated slices of rat brain. Brain Res 1975; 99: 303-318.

12. Dunlop DS, van Elden W, Plucinska I, Lajtha A. Brain slice protein degradation and development. J Neurochem 1981; 36: 258-265.

13. Austin L, Lowry OH, Brown JG, Carter JG. The turnover of protein in discrete areas of rat brain. Biochem J 1972; 126: 351-359.

14. Garlick PJ, Marshall I. A technique to measure brain protein synthesis. J Neurochem 1972; 19: 577-583.

15. Oja SS. Studies on protein metabolism in developing rat brain. Ann Acad Sci Fenn A5, 1967; 131: 1-81.

16. Seta K, Sansur M, Lajtha A. The rate of incorporation of amino acids into brain proteins during infusion in the rat. Biochim. Biophys. Acta 1973; 294: 472-480.

17. Dunlop DS, van Elden W, Lajtha A. A method for measuring brain protein synthesis rates in young and adult rats. J Neurochem 1975; 24: 337-344.

18. Shahbazian FM, Jacobs M, Lajtha A. Rates of protein synthesis in brain and other organs. Int J Dev Neurosci 1987; 5: 39-42.

19. Lajtha A, Toth J. Instability of cerebral proteins. Biochem Biophys Res Commun 1966; 23: 294-298.

20. Lajtha A, Latzkovits L, Toth J. Comparison of turnover rates of proteins of the brain, liver, and kidney in mouse in vivo following long-term labeling. Biochim Biophys Acta 1976; 425: 511-520.

21. Lajtha A, Toth J, Fujimoto K, Agrawal HC. Turnover of myelin protein in mouse brain in vivo. Biochem J 1977; 164: 323-329.

22. Shahbazian FM, Jacobs M, Lajtha A. Amino acid incorporation in relation to molecular weight of proteins in young and adult brain. Neurochem Res 1986; 11: 647-660.

23. Shahbazian FM, Jacobs M, Lajtha A. Regional and cellular differences in rat brain protein synthesis in vivo and in slices during development. Int J Dev Neurosci 1986; 4: 209-215.

24. Lajtha A, Dunlop D, Patlak C, Toth J. Compartments of protein metabolism in the developing brain. Biochim Biophys Acta 1979; 561: 491-501.

25. Dunlop DS, van Elden W, Lajtha A. Protein degradation rates in regions of the central nervous system in vivo during development. Biochem J 1978; 170: 637-642.

26. Lajtha A, Dunlop D. Turnover of protein in the nervous system. Life Sci 1981; 29: 755-767.

27. Shahbazian FM, Jacobs M, Lajtha A. Rates of amino acid incorporation into particulate proteins in vivo and in slices of young and adult brain. J Neurosci Res 1986; 15: 359-366.

28. Dunlop DS, Bodony R, Lajtha A. RNA concentration and protein synthesis in rat brain during development. Brain Res 1984; 294: 148-151.

29. Dunlop DS, Kaufman H, Lajtha A. The relation of protein synthesis to the concentrations of free and membrane-bound ribosomes in brain at different ages. Neurochem Int 1991; 19: 601-603.

30. Fando JL, Salinas M, Wasterlain CG. Age-dependent changes in brain protein synthesis in the rat. Neurochem Res 1990; 5: 373-383.

31. Ekstrom R, Liu DSH, Richardson A. Changes in brain protein synthesis during the life span of male Fischer rats. Gerontology 1980; 26: 121-128.

32. Dwyer BE, Fando JL, Wasterlain CG. Rat brain protein synthesis declines during postdevelopmental aging. J Neurochem 1983; 35: 746-749.

33. Ingvar MC, Maeder P, Sokoloff L, Smith CB. Effects of ageing on local rates of cerebral protein synthesis in Sprague-Dawley rats. Brain 1985; 108: 155-170.

34. Avola R, Condorelli DF, Ragusa N, et al. Protein synthesis rates in rat brain regions and subcellular fractions during aging. Neurochem Res 1988; 13: 337-342.

35. Lajtha A, Sershen H. Changes in the rates of protein synthesis in the brain of goldfish at various temperatures. Life Sci 1975; 17: 1861-1868.

36. Sayegh JD, Lajtha A. In vivo rates of protein synthesis in brain, muscle, and liver of five vertebrate species. Neurochem Res 1989; 14: 1165-1168.

37. Sayegh JF, Sershen H, Lajtha A. Different effects of hypothermia on amino acid incorporation and on amino acid uptake in the brain in vivo. Neurochem Res 1992; 17: 553-557.

38. Emirbekov EZ, Sershen H, Lajtha A. Lack of effects of hypothermia on cerebral amino acid uptake in vivo. Brain Res 1977; 125: 187-191.

39. Griffin WST, Woodward DJ, Chanda R. Malnutrition and brain development: cerebral weight, DNA, RNA, protein and histological correlations. J Neurochem 1977; 28: 1269-1279.

40. Winick M. Malnutrition and the developing brain. In: Plum F, editor. Res. Publ. Assoc. Nerv. Ment. Dis. NY: Raven Press, 1974; 53: 253-261.

41. Zamenhof S, van Marthens E, Grauel L. Prenatal cerebral development: Effect of restricted diet, reversal by growth hormone. Science 1971; 174: 954-955.

42. Banay-Schwartz M, Giuffrida AM, DeGuzman T, Sershen H, Lajtha A. Effect of undernutrition on cerebral protein metabolism. Exp Neurol 1979; 65: 157-168

43. Banay-Schwartz M, Zanchin G, DeGuzman T, Sershen H, Lajtha A. Decrease in cerebral protein synthesis on a low protein diet. In: Galoyan AA, editor. Problems of brain Biochemistry. Armenian Acad Sci, 1978; 13: 113-126.

44. Toth J, Lajtha A. Effect of protein-free diet on the uptake of amino acids by the brain in vivo. Exp Neurol 1980; 68: 443-452.

45. Sershen H, Lajtha A. The effect of nicotine on the metabolism of brain proteins. Neuropharmacology 1979; 18: 763-766.

46. Sershen H, Reith MEA, Lajtha A, Gennaro J, Jr. Effect of cigarette smoke on protein synthesis in brain and liver. Neuropharmacology 1981; 20: 451-456.

47. Toth E, Lajtha A. Alcohol effects on cerebral protein turnover in mice. Subst Alc Act/Misuse 1981; 2: 321-329.

48. Toth E, Lajtha A. Effect of chronic ethanol administration on brain protein breakdown in mice in vivo. Subs Alc Act/Misuse 1984; 5: 175-183.

49. Kenessey A, Banay-Schwartz M, DeGuzman T, Lajtha A. Increase in cathepsin D activity in rat brain in aging. J Neurosci Res 1989; 23: 454-456.

50. Matus A, Green GDJ. Age-related increase in a cathepsin D like protease that degrades brain microtubule-associated protein. Biochemistry 1987; 26: 8083-8086.

51. Wiederanders B, Oelke B. Accumulation of inactive cathepsin D in old rats. Mech Ageing Dev 1984; 24: 265-271.

52. Kenessey A, Banay-Schwartz M, DeGuzman T, Lajtha A. Calpain II activity and calpastatin contant in brain regions of 3- and 24-month-old rats. Neurochem Res 1990; 15: 243-249.

53. Banay-Schwartz M, DeGuzman T, Kenessey A, Palkovits M, Lajtha A. The distribution of cathepsin D activity in adult and aging human brain regions. J Neurochem 1992; 58: 2207-2211.

54. Lynch G, Larson J, Baudry M. Proteases, neuronal stability, and brain aging: An hypothesis. In: Crook T, Bartus RT, Ferris S, Gershon S editors. Treatment Development Strategies for Alzheimer's Disease. Madison, CT.: Mark Powley Assoc. Inc., 1986: 119-149.

DETERMINATION OF REGIONAL RATES OF CEREBRAL PROTEIN SYNTHESIS *IN VIVO* WITH L-[1-^{14}C]LEUCINE AS THE TRACER AMINO ACID

C. BEEBE SMITH

ABSTRACT. The quantitative autoradiographic L-[1-^{14}C]leucine method for the determination of regional rates of cerebral protein synthesis *in vivo* takes into account recycling of unlabeled leucine derived from protein degradation into the precursor pool for protein synthesis. We have evaluated the degree of recycling in whole brain by measuring the ratio of the apparent steady state leucine specific activity in the precursor pool (tRNA-bound leucine) to that in arterial plasma. In normal, conscious, adult rats this ratio (λ_{WB}) equals 0.58 indicating that 42% of leucine in the precursor amino acid pool is derived from protein breakdown. Evaluation of λ_i in local brain regions indicates that the degree of recycling does vary regionally. Local rates of leucine incorporation into protein determined with the quantitative autoradiographic technique and regional values of λ_i ranged from 11.0 in hypoglossal nucleus to 3.8 nmol/g/min in white matter. The average rate in the brain as a whole was found to be 6.1 nmol/g/min. Results of our studies of regeneration in the hypoglossal nucleus and plasticity in the developing monkey visual system suggest that chronic changes in functional activity in a pathway are more likely than acute changes to result in effects on rates of protein synthesis in structures of the pathway.

1. Introduction

In neurons the biosynthesis of proteins essential for the growth and continued maintenance of the entire cell including axons, dendrites and synaptic terminals is carried out in the perikarya. Changes in the synthesis of nerve cell proteins are probably involved in the ability of neurons to undergo long term responses to environmental changes. The processes of development and synaptogenesis, the property of neuronal plasticity, the ability of nerve cells to regenerate in response to injury, the response of target nerve cells to hormones, and probably also learning and memory are some of the conditions that may involve changes in the synthesis of proteins. It is of interest, therefore, to study the changes in protein synthesis underlying these processes, the time course of such changes, their regulation, and their localization. Such studies would of necessity be carried out in the intact animal preferably without the possible confound of anesthesia; this would require an *in vivo* method for cerebral protein synthesis and a quantitative autoradiographic technique in order to achieve regional localization.

Any method for the determination of amino acid incorporation into protein *in vivo* must apply basic principles for the assay of rates of chemical reactions with isotopes adapted for the special conditions encountered *in vivo* [1]. Briefly, the amount of labeled product specific to the metabolic

B. M. Mazoyer et al. (eds.), PET Studies on Amino Acid Metabolism and Protein Synthesis, 19–39.

step of interest (in this case the formation of labeled protein) in the selected tissue formed over an interval of time is divided by the integrated specific activity of the precursor pool over the same interval and corrected for any isotope effect [1]. This relationship can be expressed as follows:

$$R = \frac{P^*}{\Phi \int_0^T (C^*_{pp}/C_{pp})(t)\,dt} \qquad [1]$$

where R equals the rate of the overall reaction in the tissue, P^* is the amount of labeled product formed in the tissue, between zero time and time T, $(C^*_{pp}/C_{pp})(t)$ is the ratio of the concentration of labeled tracer to that of the endogenous unlabeled substrate (i.e. specific activity) in the precursor pool at any time t, and ϕ is a correction for isotope effect (if there is no isotope effect, $\phi = 1.0$). The application of these principles to *in vivo* conditions is complicated, however, because the necessary variables readily measured directly *in vitro* must be determined indirectly *in vivo*. With a kinetic modeling approach used in conjunction with the quantitative autoradiographic technique [2] to achieve spatial localization of the label in the brain, rates of protein synthesis can be determined in regions as discrete as individual cell layers and brain stem and hypothalamic nuclei. Quantitative autoradiography measures the total concentration of isotope in the tissue and not the concentrations of the specific products of the reaction of interest. Unreacted labeled precursor molecules and labeled products of other parallel biochemical reactions may also be present. In addition, proteins with very short half-lives in brain (e.g. ornithine decarboxylase) may be degraded during the course of the measurement and the labeled products removed from the tissue. Procedures must therefore be designed to avoid or correct for loss of labeled protein and for the presence of extraneous labeled compounds. Another problem is the determination of the integrated specific activity of the precursor pool. Its direct determination would require measurement of the complete time courses of the concentrations of both tracer and endogenous precursor amino acids in each region of interest. This is obviously impossible in an intact animal. Instead, the time course of the specific activity of the labeled amino acid in plasma is measured, and correction is made for the effects of the lag of the tissue behind the plasma and possible differences in the distribution spaces for the labeled and unlabeled species. These corrections are based on mathematical equations derived from a model for the behavior of the tracer amino acid in the tissue. The model must be based on the known biochemistry of the amino acid; validation of the model and underlying assumptions is essential.

2. L-[1-^{14}C]Leucine Method

2.1. THEORY

2.1.1. *Choice of Tracer.* We have attempted to deal with these problems through the choice of tracer and our experimental design [3,4,5,6]. Radioactive products of extraneous biochemical reactions were avoided by selection of L-[1-^{14}C]leucine as the labeled precursor. Carboxyl-labeled leucine, like other aliphatic, branched-chain amino acids, is either incorporated into protein or metabolized by only one pathway beginning with transamination followed by a very rapid decarboxylation [7]. Hence, the label in metabolized [1-^{14}C]leucine is converted transiently to α-[1-^{14}C]ketoisocaproic acid and then to $^{14}CO_2$, which is rapidly removed from brain by the circulation. The brain concentration of α-ketoisocaproic acid is very low (700 pmol/g) and at 2 min and later times up to 35 min following an i.v. pulse of [1-^{14}C]leucine, very little α-[1-^{14}C]ketoisocaproic acid (c. 0.7% of total brain ^{14}C) can be detected in brain [8]. Reincorporation of label from $^{14}CO_2$ is negligible because of its dilution with unlabeled CO_2 produced by glucose metabolism in brain. Therefore, virtually all ^{14}C remaining in the tissue exists either in the form of labeled protein or free leucine.

2.1.2. *Model.* We have developed a comprehensive model for the behavior of leucine and [1-^{14}C]leucine in brain [6] (Fig.1) which is based on the following assumptions. (a) The tissue region i is homogeneous with respect to concentrations of amino acids, rates of blood flow, and rates of transport, metabolism, and incorporation of amino acids into protein. (b) The

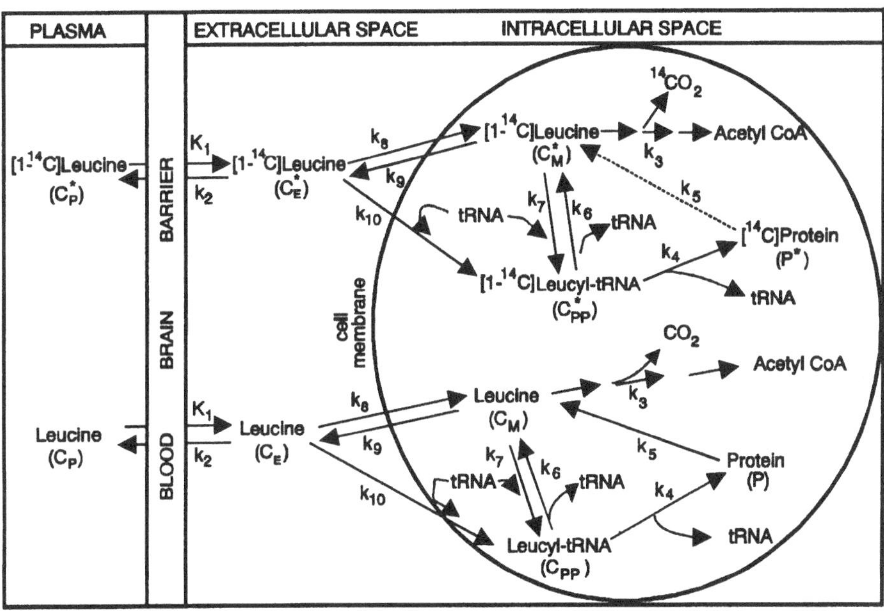

Figure 1. Diagrammatic representation of the comprehensive model for behavior of [1-^{14}C]leucine and leucine in brain. C_p, C_E, C_M, and C_{PP} represent concentrations of unlabeled leucine in plasma, extracellular space, intracellular metabolic pool, and intracellular precursor pool for protein synthesis, respectively, and C_p^*,C_E^*, C_M^* and C_{PP}^* represent concentrations of [1-^{14}C]leucine in plasma, extracellular space, intracellular metabolic pool, and intracellular precursor pool for protein synthesis, respectively, and P and P^* represent the total pool of unlabeled and ^{14}C-labeled protein in the cell, respectively. The constants K_1 - k_{10} represent the rate constants for both labeled and unlabeled leucine for carrier-mediated transport from plasma to tissue, for transport back from tissue to plasma, for metabolic degradation of leucine, for leucine incorporation into protein, for release of free leucine by protein degradation, for deacylation of leucyl-tRNA, for transfer of leucine from the intracellular metabolic leucine pool to precursor pool for protein synthesis, for transport of leucine from extracellular space to intracellular metabolic leucine pool, for transport of leucine from intracellular metabolic pool to extracellular space, and transport of leucine from extracellular space to intracellular precursor pool for protein synthesis, respectively.

concentration of unlabeled leucine in arterial plasma and rates of amino acid metabolism and protein synthesis are constant during the experimental period. (c) There is no significant loss of labeled protein nor is there recycling of [^{14}C]leucine from radioactive protein within an experimental period of 60 min, because the average half-life of proteins in brain is several days [9]; proteins with very short half-lives comprise such a small fraction of the total protein in brain that loss of label from these proteins would have a negligible effect on the average rate of synthesis. (d) Concentrations of the labeled leucine are sufficiently small that tracer theory holds, and there are no isotope effects associated with its use. The model (Fig.1) includes extracellular space, intracellular compartmentation and recycling of unlabeled, but not of labeled, leucine derived from protein degradation into the precursor pool for protein synthesis. Recycling can occur via any of several routes: within the confines of the cell or through the extracellular space. By mathematical analysis of this four-compartment model, an equation for the calculation of rates of protein synthesis can be derived [6]:

$$R_i = \frac{C_i^*(T) - C_{E_i}^*(T) - C_{M_i}^*(T) - C_{PP_i}^*(T)}{\lambda_i\left[\int_0^T \frac{C_p^*(t)}{C_p}dt - \sum_{n=1}^{3} A_n \int_0^T \frac{C_p^*(t)}{C_p} e^{-\alpha_n(T-t)}\,dt\right]}. \qquad [2]$$

where R_i is the rate of leucine incorporation into protein in tissue i; $C^*_i(T)$ is the concentration of ^{14}C in tissue i at any time, T, after introduction of the tracer into the circulation; $C^*_{Ei}(T)$, $C^*_{Mi}(T)$ and $C^*_{PPi}(T)$ represent [1-^{14}C]leucine concentrations in extracellular space, intracellular metabolizable pool, and intracellular precursor pool for protein synthesis of tissue i, respectively at any time T; $C^*_p(t)$ and C_p are the concentrations of labeled and unlabeled leucine in the arterial plasma, respectively; and t is variable time. A_n and α_n are constants composed of various combinations of the rate constants (K_1-k_{10}) of the model (Fig.1). λ_i is a constant whose exact mathematical expression comes from the solution to the system of differential equations of the model; it can be defined in terms of combinations of the rate constants.

λ_i, which has a value between 0 and 1.0, corrects for possible differences in sources of input of labeled and unlabeled leucine into the precursor pool and, therefore, also for the effects of recycling of unlabeled leucine from protein breakdown. Because leucine derived from protein breakdown during the first 60 minutes after the pulse is unlabeled, it dilutes the specific activity of the [^{14}C]leucine coming from the plasma, and this dilution must be accounted for to determine the specific activity of the precursor pool from measurements of the plasma. A value of λ_i close to 1.0 indicates that there is little recycling of unlabeled leucine from protein breakdown back into the precursor pool for protein synthesis; 1.0 - λ_i is the fraction of leucine in the precursor pool that comes from protein degradation.

2.1.3. *Experimental Design.* Experimental procedures can be designed to simplify the equation for the rate of protein synthesis in brain. Administration of [1-^{14}C]leucine as an i.v. pulse followed by a long experimental period allows most of the free [^{14}C]leucine to be cleared from the tissue by its incorporation into protein, metabolism, or transport back to plasma. The numerator of the equation is equivalent to $P^*_i(T)$, the concentration of labeled leucine incorporated into protein in tissue i during the experimental period, T. $C^*_i(T)$, the total concentration of label remaining in tissue i is measured by quantitative autoradiography; the concentrations of label remaining in the intracellular and extracellular leucine pools and in leucyl-tRNA (i.e., $C^*_{Mi}(T)$, $C^*_{Ei}(T)$, and $C^*_{PPi}(T)$, respectively, Fig. 1) must be subtracted to obtain $P^*_i(T)$. Since [1-^{14}C]leucine loses its label to

$^{14}CO_2$ during its metabolism there are no other labeled compounds in the tissue. Fixation of the brain tissue sections in formalin prior to autoradiography washes out the free amino acid pools (i.e. $C^*_{Ei}(T)$, $C^*_{Mi}(T)$), and since the pool of tRNA-bound [^{14}C]leucine ($C^*_{PPi}(T)$) is small compared to labeled protein ($P^*_i(T)$) 60 min after a pulse of [1-^{14}C]leucine, its contribution to $C^*_i(T)$ is negligible. Therefore, under these conditions, $P^*_i(T) \approx C^*_i(T)$ and can serve as the numerator of the equation. The denominator of the equation represents the integrated specific activity of [^{14}C]leucine in the precursor pool over the experimental period:

$$\int_0^T \left[C^*_{PP_i}(t)/C_{PP_i} \right] dt = \lambda_i \left[\int_0^T \frac{C^*_p(t)}{C_p} dt - \sum_{n=1}^{3} A_n \int_0^T \frac{C^*_p(t)}{C_p} e^{-\alpha_n (T-t)} dt \right] \qquad [3]$$

The three exponential terms following the minus sign are convolution integrals that correct for the lag of the precursor pool in the tissue behind that of the plasma. After a pulse, the plasma [^{14}C]leucine concentration and specific activity fall with time while the integral of the plasma specific activity continues to increase. The convolution integrals, however, decline with time, and the more rapid the turnover of free [^{14}C]leucine in the various tissue pools, the more rapid the decline. With an average half-life in gray matter of 3.6 min [10] and an experimental interval of 60 min after the pulse the contributions of the convolution integrals become negligible. Thus,

$$\lim_{T \to \infty} \int_0^T \left[C^*_{PP_i}(t)/C_{PP_i} \right] dt = \lambda_i \int_0^T \left[C^*_P(t)/C_P \right] dt. \qquad [4]$$

Therefore, by use of a pulse administration of a tracer dose of L-[1-^{14}C]leucine (e.g., less than 100 μCi contained in less than two μmols per kg of body weight to maintain tracer conditions) at zero time, an experimental period of 60 minutes, and fixation and washing of the brain sections in 10% formalin, the rather impractical Equation [2] can be reduced to the following relatively simple operational equation:

$$R_i = \frac{P^*_i(T)}{\lambda_i \int_0^T \left[C^*_P(t)/C_P \right] dt}. \qquad [5]$$

2.1.4. *Determination of* λ_i. λ_i is the only factor in the operational equation (Equation [5]) that is not measured at the same time during the procedure for the measurement of local protein synthesis and, therefore, must be evaluated in a separate series of experiments. By rearrangement of Equation [4] λ_i can be expressed as follows:

$$\lambda_i \approx \frac{\int_0^T \left[C^*_{PP_i}(t)/C_{PP_i} \right] dt}{\int_0^T \left[C^*_P(t)/C_P \right] dt}. \qquad [6]$$

Equation [6] applies only when the brain and the plasma are in a steady state with respect to unlabeled leucine and after sufficiently long periods for the convolution integrals in Equation [3] to become negligible. It is impractical to determine λ_i by measuring the integrated precursor pool and plasma specific activities because it is impossible to measure the time course of the precursor pool specific activity in the same animal. If, however, the labeled leucine is so administered that the arterial plasma specific activity achieves a constant level and remains constant long enough for the convolution integrals in Equation [3] to become negligible compared to the integrated plasma specific activity, then the precursor pool specific activity in the tissue will eventually reach a constant steady state level that will be equal to λ_i times the plasma specific activity [6]. This relationship can be expressed as follows. If the specific activity of the labeled leucine in the arterial plasma is maintained constant, then

$$\lim_{T\to\infty} \frac{\int_0^T \left[C^*_{PP_i}(t)/C_{PP_i}\right]dt}{\int_0^T \left[C^*_P(t)/C_P\right]dt} = \lim_{T\to\infty} \frac{C^*_{PP_i}(T)/C_{PP_i}}{C^*_P(T)/C_P} = \lambda_i \qquad [7]$$

The value of λ_i can be experimentally determined directly based on Equation [7] (Fig. 2). The specific activity of the labeled leucine in the arterial plasma can be maintained constant for a long period of time by means of a programmed infusion. The precursor leucine pool (C_{PP}) can be isolated from other leucine amino acid pools in the tissue by extraction of the tRNA-bound amino acids. These can be deacylated and the released leucine separated by amino acid analysis and assayed for specific activity.

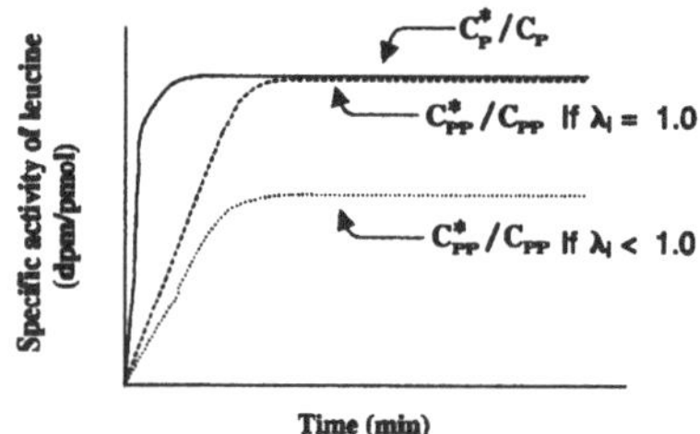

Figure 2. Determination of λ_i. Labeled leucine is administered so that a constant specific activity is maintained in the plasma over a long enough period of time for the brain to reach an apparent steady state with respect to the plasma. If the leucine specific activity in the precursor pool reaches that of the plasma then λ_i = 1.0 (dashed line); if the leucine specific activity in the precursor pool is less than that of the plasma then $\lambda_i < 1.0$

The concentration of tRNA-bound leucine in brain is too low (nM) to measure λ_i in localized brain regions by this method. The total tissue acid-soluble leucine content is, however, readily measurable (μM) in local regions in brain. We have, therefore, attempted to estimate values of λ_i from measured local values of ψ_i, the equivalent ratio of apparent steady state specific activities for the total tissue acid-soluble pool (C^*_{Ei} and C^*_{Mi}). Thus if C^*_p is held constant long enough for the tissue leucine pools to reach a steady state with the plasma, then by analogy with the precursor leucine pool for protein synthesis,

$$\lim_{T\to\infty} \frac{\left[\left(C^*_{E_i}+C^*_{M_i}\right)(T)\right]/\left[C_{E_i}+C_{M_i}\right]}{C^*_P(T)/C_P} = \Psi_i \qquad [8]$$

A quantitative relationship between the λ and ψ was determined from simultaneous measurements of both in the whole brain of animals which were administered programmed infusions to produce and maintain constant specific activities of leucine in the arterial plasma. This relationship was then used to calculate values of λ_i from measured local values of ψ_i [11].

2.2. PROCEDURES

2.2.1. *Chemicals.* Chemicals and materials were obtained from the following sources: L-[4,5-^{3}H]leucine (spec.act., 163 Ci/mmol) and L-[1-^{14}C]leucine (spec.act., 54 mCi/mmol), Amersham, Arlington Heights, IL, U.S.A.; L-[3,4,5-^{3}H]leucine (spec.act., 155 Ci/mmol), Du Pont-NEN, Wilmington, DE, U.S.A.; *Escherichia coli* tRNA, Sigma Chemical Co., St. Louis, MO, U.S.A.; vanadyl ribonucleoside complex and redistilled nucleic acid grade phenol, Bethesda Research Laboratories, Gaithersburg, MD, U.S.A.; L-norleucine, Cyclochemicals (Div. of Travenol Lab., Inc.), Los Angeles, CA, U.S.A.; 5-sulfosalicylic acid, Fluka Chemie AG, Buchs, Switzerland.

2.2.2. *Animals.* Normal, female Sprague-Dawley rats (160-250 g) were used in the studies of regeneration [4]. Rats were anesthetized with ketamine (200 mg/kg, i.m.). The hypoglossal nerve on one side was cut at the level of the anterior belly of the digastric muscle and a 3-4 mm segment of nerve was removed. Following surgery, animals were returned to their cages and allowed to recover. The rats were prepared for the determination of rates of protein synthesis by the insertion under light halothane anesthesia of polyethylene catheters into one femoral vein and artery. The rat's hindquarters were then restrained by the application of a loose-fitting plaster cast which was taped to a lead brick. After a period of 2 to 3 h for recovery from the effects of anesthesia, the procedure for the determination of local rates of protein synthesis was carried out as described below.

Studies of protein synthesis in the developing monkey visual system [12] were carried out on four full-term newborn rhesus (*Macaca mulatta*) monkeys. Three of the monkeys were subjected to unilateral tarsorrhaphy under ketamine anesthesia (0.4 mg / kg, i.m.) and allowed to recover. On the day of the [^{14}C]leucine study, monkeys were anesthetized with a mixture of halothane and nitrous oxide and polyethylene catheters were inserted into a femoral artery and a vein. Subsequently, monkeys were chaired for 2 h while recovering from the anesthesia and for the duration of the [^{14}C]leucine study.

Normal, male Sprague-Dawley rats weighing 210-350 g were used in all other studies. Under light halothane anesthesia a femoral artery and vein were catheterized, and the catheters were tunneled under the skin to exit at the nape of the neck. At least 2 h were allowed for recovery from surgery and anesthesia. Rats were allowed to move freely in their cages throughout the experimental procedure. Physiological variables, including arterial blood pressure, pH, blood gas tensions, hematocrit, plasma glucose concentration, and rectal temperature were monitored in all experimental rats; all of these variables were within the normal range for adult rats.

2.2.3. *Procedure to Determine λ_{WB}, ψ_{WB}, and λ_i.* Intravenous infusion schedules designed to achieve and maintain constant arterial plasma concentrations of labeled leucine for 90 min were determined by the method of Patlak and Pettigrew [13]. The infusion schedule consisted of an initial pulse followed by a progressively decreasing rate of infusion adjusted every 15 seconds and administered by a programmed computer-driven infusion pump (Pump 22, Harvard Apparatus,

Inc., South Natick, MA, U.S.A.). Each rat was infused with 4-17 mCi of [^{3}H]leucine in a total volume of approximately 2-3 ml over the 60 to 90 min period of the infusion. [^{3}H]Leucine was used to determine λ_i because the higher specific activity was needed to measure the leucine concentration in the very small pool of tRNA-bound amino acids in brain and in the tissue acid-soluble amino acid pools in very small (1-6 mg wet weight) regions of brain. The duration of the infusions was varied between 60 and 90 min in order to verify steady state conditions, i.e., when an apparent steady state is achieved, the ratios of pool to plasma specific activities remain constant with time. Timed arterial blood samples were drawn every 5 or 10 min during the infusion and centrifuged immediately to separate the plasma, which was then deproteinized by the addition of one third of a volume of a solution of 16% (W/V) sulfosalicylic acid containing L-norleucine as an internal standard for amino acid analyses. The deproteinized plasma samples were stored at -20°C until assayed for leucine and [^{3}H]leucine concentrations. At the end of the infusion the rats were decapitated, and the brains were quickly removed. Those brains used to determine λ_{WB} were chilled to 0°C in ice-cold 0.25 M sucrose; those brains used to determine local values of λ_i were dissected as described below on a chilled glass plate overlying crushed ice.

2.2.4. *Extraction and Purification of Aminoacyl-tRNA*. The brain from each rat was homogenized by a motor-driven loose-fitting all glass homogenizer in 10 ml of 0.25 M sucrose (0°C) containing 10 mM vanadyl ribonucleoside complex to inhibit ribonuclease, 6 mg of uncharged *Escherichia coli* tRNA as carrier, and L-norleucine (20 μM) added as an internal standard. The homogenates were centrifuged at 100,000 x *g* for 1 h to remove intact cells, cellular debris, and the subcellular organelles, and the pellets were discarded. A 100 μl volume of the soluble fraction was deproteinized by the addition of an equal volume of a solution of 8%(W/V) sulfosalicylic acid and stored at -20°C until assayed for leucine and [^{3}H]leucine concentrations in the acid-soluble pool. The cytosolic protein and RNA in the remainder of the supernatant fraction were precipitated by the addition of one tenth of a volume of a solution of 50% (W/V) trichloroacetic acid and separated by centrifugation (12,000 x *g*, 30 min). The precipitates containing the aminoacyl-tRNA were washed six times in 3% (W/V) perchloric acid to remove free amino acids, suspended in 0.3 M sodium acetate (pH 5), and extracted with an equal volume of fresh, water-saturated phenol containing 0.1% (W/V) 8-hydroxyquinoline as an antioxidant. The aqueous phase, containing the aminoacyl-tRNA, was separated, and residual phenol was removed by six extractions with anhydrous diethylether. The aminoacyl-tRNA was precipitated overnight at -20°C by addition of 2.5 volumes of ethanol containing 0.12 M potassium acetate, pH 5.5. The RNA precipitate was recovered by centrifugation (12,000 x *g*, 30 min) at 0°C; washed two times in ice-cold 1:2.5 (V/V) mixture of 0.3 M sodium acetate, pH 5, and 0.12 M potassium acetate in ethanol, pH 5.5; dissolved in 50 mM sodium carbonate (pH 10); and incubated at 37°C for 90 min to deacylate the aminoacyl-tRNA. Deacylated tRNA was precipitated overnight at -20°C by addition of 2.5 volumes of ethanol and removed by centrifugation (12,000 x *g*, 20 min). The supernatant solutions, which contained the previously tRNA-bound but now free amino acids, were dried in a stream of N_2 and redissolved in 40 μl of 0.2 M sodium citrate (pH 2.2) [6].

2.2.5. *Procedure to Dissect Brain Regions*. The pineal was first excised and brains were placed dorsal side up in a stainless steel rodent brain matrix (Activational System Inc., Warren, MI, U.S.A.), which was kept cold on crushed ice. Blades were inserted through the slots in the matrix at right angles to the sagittal axis down to the level of cerebellum. The brains were sliced at intervals of 1 mm. Brain regions were dissected from these coronal slices. The remainder of the brain was placed on the chilled glass plate, and the cerebellum was completely detached from the brain stem and set aside. The hypoglossal nuclei were dissected bilaterally from the brain stem [14]. Dissected brain regions were weighed, homogenized in 4% sulfosalicylic acid which contained 2.5 μM norleucine as an internal standard and stored at -20°C until assayed for leucine and [^{3}H]leucine concentrations. At the time of these assays the samples were thawed, vortexed and centrifuged for 30 min at 5000 x *g* at 4°C to remove the precipitated protein.

2.2.6. Assay of Specific Activity of [^{3}*H*]*Leucine*. Specific activities of [^{3}H]leucine in deproteinized plasma, acid-soluble fractions from whole brain and individual regions, and fractions derived from the deacylated tRNA-amino acids were assayed by post-column derivatization with *o*-phthaldehyde and fluorometric assay in a Beckman Model 7300 amino acid analyzer (Beckman Instrument Co, Inc., Fullerton, CA, U.S.A.). Fractions, after passage through the detector, were collected every min and assayed for ^{3}H with a TRI-CARB Liquid Scintillation Analyzer, Model 2250CA (Packard Instrument Co., Downers Grove, IL, U.S.A.). This system can measure 10-100 pmol of leucine with a 3% error. Specific activity was calculated from total ^{3}H in all fractions in the leucine peak and the total measured leucine content in the peak.

2.2.7. *Calculation of the Values of* λ_i and ψ_i. λ_i and ψ_i were evaluated based on Equations [7] and [8], respectively. The time course of the acid-soluble specific activity in arterial plasma and the specific activity of tRNA-bound leucine and acid-soluble leucine in whole brain and individual regions at the end of the experimental interval were determined as described above. The apparent steady state free leucine specific activity in the arterial plasma was calculated as the mean of the specific activities determined from 40 min to the end of the interval. Leucine from the tRNA-bound amino acid fractions was uncontaminated by leucine derived from any blood in the brain tissue because of the procedure used to separate this fraction from the free amino acids in the tissue. The specific activity of the acid-soluble leucine in the tissue, however, was corrected for contamination by leucine in the blood contained in the brain. Values reported in the literature for regional blood volume and hematocrit in rat brain [15] were used in these calculations. The equilibrium distribution of free leucine between red cells and plasma was measured and found to be 0.7; this value was used together with local tissue blood volume and hematocrit to calculate the local tissue blood leucine and [^{3}H]leucine contents from measurements in plasma. The contamination by leucine in whole blood in brain ranged from 0.6% to 1.3% of the total acid-soluble leucine in white matter regions and inferior colliculus, respectively.

2.2.8. *Determination of Local Rates of Protein Synthesis*. Animals were surgically prepared and catheterized, and their physiological states monitored as described above. The experimental period was initiated by an intravenous pulse of L-[1-^{14}C]leucine; the dose was 100 μCi/kg of body weight contained in 0.11 - 0.14 ml of physiological saline. Timed arterial samples were collected during the following 60 minutes for determination of the time courses of plasma concentrations of leucine and [^{14}C]leucine. The blood samples were immediately centrifuged and the plasma deproteinized at 4°C by the addition of an equal volume of 8% (W/V) sulfosalicylic acid. Labeled and unlabeled leucine concentrations in the acid-soluble fractions were assayed by liquid scintillation counting and by amino acid analysis, respectively. At the end of the 60 minute experimental period animals were killed by an intravenous injection of sodium pentobarbital, and the brains were rapidly removed and frozen in isopentane cooled to -40°C with dry ice. The brains were cut into sections 20 μm thick in a cryostat maintained at -20°C, and the sections were thaw-mounted on gelatin-coated slides, air-dried, then fixed and washed with 5 changes of phosphate-buffered 10% formalin for 30 minutes each, and washed in running deionized water for 30 min. The sections were then autoradiographed along with calibrated [^{14}C]methylmethacrylate standards as previously described [16]. Rates of leucine incorporation into protein in individual brain regions were determined by analysis of the autoradiograms with a computerized image processing system (MCID Imaging Research Inc. St. Catharines, Ontario, Canada). The concentration of ^{14}C in each region of interest in the autoradiograms was determined from the optical density vs. ^{14}C concentration curve for the calibrated plastic standards. Local rates of protein synthesis were calculated by means of Equation [5] with the regional values of λ_i. The rate of leucine incorporation into protein in the brain as a whole weighted for the relative masses of its component parts was determined in each animal by analysis of the autoradiograms with a Photoscan System P-1000 HS scanning densitometer (50 μm aperture) (Optronics International, Chelmsford, MA, U.S.A.); rates were calculated with the value of λ_{WB} ($\lambda_{WB} = 0.58$).

2.2.9. *Statistics*. Differences between groups (i.e., 60 vs. 90 min experiments) were analyzed by means of a one-way analysis of variance. Tissue pool to arterial plasma ratios of the apparent steady state specific activities for the acid-soluble and tRNA-bound pools were compared by means of a paired Student's t-Test. Regional values of the apparent steady state ratio of specific activity in the acid-soluble pool to that in the arterial plasma (ψ_i) were analyzed for regional homogeneity by means of a repeated measures analysis of variance. A linear least squares procedure was used to determine the coefficients of the best-fitting polynomial equation between the values of λ_{WB} and ψ_{WB}; a straight line fit was found to describe the relationship very well (Fig. 3).

3. Results

3.1. NORMAL, CONSCIOUS RATS

3.1.1. *Evaluation of* λ_{WB}. The ratios of the specific activities of [^{3}H]leucine in the tRNA-bound and in the acid-soluble pools in whole brain to that of the arterial plasma were determined after 60 and 90 min of a programmed infusion of [^{3}H]leucine that maintained a relatively constant arterial plasma [^{3}H]leucine specific activity in nine rats (Table 1) verifying that an apparent steady state had been achieved by 60 min and that all nine determinations for the tRNA-leucine pool and for the acid-soluble leucine pool can be considered as apparent steady state values. When calculated according to Equation [7], the specific activity of leucine in the precursor pool (tRNA-bound) for

TABLE 1. Steady state ratios of [^{3}H]leucine specific activities in tissue pools to plasma [^{3}H]leucine specific activity [11]

	Pool specific activity / plasma specific activity	
Experimental period	Acid-soluble pool	tRNA-bound pool
60 min	0.47 ± 0.02 (5)	0.57 ± 0.02 (5)
90 min	0.51 ± 0.02 (4)	0.59 ± 0.01 (4)
Combined 60 & 90 min	0.49 ± 0.02 (9)	0.58 ± 0.01 (9)

Values are the means ± S.E.M. for the number of rats indicated in parentheses.
* Statistically significantly different from the acid-soluble pool (paired t-Test, P < 0.0001).

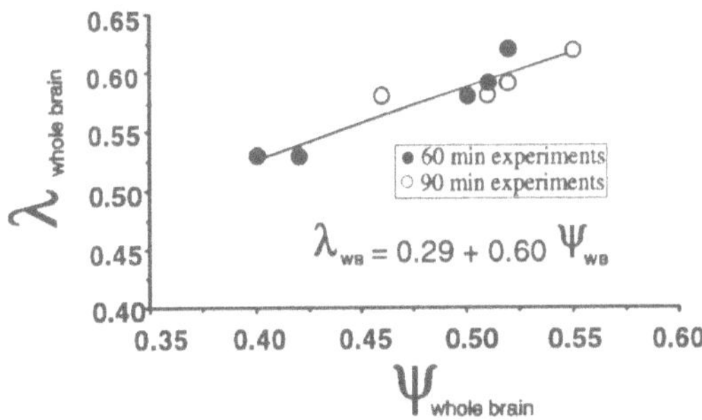

Figure 3. The relationship between experimentally determined values of ψ_{WB} and λ_{WB} in whole brain [11]. Each point represents determinations of both ratios in a single rat. Values were determined in five 60 min experiments (●) and four 90 min experiments (○). The best-fitting straight line and the equation for the line are illustrated. The correlation coefficient for the fit was 0.93 ($P < 0.0005$).

protein synthesis averaged 58% ($\lambda_{WB} = 0.58 \pm 0.01$) of that of the plasma. A paired comparison of the apparent steady state ratios of leucine pool to plasma specific activities for the tRNA-bound leucine pool (λ_{WB}) and for the acid-soluble leucine pool (ψ_{WB}) shows that λ_{WB} is an average of 20% higher than ψ_{WB} ($P < 0.0001$).Values of λ_{WB} are linearly correlated with values of ψ_{WB} ($r_{xy}= 0.93$,$P < 0.0005$) (Fig.3). Agreement between the measured values of λ_{WB} and those estimated by the linear equation (Fig.3) is excellent; no statistically significant improvement in fit was found when polynomial regressions of a higher degree were used.

3.1.2. *Evaluation of* λ_i. Seven rats in a steady state for unlabeled leucine were administered programmed infusions of [^{3}H]leucine that maintained the specific activity of [^{3}H]leucine in arterial plasma relatively constant long enough (60 min) for the acid-soluble leucine pool in brain tissue to reach a steady state with respect to plasma. In all experiments the specific activities of the acid-soluble leucine pool in all 19 regions examined were below those of the arterial plasma. The measured values of ψ_i ranged from 0.36 in the globus pallidus to 0.56 in the hypoglossal nucleus and 0.64 in the pineal. Values for λ_i (Table 2) were calculated for each brain region based on the linear regression equation (Fig. 3) fitted from the determined values of λ_{WB} and ψ_{WB}. Results of repeated measures analysis of variance indicated that values of ψ_i and λ_i are not regionally homogeneous.

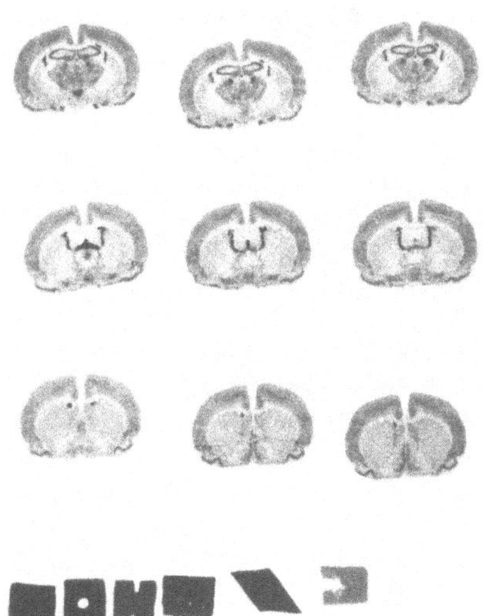

Figure 4. [1-^{14}C]Leucine autoradiogram of sections of adult rat brain and calibrated [^{14}C]methylmethacrylate standards used to quantify regional ^{14}C concentrations [10].

3.1.3. *Local Rates of Protein Synthesis.* Rates of protein synthesis in the brain as a whole, in the pineal, and in 18 brain regions were determined with the quantitative autoradiographic technique (Fig. 4); regional estimates of λ_i determined in normal conscious rats were used in the calculations (Table 2). Rates of protein synthesis

Table 2. Regional values of λ_i and rates of protein synthesis in the conscious, adult male rat [11]

Brain Region	λ_i *	Rate of leucine incorporation into protein** (nmol/g/min)
Cortex		
Anterior cingulate	0.56 ± 0.01	8.0 ± 0.2
Frontal cortex	0.57 ± 0.01	7.8 ± 0.1
Hippocampus	0.55 ± 0.01	7.1 ± 0.2
Auditory cortex	0.56 ± 0.01	8.2 ± 0.2
Visual cortex	0.57 ± 0.01	8.1 ± 0.1
Cerebellum	0.55 ± 0.01	7.5 ± 0.2
Subcortical		
Nucleus accumbens	0.56 ± 0.01	5.9 ± 0.1
Caudate putamen	0.58 ± 0.01	6.0 ± 0.1
Globus pallidus	0.51 ± 0.01	5.8 ± 0.2
Thalamus	0.55 ± 0.01	8.0 ± 0.1
Medial geniculate	0.56 ± 0.01	8.6 ± 0.2
Substantia nigra	0.52 ± 0.01	6.3 ± 0.2
Inferior colliculus	0.57 ± 0.01	8.8 ± 0.2
Hypoglossal nucleus	0.62 ± 0.01	11.0 ± 0.2
White matter		
Anterior commissure	0.54 ± 0.02	4.0 ± 0.1
Genu of corpus callosum	0.53 ± 0.01	4.0 ± 0.1
Internal capsule	0.54 ± 0.01	3.7 ± 0.1
Cerebellar white matter	0.55 ± 0.01	3.4 ± 0.1
Extracerebral		
Pineal	0.68 ± 0.02	20.5 ± 0.3
Whole brain average	0.58 ± 0.01†	6.1 ± 0.1

* Values are the means ± SEM for 7 rats in all regions except the hypoglossal nucleus, anterior commissure, and the whole brain in which determinations were made in 4, 6, and 9 rats, respectively.

**Values are the means ± SEM for 6 rats. Values were calculated with the regional values of λ_i.

† Value of λ for the brain as a whole was determined by measurement of the ratio of the apparent steady state leucine specific activity in the precursor pool (tRNA-bound) to that of the arterial plasma (Table 1).

varied throughout the brain. The values in white matter regions tended to group together and were always below those of gray matter regions. The average value in gray matter was about twice that of white matter, but individual values varied from 6 to 11 nmol leucine incorporated / g tissue / min. The average rate of protein synthesis in the brain as a whole was 6.1 nmol/g/min.

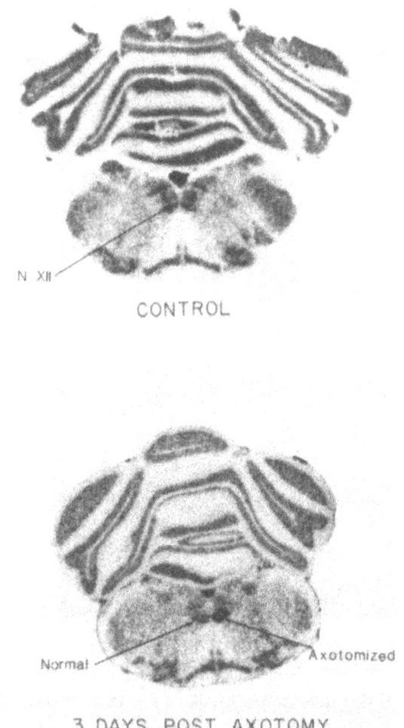

Figure 5. [^{14}C]Leucine autoradiograms from rats following right hypoglossal axotomy (bottom) and sham-operation (upper). The right side of the brain is on the right in the figure.

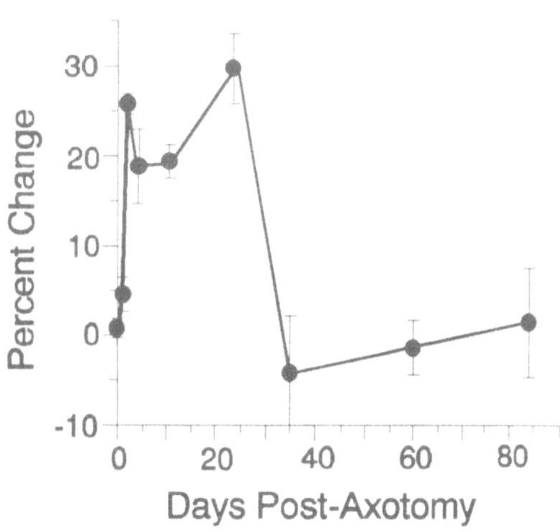

Figure 6. Time course of the changes in rates of protein synthesis in the axotomized hypoglossal nucleus. In each animal the percent change in rate of protein synthesis in the nucleus ipsilateral to the axotomy compared with the contralateral side was calculated. Each point is the mean ± S.E.M. of two rats except for the point at Day 1 which is the mean of four rats.

3.2. REGENERATION OF A CRANIAL NERVE

3.2.1. *Time course of changes in the rates of protein synthesis in the regenerating hypoglossal nucleus.* Unilateral section of the XIIth nerve in adult, female rats resulted in a 20-30% increase in rates of protein synthesis in the ipsilateral hypoglossal nucleus [4] (Fig. 5 & 6). Rates were calculated with λ_i evaluated in normal, conscious rats (Table 2). Preliminary results [17] indicate that changes in λ_i in the regenerating hypoglossal nucleus are very small (< 5.0%). Changes in rates of protein synthesis were not seen at 0.5, 3 and 8 h post-axotomy; at 24 h post-axotomy a small 5% increase was observed, and at 48 h post-axotomy protein synthesis rates were maximally increased (20-30%) over the contralateral control side and remained so through Day 24. By post-axotomy day 35, rates of protein synthesis had returned to normal.

3.3. PLASTICITY IN THE DEVELOPING MONKEY VISUAL SYSTEM

3.3.1. *Effects of acute and chronic monocular occlusion on protein synthesis in the laminae of the dorsal lateral geniculate nuclei.* The six cell laminae of the dorsal lateral geniculate nuclei are clearly visible in [^{14}C]leucine autoradiograms of a 25 day old monkey with intact binocular vision (Figure 7A.), and rates of protein synthesis in all of the laminae are more or less the same. Acute

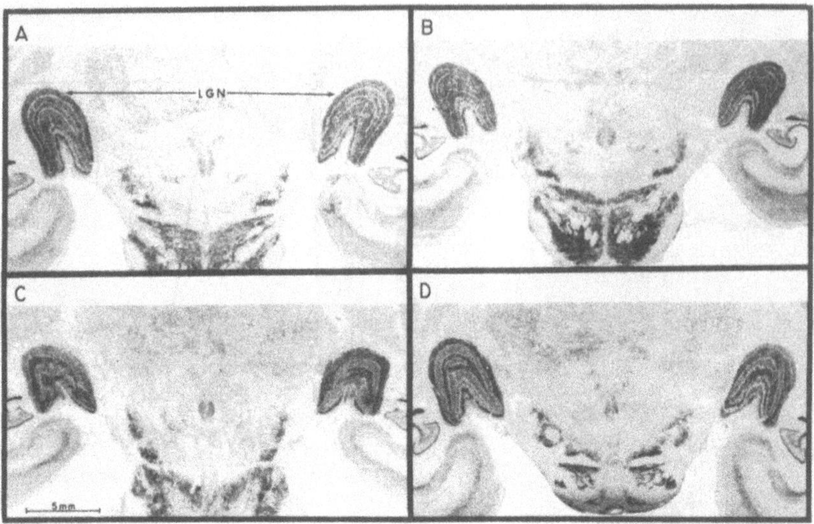

Figure 7. Autoradiograms of coronal sections of monkey lateral geniculate nuclei (LGN) obtained with the [1-^{14}C]leucine method [12]. The left side of the brain is on the left side of the autoradiograms. (A) Twenty-five day old monkey with intact binocular vision. (B) Twenty-five day old monkey with acute occlusion of the left eye. (C) Twenty-five day old monkey with chronic occlusion of the right eye initiated on the second day of life. (D) Twenty-five day old monkey with chronic occlusion of the left eye initiated on the second day of life.

monocular occlusion (Figure 7B.) had no differential effects on the rates of protein synthesis in the six laminae; the autoradiogram from the animal studied three h after unilateral lid suture was indistinguishable from that of the animal with intact binocular vision. Chronic unilateral deprivation (from Day 2 to Day 25), however, resulted in marked changes in the appearance of the autoradiograms of the lateral geniculate nuclei (Figure 7C & D). The laminae innervated by the occluded eye (1, 4, and 6 contralaterally; 2, 3, and 5, ipsilaterally) exhibited lower optical densities compared to the laminae served by the still functioning eye. The autoradiographic patterns of the lateral geniculates were inverse rather than bilaterally symmetrical. Quantification of the rates of protein synthesis, assuming that the value of λ_i was not affected under these conditions, were normal in the laminae innervated by the intact eye but were decreased by about 20-25% in the deprived laminae.

4. Discussion

4.1. VALIDITY OF THE AUTORADIOGRAHIC [1-^{14}C]LEUCINE METHOD

4.1.1. *Estimation of the specific activity of the precursor pool for protein synthesis.* A key issue for any *in vivo* method for the determination of rates of cerebral protein synthesis with a radiolabeled tracer is the estimation of the precursor pool specific activity integrated over the experimental interval in the brain region of interest. In an *in vivo* study this estimation is based on measurements of the time course of the concentrations of tracer and endogenous amino acid in the plasma and the kinetic model for the behavior of the tracer amino acid in brain. If amino acids liberated by the breakdown of unlabeled protein within the tissue are recycled into the precursor pool for protein synthesis, then the time course of the specific activity of the tracer in the plasma cannot be the sole basis from which to derive the integrated specific activity of the precursor pool in the tissue. The fractional dilution of the specific activity of the precursor pool by the recycled unlabeled amino acid derived from protein degradation must be taken into account. The results of studies of leucine [6], valine [18], and methionine [19] all show that there is significant recycling of these amino acids derived from the degradation of protein into the precursor pool for protein synthesis, and, therefore, dilution of the precursor pool specific activity cannot be ignored. Our results in adult rats show (Table1) that in the brain as a whole about 42% of leucine in the precursor pool comes from protein degradation while 58% comes from the plasma, i.e., $\lambda_{WB}= 0.58$. The four compartment model for the behavior of leucine in brain (Fig. 1) takes into account all possible routes by which recycling of leucine might occur, via the extracellular space and/or via an intracellular amino acid compartment. The steady state method used to evaluate the extent of recycling is independent of the actual relationships among the tissue compartments; the rate constants for transfer of leucine between compartments (which are in the convolution integrals) drop out of the equation (Equation [7]) under the "apparent" steady state conditions. (The steady state is "apparent" because it pertains only to the free leucine pools and not the pool of leucine incorporated into protein.)

Equation [5] for the determination of rates of cerebral protein synthesis is valid when an i.v. pulse of L-[1-^{14}C]leucine is followed by a long experimental interval and the quantitative autoradiographic technique is applied to formalin-fixed and washed tissue sections provided that the value of λ_i is known. The value of λ_{WB} applies only to the brain as a whole in normal, conscious rats. In order to take advantage of the anatomic resolution of the autoradiographic technique and to apply the technique to neurobiological questions the possibility that λ_i varies in different brain regions and in different biological states must be examined.

The linear relationship between λ_{WB} for the true precursor pool and ψ_{WB} for the acid-soluble pool (Fig. 3) provided us with a rational approach for estimating regional values of λ_i. Direct chemical determination of local values of λ_i, as was done for whole brain, is beyond the limits of our technique for assaying the specific activity of the tRNA-bound leucine. The leucine content of the acid-soluble pool in brain is approximately three orders of magnitude greater than that of the tRNA-bound leucine. We have, therefore, measured the steady state ratio of the tissue : plasma specific activities of leucine in the regional acid-soluble pools and used the linear relationship between the two pools, acid-soluble and tRNA-bound, in whole brain, to estimate the local values of λ_i. We have assumed that the linear relationship between λ_{WB} and ψ_{WB}, determined for the whole brain, applies to local regions and is relatively constant from region. Though it is possible that this relationship may not be exactly the same for all brain regions, it may give us the closest estimate of actual regional values of λ_i (Table 2).

The results show that in normal, conscious rats there is some regional variation in the extent of recycling of leucine derived from protein degradation into the precursor pool for protein synthesis. In white matter 46% of the leucine in the precursor pool is derived from products of protein degradation while in neuronal cell body-rich regions, such as the hypoglossal nucleus, 38% of the leucine is recycled. In some subcortical brain regions, such as globus pallidus and substantia nigra, with mixed gray and white matter, recycling is as high as 50 and 48%, respectively. In the pineal, which is outside the blood-brain barrier, only 33% of leucine in the precursor pool is derived from products of protein degradation. In most regions of gray matter examined, other than the hypoglossal nucleus, however, the value of λ_i is within 5% of the λ_i value for the brain as a whole.

Some regional variation in the value of λ_i is not surprising. The site of protein synthesis in neurons is the cell body which supplies the entire axon and dendritic arbor with proteins to maintain their structure and function. In regions rich in neuronal cell bodies, such as the hypoglossal nucleus which has the highest value of λ_i in the brain regions examined (Table 2), much of the protein synthesized by the neurons leaves the cell bodies via axonal transport. Only a fraction of the synthesized protein remains available in the cell body region for degradation and recycling. The sites of biosynthesis of peptide hormones, such as the paraventricular and supraoptic nuclei in the hypothalamus, might also be expected to have higher values of λ_i because a large fraction of the protein synthesized in such cells is exported. These regions of brain have some of the highest rates of protein synthesis [5,10]. Determination of λ_i in these very small nuclei was beyond the limits of our ability to dissect them accurately and to measure reproducibly their leucine concentrations. In regions rich in glial cells and interneurons, such as the substantia nigra, globus pallidus, hippocampus, cerebellum and white matter regions, protein synthesized in the cells is not so extensively exported and remains available for degradation at the site of protein synthesis resulting in lower values of λ_i, i.e., higher degrees of recycling. The highest value of λ_i was found in the pineal which is a highly vascularized structure without an effective blood-brain barrier; it is then to be expected that delivery of leucine via the circulation and plasma:tissue transport would be greater than in brain tissue transport-limited by the blood:brain barrier.

4.1.2. *Regional rates of protein synthesis.* The regional cerebral rates of leucine incorporation into protein presented in Table 2 are unique in that they have been calculated with regional correction for recycling (λ_i). The only other published study in which leucine was the tracer amino acid and an attempt was made to determine "λ" was that of Hargreaves-Wall et al. [20] in which the rate of protein synthesis in a hemisphere was determined to be 0.62 nmol leucine incorporated/g tissue/min. This value, which is 10% of the value for whole brain that we observed (Table 2), was determined in adult rats under ketamine anesthesia with an internal carotid artery perfusion technique. While it is possible that the difference between the results of Hargreaves-Wall et al. [20] and ours is due to an effect of ketamine, we have not found such changes in cerebral protein synthesis with ketamine anesthesia [21]). Perhaps more pertinent is the question of whether or not perfusion of the brain at a flow rate of 1.0 ml/min with red blood cells reconstituted in an oxygenated buffer is adequate to maintain the brain in a normal state over a 30 min period. In a study of conscious, adult rats [8] the rate of leucine incorporation into cerebral protein was reported to be 3.2 nmol/g/min, a value about 1/2 the rate that we find (Table 2). These investigators [8] specify that the rate applies to incorporation of leucine derived from an "exogenous source", presumably the plasma; and, if so then only minimal rates of protein synthesis uncorrected for recycled leucine derived from protein degradation were determined.

Amino acids other than leucine have been used as tracers for protein synthesis. In studies with the [1-^{14}C]valine flooding technique, in which the precursor pool specific activity is based on measurement of the tissue acid-soluble pool specific activity, Dunlop et al. [22] reported a rate for the brain as a whole of 2.8 nmol valine incorporated/mg protein/h which is approximately equivalent to 4.7 nmol valine/g tissue/min (100 mg protein/g tissue assumed). The value for valine incorporation can be compared with its equivalent value for leucine incorporation by dividing the valine rate by 0.7, the ratio of valine to leucine in whole rat brain protein [23]; the equivalent value of 6.7 nmol leucine/g/min thus calculated from the rate of valine incorporation is in reasonable agreement with our value for whole brain. In two published studies of regional protein synthesis in awake, behaving rats, either [1-^{14}C]valine [24] or [^{35}S]methionine [25] were used as tracers. Both of these published studies failed to take into account recycling of the tracer amino acid.

Regional evaluation of λ_i for leucine enables local rates of cerebral protein synthesis to be determined with a fully quantitative autoradiographic [1-^{14}C]leucine method, at least in conscious adult rats. The stability of λ_i in the brain as a whole under different physiological and pharmacological experimental conditions is currently under study. The effect of recycling of amino acids derived from protein degradation on the specific activity of the precursor pool is, at least in part, a function of the rate of protein degradation in the tissue. Evaluation of λ_i, therefore, in addition to allowing more accurate determinations of rates of protein synthesis, also provides us with an indicator of the rate of protein degradation in the tissue.

4.2. EFFECTS OF FUNCTIONAL STATE ON RATES OF PROTEIN SYNTHESIS

4.2.1. *Regeneration of a cranial nerve.* Unilateral axotomy of the hypoglossal nerve in adult, female rats resulted in increased rates of protein synthesis in the hypoglossal nucleus ipsilateral to the cut. Protein synthesis rates in the nucleus were not increased until two days after the axotomy and remained elevated until some time between Day 24 and Day 35, more than four days after a functional connection was established with the tongue. The continued high rates of protein synthesis may be associated with reestablishment of the afferent synapses in the nucleus which occurs after the completion of a functional connection [26, 27]. The anabolic state of the hypoglossal cells as assessed in our studies may be biphasic with the initial response reflecting regeneration of the injured hypoglossal axons, and the second phase reflecting repair and reformation of afferent synaptic junctions. It is important to note that our studies on protein synthesis apply to the hypoglossal nucleus as a whole. Which individual cell types are involved cannot be determined from the low resolution autoradiograms (Fig. 5). Morphological studies have shown that microglia, oligodendrocytes, and astrocytes within the hypoglossal nucleus all react after axotomy [28,29]. The temporal correspondence between the changes in protein synthesis and morphologic changes is best for the neurons but the influence of changes in glial cells cannot be ruled out.

4.2.2. *Plasticity in the developing monkey visual system.* In the newborn rhesus monkey chronic monocular occlusion (Day 2 to Day 25) resulted in marked reductions in protein synthesis in the laminae of the dLGN served by the deprived eye [12]. These changes in rates of protein synthesis may underlie the widening of the nondeprived ocular dominance columns in layer IV of striate cortex (Fig. 8C) at the expense of the columns for the deprived eye [30]. We hypothesized that the reorganization of the terminals in layer IV of striate cortex which takes place in response to the deprivation would be reflected in a change in the protein synthesis rate in the cell bodies of origin of these terminals which are located in the six cell laminae of the dLGN. Either the nondeprived cells actively grow into the adjacent columns and make synaptic connections with cells normally responsive to the deprived eye or the deprived cells cease to maintain their growth rates and fail to compete effectively for synaptic connections in striate cortex. Our results suggest that the loss of ocular dominance columns in striate cortex is the consequence of depressed protein synthesis and deficient axonal growth in the developing geniculocortical pathway for the deprived eye.

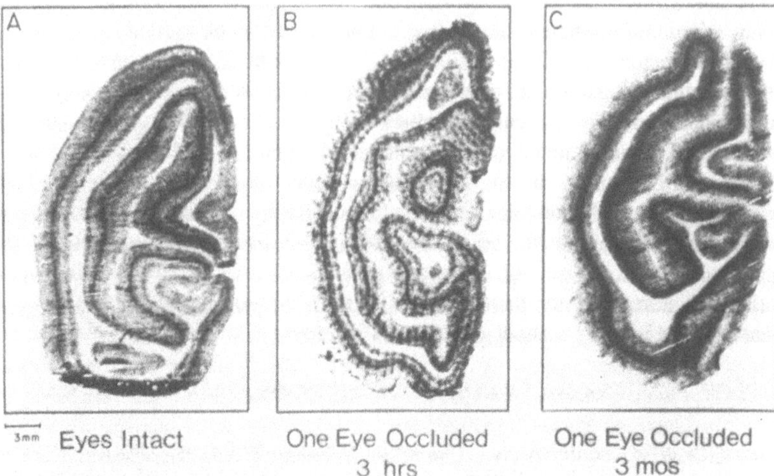

Figure 8. [^{14}C]Deoxyglucose autoradiograms of coronal sections of neonatal rhesus monkey striate cortex [30]. Panels A-C: A. studied on the first day of life with intact binocular vision; B. studied on the first day of life 3 hours after occluding one eye; and C. studied at 3 months with monocular occlusion having been initiated on the first day of life.

Effects of acute v. chronic functional activation. In our studies of both regeneration and plasticity effects on protein synthesis rates occurred in response to chronic rather than acute changes in input. We know that the acute changes in input reached the pertinent structures because other studies have shown that rates of glucose utilization were affected. For example, acute monocular occlusion in monkeys resulted in striking effects on rates of energy metabolism in striate cortex (Fig.8) [30]; the [^{14}C]deoxyglucose autoradiogram of striate cortex of an acutely occluded animal (Fig. 8B) shows a distinct pattern of ocular dominance columns in contrast to the autoradiogram from the animal with binocular vision (Fig.8A). The [^{14}C]leucine autoradiogram of dLGN from an acutely occluded newborn monkey was indistinguishable from that of a monkey with intact binocular vision. Other evidence that rates of protein synthesis are unaffected by acute changes in input comes from our studies of direct electrical stimulation of a peripheral nerve [31]. Unilateral electrical stimulation (10 Hz) of sciatic nerve in rats had no acute affect on either rates of protein synthesis or on the values of λ_i in either the dorsal root ganglia or the spinal cord . However, electrical stimulation of the sciatic nerve under the same conditions as those used in our studies of protein synthesis resulted in 100% increases in glucose utilization in the dorsal horn ipsilateral to the stimulation [32] demonstrating that stimulation did activate the tissue. Our results suggest that chronic changes in functional activity in a pathway are more likely than acute changes to result in effects on rates of protein synthesis in structures of the pathway.

Acknowledgments

I wish to thank J.D. Brown for preparing the photographs.

References

1. Sokoloff L, Smith CB. Basic principles underlying radioisotopic methods for assay of biochemical processes *in vivo*. In: Lambrecht RM, Rescigno A, editors. Tracer kinetics and physiologic modeling. Berlin:Springer, 1983: 202-234.

2. Reivich M, Jehle JW, Sokoloff L, Kety S. Measurement of regional cerebral blood flow with antpyrine-^{14}C in awake cats. J Appl Physiol 1969; 27:296-300.

3. Smith CB, Davidsen L, Deibler G, Patlak C, Pettigrew K Sokoloff L. A method for the determination of local rates of protein synthesis in brain. Trans Am Soc Neurochem 1980;11:94.

4. Smith CB, Crane AM, Kadekaro M, Agranoff B, Sokoloff L. Stimulation of protein synthesis and glucose utilization in the hypoglossal nucleus induced by axotomy. J Neurosci 1984; 4:2489-2496.

5. Ingvar MC, Maeder P, Sokoloff L, Smith CB. Effects of aging on local rates of cerebral protein synthesis. Brain 1985; 108:155-170.

6. Smith C Beebe, Deibler GE, Eng N, Schmidt K, Sokoloff L. Measurement of local cerebral protein synthesis *in vivo*: Influence of recycling of amino acids derived from protein degradation. Proc Natl Acad Sci, USA 1988; 85:9341-9345.

7. Banker G, Cotman CW. Characteristics of different amino acids as protein precursors in mouse brain: advantages of certain carboxyl-labeled amino acids. Arch Biochem Biophys 1971; 142:565-573.

8. Keen RE, Barrio JR, Huang S-C, Hawkins RA, Phelps ME. *In vivo* cerebral protein synthesis rates with leucyl-transfer RNA used as a precursor pool: determination of biochemical parameters to structure tracer kinetic models for positron emission tomography. J Cereb Blood Flow & Metab 1989; 9:429-445.

9. Lajtha A, Latzkovits L, Toth J. Comparison of turnover rates of proteins of the brain, liver and kidney in mouse *in vivo* following long term labeling. Biochim Biophys Acta 1976; 425:511-520.

10. Smith, C Beebe. The measurement of regional rates of cerebral protein synthesis *in vivo*. Neurochem Res 1991; 9:1037-1045.

11. Sun Y, Deibler GE, Sokoloff L, Smith C Beebe. Determination of regional rates of cerebral protein synthesis adjusted for regional differences in recycling of leucine derived from protein degradation into the precursor pool in conscious adult rats. J Neurochem (in press).

12. Kennedy C, Suda S, Smith CB, Miyaoka M, Ito M,. Sokoloff L. Changes in protein synthesis underlying functional plasticity in immature monkey visual system. Proc Natl Acad Sci, USA 1981; 78:3950-3953.

13. Patlak CS, Pettigrew KD. A method to obtain infusion schedules for prescribed blood concentration time courses. J Appl Physiol 1976; 40: 458-463.

14. Yu WA. Dissection of motor nuclei of trigeminal, facial, and hypoglossal nerves from fresh rat brain. In: Shahar A, Velis J D, Vernadakis A, Haber B, editors. A dissection and tissue culture manual of the nervous system. New York: Alan R. Liss, 1989:30-39.

15. Cremer JE, Cunningham VJ, Seville MP. Relationships between extraction and metabolism of glucose, blood flow, and tissue blood volume in regions of rat brain. J Cereb Blood Flow & Metab 1983; 3:291-302.

16. Sokoloff L, Reivich M, Kennedy C, Des Rosiers MH, Patlak CS, Pettigrew KD,et al. The [^{14}C]deoxyglucose method for the measurement of local cerebral glucose utilization: theory, procedure, and normal values in the conscious and anesthetized albino rat. J Neurochem 1977; 28: 897-916.

17. Sun Y, Sokoloff L, Smith C Beebe. Effects of axotomy on protein synthesis and degradation in the rat hypoglossal nucleus. Soc Neurosci Abstracts 1991; 17:48.

18. Smith C Beebe, Sun Y, Deibler GE, Sokoloff L. Effect of loading doses of L-valine on relative contributions of valine derived from protein degradation and plasma to the precursor pool for protein synthesis in rat brain J Neurochem 1991; 57:1540-1547.

19. Bobillier P this volume

20. Hargreaves-Wall KM, Bucink JL, Pardridge WM. Measurement of free intracellular and transfer RNA specific activity and protein synthesis in rat brain *in vivo*. J Cerebr Blood Flow & Metab. 1990; 10:162-169.

21. Eintrei C, Smith CB, Sokoloff L. Effects of thiopental and ketamine anesthesia on local rates of protein synthesis in rat brain. J Cerebr Blood Flow & Metab. 1989; 9, Suppl. 1: S755.

22. Dunlop DS, van Elden W, Lajtha A. A method for measuring brain protein synthesis rates in young and adult rats. J Neurochem 1975; 24:337-344.

23. Roberts S. Protein Synthesis. In: Lajtha A, editor. Handbook of neurochemistry, vol 5A. New York: Plenum Press, 1971:1-48.

24. Kirikae M, Diksic M, Yamamoto YL. The transfer coefficients for L-valine and the rate of incorporation of L-[1-^{14}C]valine into proteins in normal adult rat brain. J. Cerebr. Blood Flow and Metab. 1988; 8: 598-605.

25. Lestage P, Gonon M, Lepetit P, Vitte PA, Debilly, G, Rossatto C, et al. An *in vivo* kinetic model with L-[^{35}S]methionine for the determination of local cerebral rates for methionine incorporation into protein in the rat. J Neurochem 1987; 48:352-363.

26. Sumner BEH. A quantitative analysis of the response of presynaptic boutons to postsynaptic motor neuron axotomy. Exp Neurol 1975; 46:605-615.

27. Rotter A, Birdsall NJM, Burgen ASV, Field PM, Smolen A, Raisman G. Muscarinic receptors in the central nervous system of the rat. IV. A comparison of the effects of axotomy and deafferentation on the binding of [^{3}H]propylbenzilylcholine mustard and associated synaptic changes in the hypoglossal and pontine nuclei. Brain Res Rev 1979; 1:207-224.

28. Watson WE. An autoradiographic study of the incorporation of nucleic-acid precursors by neurones and glia during nerve regeneration. J Physiol (Lond) 1965; 180:741-753.

29. Watson WE. Some quantitative observations upon the responses of neuroglial cells which follow axotomy of adjacent neurones. J Physiol (Lond) 1972; 225:415-435.

30. Des Rosiers MH, Sakurada O, Jehle J, Shinohara M, Kennedy C, Sokoloff L. Functional plasticity in the immature striate cortex of the monkey shown by the [^{14}C]deoxyglucose method. Science 1978; 200:447-449.

31. Smith C Beebe, Sun Y, Kadekaro M, Deibler GE, Sokoloff L. Electrical stimulation of sciatic nerve: effects on protein synthesis. Trans Amer Soc Neurochem 1992; 23:307.

32. Kadekaro M, Crane A, Sokoloff L. Differential effects of electrical stimulation of sciatic nerve on metabolic activity in spinal cord and dorsal root ganglion in the rat. Proc Natl Acad Sci, USA 1985; 82:6010-6013.

METHIONINE METABOLISM IN RAT BRAIN

P. BOBILLIER, E. GRANGE, A. GHARIB, M. LECLERC, N. SARDA
and P. LEPETIT

1. Introduction

A quantitative autoradiographic method has been developed for the in vivo measurement of local cerebral rates for methionine incorporation into proteins in the free-moving rats [1].The accurate measurement of brain protein synthesis, using this model, depends on two main assumptions. The first one is that methionine enters the precursor pool for protein synthesis from the plasma and that unlabeled methionine derived from steady-state protein degradation does not recycle into this pool. The second one is that metabolism of methionine through the transmethylation-transsulfuration pathway is negligible.

The main objectives of the work were -1/ to evaluate the degree of recycling by measuring the ratio of steady-state specific activity of L-^{35}S methionine in the tRNA-bound pool to that of plasma -2/ to analyze, during continuous infusion of L-^{35}S methionine, the end products of both incorporation into proteins and metabolism -3/ to estimate the influence of methionine recycling on protein synthesis rates mesured in different experimental conditions.

In immobilized rats, at normal plasma amino acid level, the relative contribution of methionine derived from protein degradation to the precursor pool for protein synthesis was 26%. The utilization of the metabolic pool of the free tissue methionine for protein synthesis was 90%. In other experiments, about 60% increase of local cerebral rates for methionine incorporation into protein has been measured by quantitative autoradiography during acute immobilization stress or after chronic adrenalectomy. Preliminary results indicate that, under these experimental conditions, the index of the degree of recycling of methionine derived from protein degradation into the precursor pool for protein synthesis does not vary by more than 10%, in whole brain or in individual regions such as frontal cortex, as compared to free-moving conditions.

In conclusion L-^{35}S methionine is the amino acid which allows, to date, to measure, with minimal underestimation, rates of brain protein synthesis in vivo.

2. Materials and Methods

2.1. ANIMALS

All studies were carried out on young adult male Sprague Dawley rats (IFFA CREDO, l'Arbresle, France) weighing 200-220 g. They were allowed food and water ad libitum and

B. M. Mazoyer et al. (eds.), PET Studies on Amino Acid Metabolism and Protein Synthesis, 41–52.

were kept on a 12 h light/dark cycle (lights on at 07:00 a.m.). Under anesthesia, teflon catheters were implanted in the inferior vena cava and abdominal aorta as described previously [2]. During surgery some animals were subjected to bilateral adrenalectomy. The rats were then replaced in their habitual surrounding until experiments were performed 7 days later.

2.2. L-^{35}S METHIONINE ADMINISTRATION

All administrations of L-^{35}S methionine (specific activity > 1000 Ci/mmol, Amersham) were performed between 10:00 and 11:00 a.m. In a first experiment, animals were partially immobilized. They received i. v. infusions of a saline solution of L-^{35}S methionine, at a constant rate of 1 ml (500 µCi)/h. Arterial blood samples were collected at various times during the infusion. Plasma (200 µl) was separated by centrifugation of blood, combined with 2% (vol/vol) thiodiglycol and stored at -60°C until further analysis. After 1 h of infusion, rats were decapited and brains were rapidly removed and frozen in liquid nitrogen, the process taking less than 45 sec.

In a second experiment, normal free-moving rats and adrenalectomized free-moving rats received an i. v. pulse of L-^{35}S methionine (200 µCi). Sequential arterial blood samples were then collected to determine free plasma ^{35}S concentration. Sixty min later, animals were killed by an overdose of pentobarbital and brains were stored at -60°C.

2.3. BIOCHEMICAL ANALYSIS

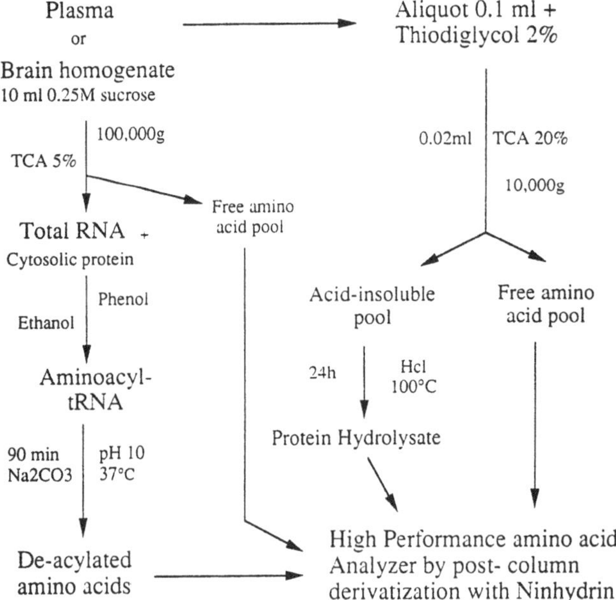

Figure 1. Summary of the methods used in the analysis of plasma and brain amino acids pools.

All steps of the extraction, purification and hydrolysis of aminoacyl-tRNA were performed according to previous procedure [3]. They are summarized in Fig. 1. Amino acids in plasma and brain samples were assayed with conventional ion-exchange chromatography (Kontron Liquimat III or Beckman 7300 analyzer).

The concentrations of S-adenosyl-L-methionine and S-adenosyl-L-homocysteine were determined by reverse-phase HPLC [4]. All peak fractions were collected and measured for ^{35}S by liquid scintillation spectroscopy.

2.4 DETERMINATION OF METHIONINE INCORPORATION INTO PROTEINS

Plasma and brain homogenate were used for determination of the total acid-soluble and acid-insoluble ^{35}S [5], and assayed for protein content [6]. Pellets were washed with cold 10% trichloroacetic acid and then hydrolyzed with 6 mol/l HCl. The protein hydrolysate was analyzed by ion-exchange chromatography as described above.

3. Results

3.1. STEADY-STATE OF METHIONINE IN PLASMA AND BRAIN

In immobilized rats, the specific activity of plasma L-^{35}S methionine was in the same range from 40-60 min indicating that plasma tracer had reached a level within 10% of steady-state during this period (Fig. 2). At normal plasma methionine concentrations, the values of the ratios of specific activities of L-^{35}S methionine in the brain acid-soluble and tRNA-bound pools to that of the arterial plasma were determined after 60 min of a rate constant infusion of L-^{35}S methionine. Both values were similar and averaged 0.77 and 0.74, respectively (Table 1 and Fig. 3). The mean ± SEM value of methionine concentrations in the methionyl-tRNA fractions was estimated to be 40 ± 7 pmol/g of brain (n = 7 rats). This value was probably underestimated since there was no estimation of the recovery.

Table 1. Comparison of the ratios of specific activities of L-^{35}S methionine in acid-soluble tissue pools to plasma in different experimental conditions.

Structure	Free-moving rats (3)	Immobilized[a] rats (6)	Adrenalectomized rats
Whole brain	0.88 ± 0.14	0.77 ± 0.10	1.00 - 1.02
Frontal cortex	0.89 ± 0.13	—	0.94 - 1.02

In immobilized rats the values were determined after 60 min constant infusions of L-^{35}S methionine. [a]Data from [7], with permission. In free-moving and adrenalectomized rats the values were determined 60 min after a pulse injection of L-^{35}S methionine. Values are the means ± SD for the number of animals shown in parentheses.

In free-moving and adrenalectomized rats the ratios of specific activities of L-^{35}S methionine in the tissue acid-soluble pools to that of the plasma have been measured at a

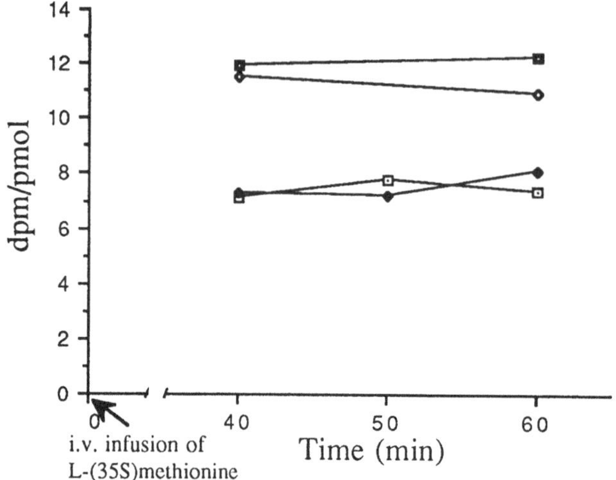

Figure 2. Time course of the specific activity of L-^{35}S methionine in arterial plasma. Rats were immobilized before starting the infusion of tracer at a constant rate of 1 ml (0.5 mCi)/h. Data from [7], with permision.

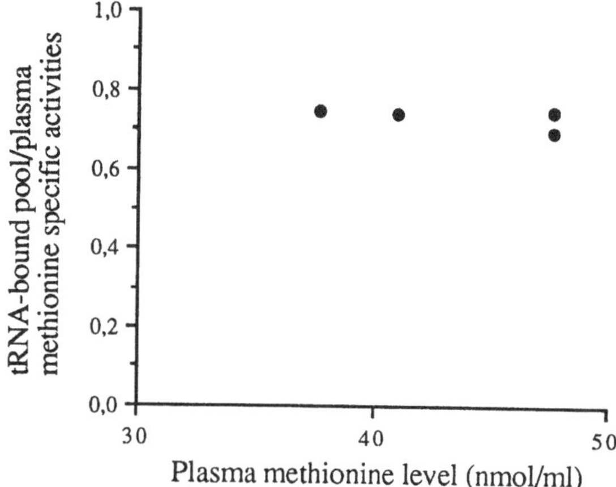

Figure 3. Ratios of the specific activities of L-^{35}S methionine to plasma L-^{35}S methionine determined after 60 min of L-^{35}S methionine. The mean value ± SD is 0.74 ± 0.02. Data from [7], with permission.

time where equilibrium should have been reached during the 60 min of tracer circulation. The values averaged 0.88 and 1, respectively (Table 1). There was no variation of these values between whole brain and frontal cortex.

3.2. ANALYSIS OF PLASMA AND BRAIN ^{35}S RADIOACTIVITY

In plasma, 60% of the total radioactivity was detected in proteins (Fig. 4). Most of the acid-soluble radioactivity was found in 2 peaks corresponding to methionine and cystine. An equal amount of radioactivity was distributed between these 2 compounds and the sum represented 80% of the total acid-soluble radioactivity.

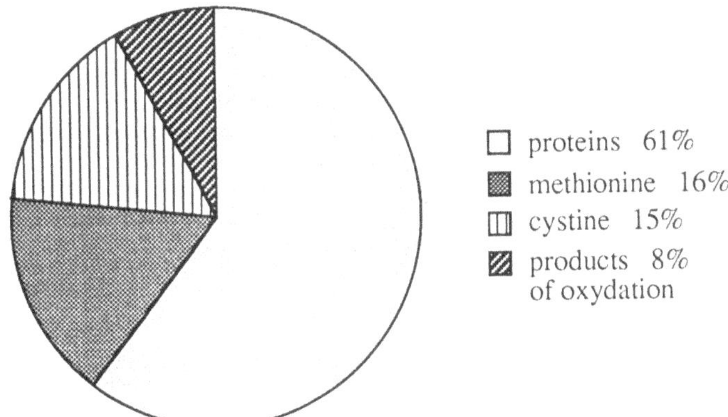

Figure 4. ^{35}Sulfur activity in plasma acid-soluble and acid-insoluble fractions 60 min after continuous infusion of L-^{35}S methionine in immobilized rats. Results are expressed as mean % of the total radioactivity in the fractions obtained from 7 animals.

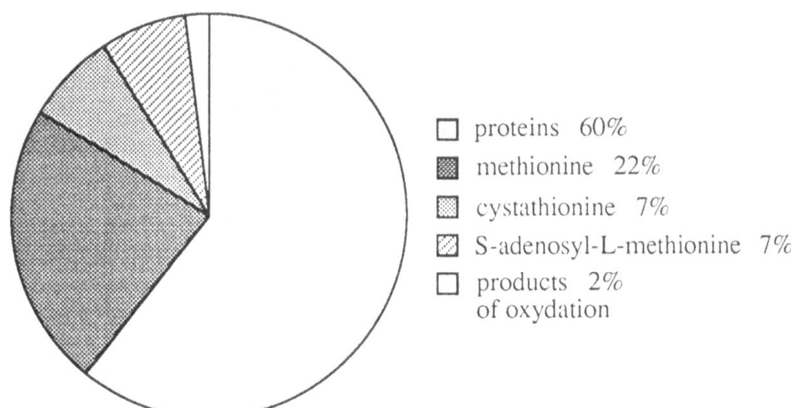

Figure 5. ^{35}Sulfur activity in brain acid-soluble and acid-insoluble fractions 60 min after continuous infusion of L-^{35}S methionine in immobilized rats. Results are expressed as mean % of the total radioactivity in the fractions obtained from 7 animals

In brain 41% of the radioactivity was in the acid-soluble pool and 59% was in proteins (Fig. 5). The majority (56%) of the acid soluble radioactivity was methionine and 36%

went equally to two ^{35}S metabolic products of methionine: S-adenosyl-L-methionine and cystathionine.

3.3. ESTIMATION OF THE RATES OF METHIONINE INCORPORATION INTO BRAIN PROTEINS

The content of methionine in proteins expressed as nmol/mg of protein was 207 for total brain proteins. Protein concentration expressed as mg of protein/g of brain was 89.

In immobilized rats, the turnover rates were then 2.82 nmol/g of tissue/min for total brain proteins or 0.92 %/h, as calculated from the ratio of increase specific activity of methionine in proteins (0.07 dpm/pmol) to the average specific activity of methionine from tRNA pools. The hydrolysis of proteins contained in the acid-insoluble fractions of brain homogenate showed that, at 60 min of infusion, 95% of the radioactivity was recovered as methionine and 4% as cysteine. Therefore brain protein synthesis may be estimated directly from the accumulation of ^{35}S in the acid-insoluble protein pools divided by the average specific activity of methionine in tRNA pools. The values averaged 2.48 ± 0.22 nmol/g of tissue/min (Table 2). They were not different at 40 min (2.58 ± 0.24), indicating that a steady-state has been reached between the plasma and cellular pools [7].

Table 2. Estimation of brain protein synthesis in different experimental conditions.

Structures	Free-moving rats (3)	Immobilized[a] rats (6)	Adrenalectomized rats
Whole brain	1.26 ± 0.13	2.48 ± 0.22	2.00 - 2.26
Frontal cortex	1.14 ± 0.09	—	1.90 - 1.99

Values of tissue rates of methionine incorporation into proteins are the means ± SD and expressed as nmol/g/min. (n), number of animals. In immobilized rats protein synthesis was determined after 60 min continuous infusion of L-^{35}S methionine. [a]Data from [7], with permission. In free-moving and adrenalectomized rats synthesis was determined 60 min after a pulse injection of tracer and were not corrected for dilution of the precursor pool.

In free moving and adrenalectomized rats protein synthesis was estimated from the tissue accumulation of ^{35}S divided by the denominator of the operational equation of the L-^{35}S methionine method [1]. The values, for whole brain, averaged 1.26 ± 0.13 and 2.13 nmol/g of tissue/min, respectively.

4. Discussion

The primary aim of this study was to evaluate the relative contribution of methionine derived from plasma and protein degradation to the precursor pool for protein synthesis in conscious rat brain. For this purpose, we have measured the ratio of the steady-state specific activites of the amino acid in the precursor pool for protein synthesis and in the arterial plasma. It has a theoritical range of value from 0 to 1. When the amount of amino acid entering the precursor pool from plasma is much greater than that derived from protein degradation the ratio approaches its maximum value of 1. An intermediate value corresponds to the situation where equal amounts of amino acids for protein synthesis originate from plasma and metabolic pools. The value of this parameter allows, therefore, to correct the in vivo rates of protein synthesis which are determined from measurement of the integrated specific activity of the plasma labeled amino acid.

Our results indicate that there is a significant dilution of L-^{35}S methionine specific activity in brain methionyl-tRNA and free methionine compared to plasma. In brain of immobilized rats, the dilution by unlabeled methionine derived from protein degradation approximates to 26% and 23% of the brain methionyl-tRNA and free methionine pools, respectively. Although brain cells appear to derive their methionine for protein synthesis preferentially from plasma, i.e. 74%, the determination of rates of cerebral protein synthesis in vivo with L-^{35}S methionine that depends on estimation of precursor pool specific activity in tissue from measurement in plasma must lead to underestimated values.

Assuming that there is no source of L-^{35}S methionine in brain other than plasma the steady-state ratio of specific activity of L-^{35}S methionine in the tRNA-bound pool to that of the arterial plasma has a theoretical upper limit of unity. Experimentally, however, this parameter cannot be estimated without an error component. This probably explains the preliminary conclusion that plasma was the primary source of methionine entering the precursor pool for protein synthesis [8, 9].

Previous studies [3, 10] showed that, in rats with normal plasma amino acid concentration, the relative contributions of leucine and valine from protein degradation to the precursor pool for protein synthesis were 42% and 63% respectively. Our finding together with these results indicate that there is considerable differences in the degree of recycling between these three essential amino acids. The comparison of the rates of amino acid influx into brain from plasma [11] does not bring a likely explanation for these differences. For essential amino acids the rate of influx approximates to the rate of protein synthesis and, at steady-state, proportionnally equal amounts of amino acids would be released from protein degradation. It is, therefore, expected that the tissue/plasma ratio of the three labeled amino acids would be in the same range, which is not the case. The causes of the differences in amino acid recycling should be looked for rather in the intrinsic chemistry, compartmentalization or exchange of each amino acid in the neuron and glial cells.

The amount of methionyl-tRNA recovered from whole brain represents about 1/1000 of the tissue acid-soluble methionine pool. Evaluation of possible differences in the degre of methionine recycling, under different experimental conditions, would then be a difficult task.We demonstrate, in the present study, that the specific activity of brain methionyl-tRNA equals that of the free methionine pool. It is, therefore, possible to use the steady-state tissue/plasma ratios of free L-^{35}S methionine as an index of the degree of recycling of methionine derived from protein degradation. Preliminary mesurements of these ratios, at steady-state, indicate that the rates of protein synthesis measured in free-moving rats (Table 3) are underestimated by 12% whereas those measured in adrenalectomized rats do not need to be corrected. In both experimental conditions, the values of these ratios are similar in whole brain and in frontal cortex. This suggests that the degree of methionine recycling should be relatively stable between brain regions.

The comparison of the tissue/plasma ratios of free L-^{35}S methionine (Table 1) shows that the values do not vary by more than 10% between physiological (free-moving rats) and experimental (immobilized and adrenalectomized rats) conditions, a value which is in the range of experimental errors for the determination of the rate of cerebral protein synthesis. In previous studies, when protein synthesis rates where measured with the assumption of no recycling of methionine derived from protein degradation into the precursor pool for protein synthesis in both control and experimental rats, the variations of brain protein synthesis reported then [12-18] are likely to be only slightly underestimated or overestimated.

The rate synthesis of S-adenosine-L-methionine, calculated using the brain concentrations of S-adenosine-L-methionine measured in immobilized rats (mean ± SEM : 28 ± 1 mmol/g. n=8) and the rate constant (0.01 min^{-1}) of S-adenosine-L-methionine synthesis [19], would be 280 pmol/g of tissue/min. The relative rates of methionine flux into the transmethylation pathway may thus represent 10% of the rate of methionine incorporation into brain proteins.

The results of the evaluation of the rate of formation of sulfur labeled metabolites 1 h after an infusion of L-^{35}S methionine confirm that, in brain as a whole, the flux of methionine into the transmethylation-transsulfuration pathways is quantitatively much less important than the flux of methionine into proteins. Given the low cystathionase activity in brain [20], a low amount of ^{35}S cysteine is expected to be formed from ^{35}S cystathionine, which is the case since ^{35}S cysteine could be detected only after its accumulation in proteins. Thus, a kinetic three-compartment model is sufficient to describe the behavior of methionine and L-^{35}S methionine in brain.

In contrast to brain, the peripheral transmethylation-transsulfuration pathway is much more active. This is demonstrated by the fact that a substantial amount of ^{35}S is recovered as cysteine and cysteic acid after hydrolysis of plasma proteins (20% and 2% of the total radioactivity, respectively). The failure to detect (^{35}S)cysteine in brain, in spite of equal amounts of circulating L-^{35}S methionine and ^{35}S cysteine, confirms that cysteine crosses

Table 3. Effect of immobilization stress on local rates of protein synthesis in the rat.

Brain region	Methionine incorporation into proteins (nmol/g/min)		
	Free-moving (5)	Immobilized (5)	percent effect
Piriform cortex	1.81 ± 0.09	2.70 ± 0.10	+49
Parietal cortex	0.89 ± 0.03	1.53 ± 0.09	+72
Lateral septum	0.82 ± 0.04	1.29 ± 0.04	+57
Hippocampus CA3 layer	2.22 ± 0.10	3.43 ± 0.05	+55
Medial habenula	2.84 ± 0.08	4.31 ± 0.14	+52
Anteroventral thalamic nucleus	1.15 ± 0.04	2.05 ± 0.08	+78
Magnocellular paraventricular hypothalamic nucleus	3.55 ± 0.14	6.04 ± 0.14	+70
Substantia nigra compacta	1.66 ± 0.07	2.78 ± 0.05	+67
Dorsal raphe nucleus	1.52 ± 0.04	2.81 ± 0.04	+85
Inferior olive	1.69 ± 0.11	2.70 ± 0.09	+60
Corpus callosum	0.33 ± 0.02	0.56 ± 0.05	+70

Values are the means ± SEM for the number of animals indicated in parentheses. They were determined by quantitative autoradiography, using the L-^{35}S methionine model. Data from [23]. Values were corrected by using the steady-state tissue/plasma ratio of L-^{35}S methionine: 0.88 and 0.77 for free-moving and immobilized rats, respectively.

poorly the blood-brain barrier. It is, therefore, necessary to substract plasma labeled cysteine from the ^{35}S plasma radioactivity to improve the arterial input function of the model as previously stated [1].

The good correspondance of the present results with those which were predicted on the basis of previous investigations [1] validates the utilization of the operational equation of the L-^{35}S methionine model. Moreover, in free-moving rats, the rate of methionine incorporation into the frontal cortex measured by biochemistry (Table 2) is similar to that usually determined by quantitative autoradiography both after removing the free ^{35}S by formalin fixation of tissue sections or by computing the correction (about 1 nmol/g/min). In free-moving rats the actual rate of methionine incorporation into brain protein may be calculated with the values of the tissue/plasma ratio (0.88) or the tRNA/plasma ratio (0.74) of the L-^{35}S methionine measured in restrained rats. The estimated values, i.e. 1.43-1.70 nmol/g of tissue/min (0.47-0.57 %/h) are underestimated and overestimated, respectively. They probably delimit the range within which the rate of methionine incorporation lies. For comparison, rates of brain protein synthesis in the rat, reported in the literature, vary from 0.3 to 0.8 %/h [21]. The rate of methionine influx into brain of 1.56 nmol/g/min [22] is also in good agreement with our data.

Studies with the quantitative autoradiographic L-^{35}S methionine method have shown that, in the rat, acute immobilization stress and chronic bilateral adrenalectomy are associated with increases of regional rates of methionine incorporation into brain proteins (Tables 3 and 4). The magnitude of the changes seems to be regionally selective, varying from +52% (medial habenula) to +85% (dorsal raphe nucleus) in restrained rats and +32% (paraventricular hypothalamic nucleus) to +72% (corpus callosum) in adrenalectomized rats. All the main areas of brain examined are affected. Indeed, the present biochemical

Table 4. Effect of chronic bilateral adrenalectomy on local rates of protein synthesis in the rat.

	Methionine incorporation into proteins (nmol/g/min)		
Brain region	Free-moving (6)	Adrenalectomized (7)	percent effect
Entorhinal cortex	1.15 ± 0.05	1.71 ± 0.05	+49
Parietal cortex	1.05 ± 0.04	1.61 ± 0.09	+53
Lateral septum	1.01 ± 0.06	1.56 ± 0.08	+54
Subiculum	1.09 ± 0.04	1.53 ± 0.07	+40
Anteroventral thalamic nucleus	1.26 ± 0.06	1.88 ± 0.07	+49
Magnocellular paraventricular hypothalamic nucleus	3.66 ± 0.06	4.83 ± 0.19	+32
Dorsal raphe nuleus	1.59 ± 0.04	2.16 ± 0.08	+36
Sensory trigeminal nucleus	1.36 ± 0.03	1.93 ± 0.05	+42
Facial nucleus	1.38 ± 0.04	2.14 ± 0.10	+55
Corpus callosum	0.43 ± 0.04	0.74 ± 0.06	+72

Values are the means ± SEM for the number of animals indicated in parentheses. They were determined by quantitative autoradiography, using the L-^{35}S methionine model. Values were corrected by using the steady-state tissue/plasma ratio of L-^{35}S methionine: 0.88 and 1 for free-moving and adrenalectomized rats, respectively.

work confirms that the two experimental conditions both induce an incease of protein synthesis of similar mean amplitude in the brain as a whole.

In conclusion, our results demonstrate that there is a significant recycling of unlabeled methionine derived from protein degradation in restrained rat brains. Therefore the cerebral rates of protein synthesis measured in vivo with L-^{35}S methionine would lead to underestimated values. Nevertheless, preliminay studies under physiological or pathological conditions suggest that the variations in methionine recycling are of small amplitude and would not significantly affect the experimental effect measured by quantitative autoradiography. To improve the accurancy of the L-^{35}S methionine method it is possible to estimate the actual rates of protein synthesis by using an index of the degree of methionine recycling. Indeed, L-^{35}S methionine presents radiobiochemical properties required to measure in vivo local rates of cerebral protein synthesis, with minimal underestimation, for three reasons: -1/ The rate of methionine incorporation into proteins is ten fold higher than the rate of methionine utilization through the transmethylation pathway. -2/ Only two ^{35}S metabolites are retained in the brain, representing about 8% of the total radioactivity, 90 min after pulse labeling of the tracer. -3/ At normal plasma amino acid

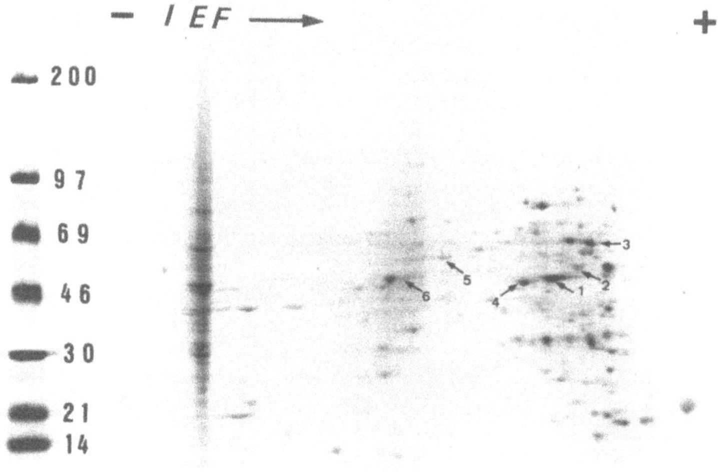

Figure 6. Two dimensional gel autoradiograph of L-^{35}S methionine-labelled proteins from the cerebellum of conscious freely moving rat injected intracisternaly with 100 µCi of L-^{35}S methionine, 90 min before sacrifice. Proteins were separated on immobilized pH gradient gels (pH 3 - 10.5) for the first dimension and SDS-PAGE pore gradient gels (8% - 18%) for the second dimension. The following proteins have been identified either by comigration or immunoblotting: 1/ actin ; 2/ gamma enolase ; 3/ ß tubulin ; 4/ creatine phosphokinase ; 5/ alpha enolase ; 6/ glutamine synthetase. The high optical density of actine, creatine phosphokinase and tubulin reflects their high turnover.

concentrations, methionine is the amino acid which, to date, presents the lowest degree of recycling into the precursor pool for protein synthesis in conscious rats. Finally, the high specific activity of L-^{35}S methionine allows the analysis of individual proteins (Fig. 6). This latter approach should be particularly useful to study the differential regulations of proteins in brain regions of interest previously visualized on macroautoradiographs.

5. Acknowledgements

The authors gratefully thank N.Gay, J. Allard and J. Guillaud for helpful technique assistance. This work was supported by CNRS, DRET (grant 90123) and INSERM (grant 896002).

6. References

1. Lestage P, Gonon M, Lepetit P et al. An in vivo kinetic model with L-(^{35}S)methionine for the determination of local cerebral rates for methionine incorporation into protein in the rat. J Neurochem 1987; 48: 352-363.
2. Lestage P, Vitte PA, Rolinat J P, Minot R, Broussolle E, Bobillier P. A chronic arterial and veinous cannulation method for freely moving rats. J Neuroscience Methods 1985; 13: 213-222.
3. Smith CB, Deibler GE, Eng N, Schmidt K, Sokoloff L. Measurement of local cerebral protein synthesis in vivo: influence of recycling of amino acids derived from protein degradation. Proc Natl Acad Sci 1988; 9341: 9345.
4. Gharib A, Sarda N, Chabannes B, Cronenberger L, Pacheco H. The regional concentrations of S-adenosyl-L-methionine, S-adenosyl-L-homocysteine and adenosine in rat brain. J Neurochem 1982; 38: 810-815.
5. DunlopDS, Van Elden W, Lajtha A. A method for measuring brain protein synthesis rate in young and adult rats. J Neurochem 1975; 24: 337-344.
6. Bradford MM. A rapid and sensitive method for quantitation of microgram quantities of protein utilizing the principle of protein dye-binding. Analyt Biochem 1976; 72: 248-254.
7. Grange E, Gharib A, Lepetit P, Guillaud J, Sarda N, Bobillier P. Brain protein synthesis in the conscious rat using L-^{35}S methionine: relationship of methionine specific activity between plasma and precursor compartment and evaluation of methionine metabolic pathways. J Neurochem 1992; in press.
8. Grange E, Gharib A, Lepetit P, Guillaud J, Sarda N, Bobillier P. Biosynthèse des protéines cérébrales chez le rat vigile: équilibre de la L-méthionine entre le plasma et le compartiment précurseur direct. C R Acad Sci (Paris) 1991; 312: 255-260.
9. Grange E, Gharib A, Lepetit P, Sarda N, Bobillier P. Brain protein synthesis in conscious rats: equilibration of L-(35S)methionine specific activity between plasma and methionyl-tRNA. J Cereb Blood Flow Metab 1991; 11 suppl2: S356.
10. Smith CB, Sun Y, Deibler G E, Sokoloff L. Effect of flooding with valine on recycling of valine from protein breakdown into precursor pool for protein synthesis. J Cereb Blood Flow Metab 1991; 11 (suppl 2): S 583.
11. Smith QR, Momma S, Aoyagi M, Rapoport S. I. Kinetics of neutral amino acid transport across the blood-brain barrier.J Neurochem 1987; 49: 1651-1658.

12. Lepetit P, Lestage P, Gauquelin G et al. Differential effects of chronic dehydratation on protein synthesis in neurons of the rat hypothalamus. Neurosci Lett 1985; 62: 13-18.
13. Rossatto C, Lestage P, Dalery J, Bobillier P. Effets d'une administration aigue ou chronique de desmethylimipramine sur les taux locaux d'incorporation de méthionine dans les protéines cérébrales chez le rat libre de se mouvoir. C R Acad Sci (Paris) 1986; 303: 761-764.
14. Lepetit P, Lestage P, Jouvet M, Bobillier P. Localisation of cerebral protein synthesis alterations in response to water deprivation in rats. Neuroendocrinology 1988; 48: 271-279.
15. Lepetit P, Touret M, Grange E, Gay N, Bobillier P. Decreased protein synthesis in hypothalamic nuclei following L-5-hydroxytryptophan in intact and p-chlorophenylalanine pretreated rats. Neurosci Lett 1991; 122: 218-220.
16. Lepetit P, Touret M, Grange E, Gay N, Bobillier P. Inhibition of methionine incorporation into brain proteins after the systemic administration of p-chlorophenylalanine and L-5-hydroxytryptophan. Europ J Pharmac 1991; 209: 207-212.
17. Lepetit P, Grange E, Gay N, Bobillier P. Progressive increases of protein synthesis in the circumventricular organs of the rats following chronic dehydratation. Prog Brain Res 1992; 91: 435-438.
18. Lepetit P, Grange E, Gay N, Bobillier P. Comparaison of the effects of chronic water deprivation and hypertonic saline ingestion on cerebral protein synthesis in rats. Brain Res 1992; in press.
19. Lakher M, Wurtman RJ, Blusztajn J et al. Brain phosphatidylcholine pools as possible sources of free choline for acetylcholine synthesis. in Borchardt RT, Creveling CR and Ueland PM editors. Biological Methylation and Drug Design. Clifton New Jersey: Humana press, 1986: 101-110.
20. Finkelstein JD. Methionine metabolism in mammals. J Nutr Biochem 1990; 1: 228-237.
21. Dunlop DS. Measuring protein synthesis and degradation rates in CNS tissue. in: Marks N and Rodnight R, editors. Research Methods in Neurochemistry. New York: Plenum Press, 1978: Vol. 4: 91-141.
22. Baños G, Daniel PM, Moorhouse SR, Pratt OE. The influx of amino acids into the brain of the rat in vivo: the essential compared with some non essential amino acids. Proc Roy Soc Lond [Biol] 1973; 183: 59-70.
23. Lestage P. Mesure ex vivo de la synthèse protéique cérébrale par radioautographie quantitative de la L-^{35}S méthionine: étude chez le rat libre de se mouvoir ou immobilisé. Thèse de doctorat de Neurosciences, Lyon 1987.

PROTEIN SYNTHESIS STUDIES IN RATS WITH METHIONINE

A.M. PLANAS, C. PRENANT, B.M. MAZOYER, S. CHADAN, D. COMAR,
L. DiGIAMBERARDINO

ABSTRACT. 1) Total brain radioactivity was found to be regionally correlated (r = 0.97) with radioactivity incorporated into proteins following a bolus injection of [^{14}C-methyl]methionine. This suggests that regional differences in total label accumulation correspond to differences in the incorporation of label into proteins. 2) Under steady-state conditions (i.e., during continuous infusion of [^{14}C-methyl]methionine), regional brain specific activity (SA) was found to be lower than the plasma SA. Brain SA was diluted by an endogenous source of free methionine likely to be from protein breakdown. Assuming that all endogenous brain methionine can contribute to protein synthesis, measuring labelled methionine incorporation (without accounting for tissue SA), would lead to underestimated rates. 3) In contrast to these results, similar studies carried out on the rat heart have shown a ratio of heart to plasma SA equivalent to unity. According to this, methionine recycling did not became apparent in the heart during the experimental time and under our experimental conditions.

1. Introduction

Accumulating evidence questions the validity of using [^{11}C-methyl]methionine incorporation into the tissue to estimate the rate of protein synthesis by PET. Up to present there is no satisfactory model for quantifying protein synthesis rates in vivo using labelled methionine, although attempts with three or more compartmental models have been made (1-3). Two basic processes may contribute to the difference between the rate of total label methionine accumulation in the tissue and the rate of protein synthesis. One is the metabolism of [^{11}C-methyl]methionine, occuring in both plasma and tissue (4-11), and the other is the endogenous recycling of amino acids in the tissue (12-13).

Regarding the metabolism of labelled methionine, even though this point has been extensively discussed in the literature in favour of using other labelled amino acids (4, 5, 8), [^{11}C-methyl]methionine has been widely used in PET studies (14-18) and is now currently utilized in the field of oncology (1, 19-24). We have carried out a study using [^{14}C-methyl]methionine with the aim to finding out whether a relationship exists between total label accumulation in brain tissue and incorporation of label into brain proteins. In particular, we were interested in looking at the relationship between regional differences in label accumulation and the corresponding regional incorporation into proteins.

We have also considered a second problem related to the estimation of brain protein synthesis rates: amino acid recycling within the brain tissue. It has been suggested in the literature (8, 12, 13) that protein breakdown could be an internal source of leucine for brain

B. M. Mazoyer et al. (eds.), PET Studies on Amino Acid Metabolism and Protein Synthesis, 53–68.

protein synthesis. We have carried out a study to find out evidences indicating whether or not methionine recycling takes place in eight different regions of the rat brain. In addition, we have carried out equivalent studies on the heart tissue. In the brain and for all the regions considered, we have found a ratio of brain to plasma lower than unity. In contrast, the heart to plasma SA ratio was equal to unity. These results indicate that, in the brain tissue but not in the heart, there is an endogenous source of brain free methionine which does not readily exchange with plasma during the experimental time and may contribute to protein synthesis. Assuming that all brain free methionine can be incorporated into proteins, then regional brain SA must be taken into account to avoid underestimating the rates of methionine incorporation into brain proteins.

2. Materials and Methods

2.1 MATERIALS

[^{14}C-methyl]methionine (50 mCi/mmol) was purshased from New England Nuclear (DuPont de Nemours, France). Otherwise stated, all chemicals used were from Sigma (Sigma chimie S.a.r.l., France).

2.2 ANIMALS

Male Sprague-Dawley rats weighing around 200-250 g and having free access to food and tap water had a vein and an artery of the tail cannulated under halothane anesthesia (3%) and were allowed to recover for about 2 h before the beginning of the experiment. During this time rats were kept under mild restrain. ^{14}C-methionine was i.v. given to awake rats.

2.3 BRAIN AUTORADIOGRAPHY

We have carried out an autoradiographic study using ^{14}C-methionine. Basically, at 45 min following an i.v. bolus injection of [^{14}C-methyl]methionine (200 μCi/ml/kg), rats (n = 8) were decapitated, the brain removed from the skull and frozen into isopentane cooled down to -70 ^{o}C. Cryostat coronal sections (20μm) were made through the brain in sets of three consecutive sections. The first was directly exposed to the autoradiographic film (Hyperfilm ß-max, Amersham) to determine total radiolabel accumulation. The second section of each set had a treatment with acid (trichloroacetic acid washes at either 5 or 10% concentration for 30 min and overnight, respectively; followed by fixing and washing the sections according to ref. 25) allowing to washing out of the brain sections those labelled products not incorporated into proteins. This treated slice, second of each set, was exposed next to the first slice on the same film and for the same length of time. Calibrated ^{14}C-standards (Amersham) were exposed to the film together with the cryostat sections to allow for quantitation of tissue label in terms of nCi/g of tissue. Finally, the third section of each set was stained with thionine for microscopic examination. The amount of label associated to different regions of interest (n = 19) for each of the two sections per set was determined using an image analyser system (IMSTAR, France).

2.4 INFUSION OF [^{14}C-METHYL]METHIONINE

2.4.1 *Working hypothesis.* As a starting hypothesis we considered that if plasma methionine was the only source of brain methionine then brain - plasma SA should be equivalent, under steady-state conditions for both endogenous and labelled methionine. In contrast, should there be an internal source of unlabelled methionine within the brain, the corresponding SA would be diluted by the endogenous methionine contribution resulting in a brain SA lower than the plasma SA (12-13). Experimentally, to test this hypothesis a constant [^{14}C-methyl]methionine concentration should be kept in both plasma and brain to fulfil the steady-state requirement, under the assumption that the endogenous content of methionine will remain constant in plama and brain during the experimental time.

2.4.2. *Estimation rates of methionine incorporation into proteins.* Considering plasma methionine alone as the source of brain methionine, the apparent rate of methionine incorporation into proteins, R_a, can be calculated as follows:

$$Ra = \frac{P^*(T)}{\int_0^T [(Mp^* / Mp)\, t]\, dt}$$

where, P* (nCi/g of tissue) is the amount of label incorporated into proteins at a time T (min), and the denominator is the time integral of the plasma SA (Mp* is the plasma label methionine concentration, nCi/ml; and Mp is the plasma endogenous methionine content, nmol/ml).

Assuming that the relationship between tissue free methionine SA and methionyl-tRNA SA is closed to unity under our experimental conditions, the real rate of methionine incorporation into proteins can be calculated according to Hargreaves-Wall et al. (13):

$$Ri = \frac{P^* / T}{Mt^* / Mt}$$

where, Mt* is the tissue label methionine concentration (nCi/g of tissue), and Mt is the endogenous tissue methionine (nmol/g of tissue).

Both rates, R_a and R_i, were estimated for each particular brain region.

2.4.3. *Infusion protocol.* Labelled methionine is rapidly eliminated from plasma after an i.v. bolus injection. Elimination can be compensated by starting an infusion of [^{14}C-methyl]methionine simultaneously to the bolus injection. This should be a programmed infusion administering concentrations of [^{14}C-methyl]methionine variable with time to maintain its level constant in the plasma. We have worked out the entry function according to the technique of Patlak and Pettigrew (26) which allowed to obtain a level of [^{14}C-methyl]methionine in the plasma remaining constant from about 5 to 60 min with continuous programmed i.v. infusion (see ref. 27). The infusion protocol was carried out on awake rats and lasted for either 15 min (n = 3), 30 min (n = 4) or 60 min (n = 3). The total amount of label that rats received varied from 100 μCi after 15 min infusion to 200 μCi after 1 h. During the experiment, multiple timed arterial blood samples were withdrawn. At the end of the infusion period, rats were decapitated, brains were dissected out into different regions, which were weighed and frozen into liquid nitrogen until further analyses. In some of these infused rats and in other rats receiving a similar [^{14}C-methyl]methionine infusion protocol, the heart was removed and a portion was weighed and frozen into liquid nitrogen. For these experiments the infusions lasted for either 15 min (n = 3); 30 min (n = 2) or 60 min (n = 3).

2.4.4 *Processing plasma and tissue samples.* After [^{14}C-methyl]methionine infusion, plasma and tissues (either brain regions or heart fractions) were processed as follows: proteins were precipitated out in acid and ^{14}C-methionine was separated from labelled metabolites using HPLC (27). Basically, this biochemical analysis allowed to fractionate total radioactivity in the sample (plasma or tissue) into three different fractions: free unmetabolized [^{14}C-methyl]methionine, labelled metabolites and labelled proteins. In addition, the endogenous content of methionine was determined in plasma and in tissues. Essentially, the HPLC analysis was carried out using a reverse phase column and fluorimetric detection after precolumn derivatisation using o-phthaldialdehyde. 0.5 min fractions were collected out of the column and the radioactivity associated to peaks was determined by liquid scintillation. The tissue protein content was measured by a conventional spectrophotometric method (28). Biochemical analyses were carried out on the following brain regions after manual dissection: cerebellum (CE), medulla (ME), striatum (STR), hippocampus (HIP), hypothalamus (HY), thalamus (TH), cortex (CTX) and white matter (WM).

2.5 ELECTROPHORESIS

The postcingulate cortex was dissected out at different times following administration of [^{14}C-methyl]methionine kept frozen for later electrophoretic analysis. The tissue was homogenized in Laëmmli buffer (tris-HCl 60mM pH 6.8, EDTA 3mM, SDS 20% w/v, glycerol 10% w/v, ß-mercaptoethanol 5% w/v; 1mg tissue / 100µl buffer), heated at 100 °C for 4 min, and aliquots (1.8 mg of protein) were run on one dimensional (1D) SDS-PAGE according to the method of Laëmmli (29). Electrophoresis was carried out on a linear gradient (5-15%) polyacrylamide gel (15 cm height x 1.5 mm thickness) with a stacking gel (20% acrylamide, 0.125 M tris pH 6.8, 0.1 % SDS) at 10mA for 15 h. The electrophoresis buffer was: 0.05M tris-HCl, 0.38 M glycine, 0.1 % SDS pH 8.3. The SDS gel was stained with coomassie blue, vacuum dried and exposed to an autoradiography film (X-OMAT, Kodak) for 5 months.

2.6 STATISTICAL ANALYSES

Linear regression, analyses of variance (ANOVA) and t-tests were carried out. Two-way ANOVA with a repeated measures design was applied to test for significant differences between times and between brain regions after the infusion. Particular regional differences were tested using pooled pairwaise Student t-test with the Bonferroni probability. Analyses were carried out using the BMDP statistical package (BMDP Statistical Software, Inc., Los Angeles,CA, U.S.A.). Data are presented as mean ± SD.

3. Results and Discussion

3.1 BRAIN AUTORADIOGRAPHY

Images from the autoradiography films were digitalized using an image analizer system. Results are presented in fig. 1 for coronal sections of the brain at the level of cortex and striatum to illustrate the good similarity between the image on the left, showing total label, and the image on the right, showing the label incorporated into proteins only (since the rest of label was washed out of the tissue). The label concentration (nCi/g of tissue) for 19 different brain regions was determined using the curve of grey level by label concentration obtained from the calibrated standards. Results are shown in fig. 2 where brain regions are ranked according to increasing values of total label accumulation. The % of total label incorporated into proteins at 45 min varied from 40 to 75 %. This is comparable to results obtained after a bolus injection of ^{14}C-methionine, dissection of different brain regions, homogenisation and acid precipitation of proteins (data not shown). The regional pattern of label incorporation into proteins is very similar to that of total label accumulation. Values of free label, which were obtained by numerical substraction of data measured from each of the two sections, did not vary significantly from region to region.

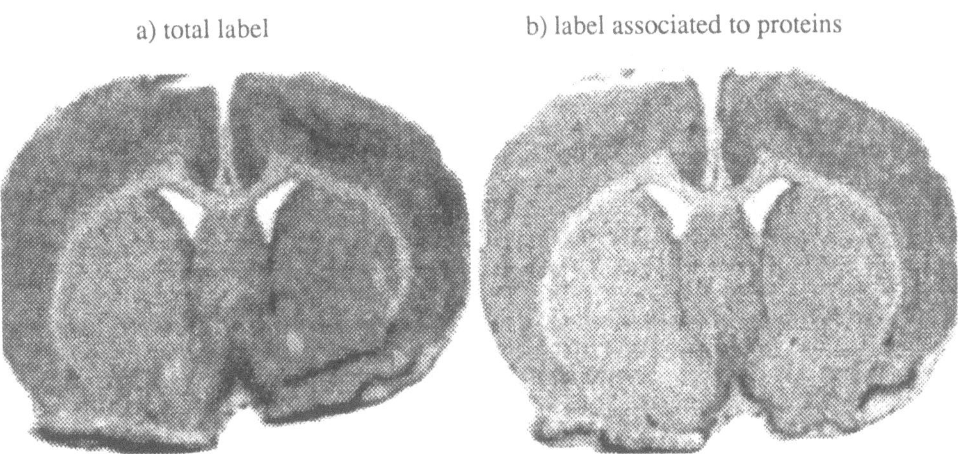

FIGURE 1. Autoradiography images at the level of the striatum. The rat was sacrified 45 min after the i.v. injection of a bolus of ^{14}C-methionine. a) The brain section was directly exposed to the autoradiographic film so the image accounts for total tissue radioactivity. b) The brain section was treated with acid according to the technique of Yoshimine et al. (25) to wash out label not incorporated into proteins. Films were exposed for 14 days.

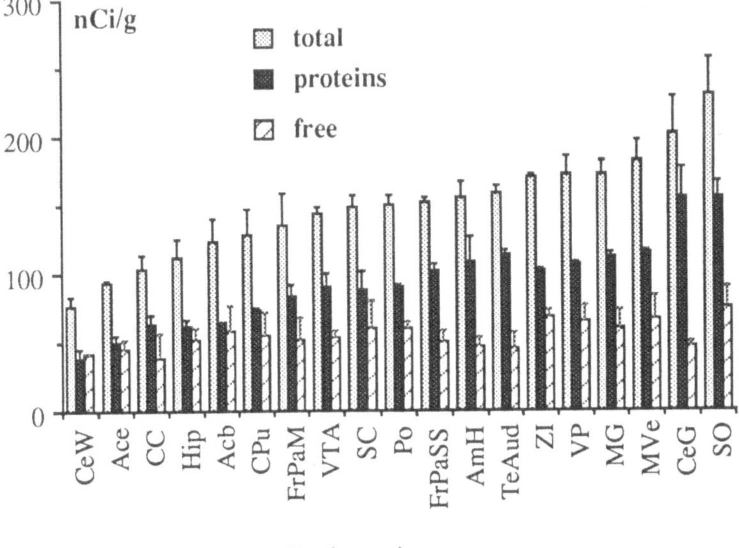

FIGURE 2. Mean (n = 8) label distribution (nCi/g of tissue) in different rat brain regions. Regions are ranked according to increasing values of total radioactivity. Radioactivity was determined using quantitative autoradiography (IMSTAR, Paris, France). Measurements of total label and of label associated to proteins were independently carried out on consecutive brain sections taken from the same rat and

same autoradiographic film (Hyperfilm ß-max, Amersham) for the same period of time (14 days). For each region, one of these sections was exposed to the film directly, whilst the other was washed with acid to precipitate proteins (25). Free label is the result of the numerical substraction of the two measurements: total label - label in proteins. The brain regions studied were: (CeW) cerebellar white matter; (Ace) anterior commissure; (CC) corpus callosum; (Hip) dentate gyrus of the hippocampus; (Acb) accumbens nucleus; (CPu) caudate putamen; (FrPaM) frontoparietal cortex, motor area; (VTA) ventral tegmental area; (SC) superior colliculus; (Po) posterior thalamic nucleus; (FrPaSS) frontoparietal cortex, somatosensory area; (AmH) hippocampal Ammon's horn, field CA1; (TeAud) temporal cortex, auditory area; (ZI) zona incerta; (VP) ventro posterior thalamic nucleus; (MG) medial geniculate nucleus; (MVe) medial vestibular nucleus; (CeG) cerebellar grey matter; (SO) superior olive.

Studying the relationship between total label accumulation and label incorporated into proteins we found a very good correlation ($r = 0.97$) indicating that the main source of region-to-region differences in label accumulation is the regional incorporation of label into proteins. Results from this study suggests that, in normal conditions, autoradiographic or PET images (obtained from 45 min or later after i.v. injection of labelled methionine) would reflect the incorporation of label into tissue proteins.

3.2 BRAIN ELECTROPHORESIS

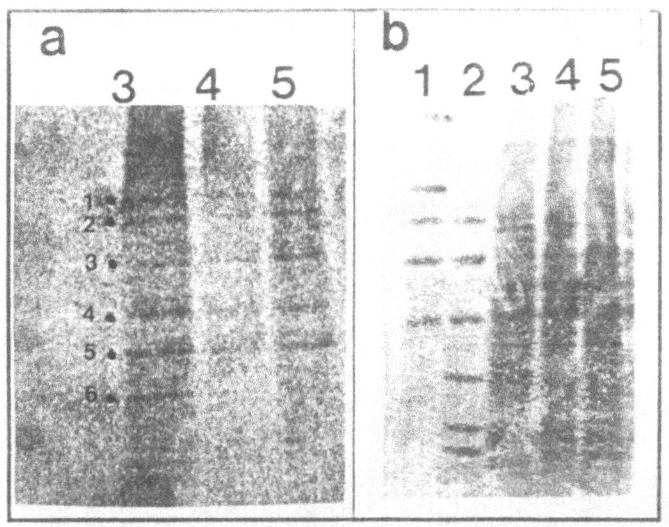

FIGURE 3. Autoradiography (a) of coomassie blue stained 1D polyacrilamide gel (b). Wells are indicated on the top of each figure. Wells 3, 4 and 5 in (a) and (b) correspond to total protein of postcingulate cortex samples. Rats were given ^{14}C-methionine either as a 60 min infusion in well 3; or as a bolus injection sacrifying the rat 15 min later in well 4, or 24 h later in well 5. (a) The vertical points on the left of the figure mark the position of the radioactive proteins: 1 = 100kD, 2 = 95 kD, 3 = 68 kD, 4 = 56 kD, 5 = 43 kD, 6 = 36 kD. Autoradiograms were obtained after exposing the film (X-OMAT, Kodak, France) for 5 months at -80°C with a lightenning plus screen (Dupont Cronex, France). b) The gel was run according to the method in ref. 29. Bands in wells 1 and 2 correspond to molecular weight standards (BioRad, France) with the following molecular weights, from the top to the bottom: 200, 116, 97, 66, 43, 31, 21 and 14 kD.

Label is incorporated into the different brain proteins which are synthesized during the experimental time. We were interested in looking at the contribution of particular labelled proteins to the whole label. We carried out monodimensional electrophoresis on polyacrylamide gel followed by autoradiography. This technique was applied, not with the aim of identifying the different labelled proteins, but to find out whether most of the label was distributed among many proteins, or whether it was restricted in only a few. As shown in figure 3, we have found a few major labelled bands of molecular weights of 100, 95, 68, 56, 43 and 36 kD. Labelled proteins with molecular weights of 68, 56 and 43 kD may correspond to neurofilaments (NF-L), tubulin and actin, respectively. These results showed mainly the incorporation of labelled methionine into the structural proteins of the brain, which are quantitatively the most abundant. Qualitatively, the pattern of labelled bands did not appear to be very different from 15 min (fig 3a, well 4) to 24 h (fig 3a, well 5). For a more extensive study on the time course and identity of specific labelled proteins using ^{14}C-valine we refer you to Shahbazian et al. (30).

3.3 BRAIN AND PLASMA [^{14}C-METHYL]METHIONINE STEADY-STATE

3.3.1 *Endogenous measurements*

The mean endogenous content of methionine in the plasma was 41 ± 6 nmol/ml for 10 rats. The corresponding average of the mean regional (n = 8 regions) brain methionine was 26 ± 8 nmol/g. Statistical analyses showed significant differences in the methionine content for the brain regions studied ($F = 5.2$; $df = 7, 63$; $p < 0.001$). The regions showing significant differences were white matter ($p < 0.01$), medulla ($p < 0.005$) and cerebellum ($p < 0.001$) all showing methionine levels smaller than the striatum.

The corresponding brain protein content was measured in the different regions studied and was found to vary from 75 ± 15 mg/g in the hypothalamus to 113 ± 9 mg/g in the cortex. No regional correlation was found between the content of methionine and that of protein per g of tissue.

3.3.2 *Plasma radioactivity*

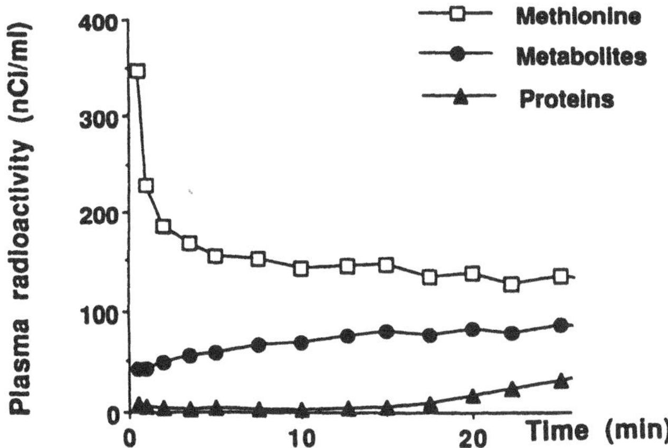

FIGURE 4. Typical time course of plasma radioactivity during 30 min infusion of ^{14}C-methionine. Values correspond to unmetabolized free ^{14}C-methionine, ^{14}C-metabolites or ^{14}C-labelled proteins, expressed as nCi/ml of plasma. Programmed infusions were carried out using the method of Patlak and Pettigrew (26).

The concentration of labelled methionine was maintained constant in the plasma from about 5 to 60 min after the beginning of the infusion. Fig 4 shows the time course of plasma radioactivity including ^{14}C-methionine, labelled metabolites and labelled proteins from a representative 30 min infusion experiment. Labelled metabolites increased slightly with time and a labelled protein fraction was detected increasing linearly with time, from about 15 min after the beginning of the experiment,.

3.3.3 *Brain radioactivity*

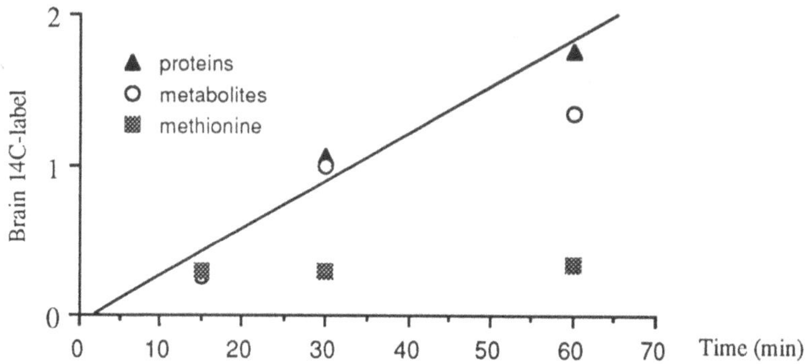

FIGURE 5. ^{14}C-label distribution into brain fractions. Values are the mean of 8 brain regions for 10 rats. Values are nCi/g normalised, for each rat, by the ^{14}C-methionine plasma concentration (nCi/ml).

Figure 5 shows the results of radiolabel accumulation in the brain tissue as the average of the different regions studied at 15, 30 and 60 min. Results are presented normalized by the plasma ^{14}C-methionine concentration for comparison purposes. Total label increased with time together with label incorporated into proteins which seemed to follow a linear function ($r = 0.9$). The level of labelled metabolites also increased from 15 to 60 min (times 5), although this increase was lower than that of labelled proteins (times 6). In contrast, the level of unmetabolized ^{14}C-methionine was not significantly different for the studied times.

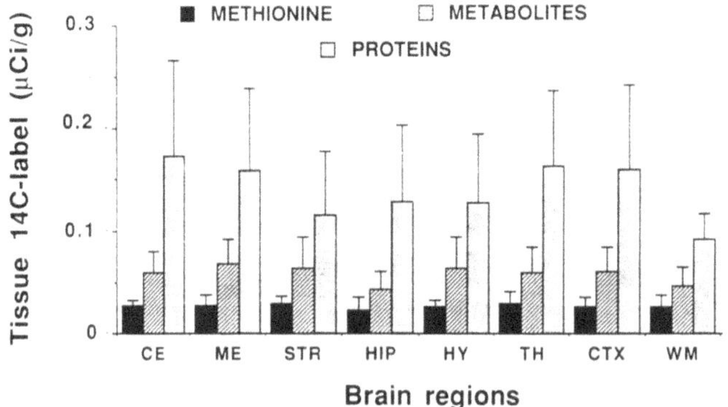

FIGURE 6. Distribution of label between labelled proteins, labelled metabolites and unmetabolized ^{14}C-methionine for different brain regions, after 1h infusion of [^{14}C-methyl]methionine to the rats. Results are presented as mean µCi/g ± S.D (n = 3). Brain regions are: cerebellum (CE), medulla (ME), striatum (STR), hippocampus (HIP), hypothalamus (HY), thalamus (TH), cortex (CTX) and white matter (WM).

At 60 min, total radioactivity accumulated in the brain regions was found to vary from region to region (fig 6). At this time, label incorporated into proteins was found to be significantly different between brain regions ($p < 0.03$), whilst no regional significant variation was observed for labelled metabolites and labelled unmetabolized methionine.

These results are in agreement with the results we found by autoradiography showing a good regional correlation between total radiactivity accumulation and label incorporation into proteins. In addition, the fact that the concentration of unmetabolized ^{14}C-methionine did not vary from 15 to 60 min shows that a steady-state situation for ^{14}C-methionine was kept for the different brain regions during the infusion. A steady-state situation was also maintained in the plasma starting from about 5 min after the beginning of the infusion (see section 3.3.2).

3.3.4 *Brain to plasma radioactivity*

Considering labelled methionine and metabolites, fig 7 shows the results of typical HPLC analyses comparing the percent plasma [^{14}C-methyl]methionine at 3 and 55 min with the corresponding percent in the tissue at 55 min. Comparatively, more metabolites were acumulated in the brain tissue than in the plasma one hour after the beginning of the infusion.

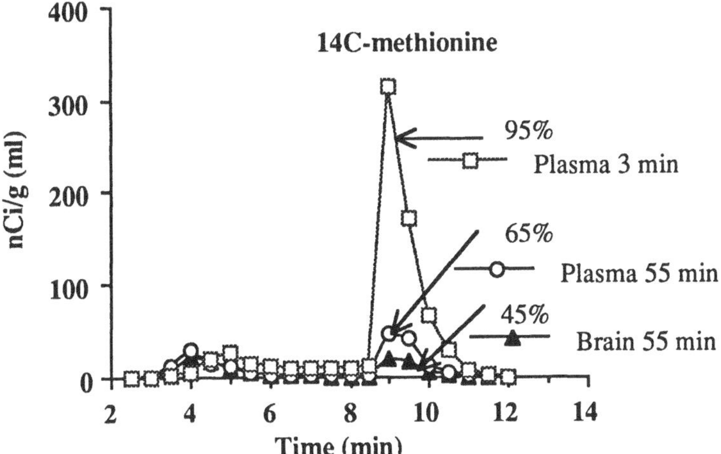

FIGURE 7. Radioactive fractions collected out of the HPLC system (see Methods). The radioactivity associated to each 0.5 min-fraction was determined by liquid scintillation. Results of plasma withdrawn a few minutes after starting the [^{14}C-methyl]methionine infusion are compared with results of plasma withdrawn one hour later. Plasma results are compared to brain results from tissue sampled one hour after the begining of the infusion.

3.3.5 *Brain to plasma specific activity*

According to the results presented above, the plasma and brain SA remained constant from 15 to 60 min. The plasma SA time course was followed for each rat. In contrast, the brain time course resulted from rats infused for different periods of time. The brain-to-plasma SA ratios were calculated for the different brain regions and are presented in table 1. The mean brain-to-plasma ratio was found at 0.50 ± 0.16. Although global significant differences were found between regions using the ANOVA test ($F = 3$; $df = 7, 63$; $p < 0.01$), the comparison between individual regions was not able to show specific regional differences. Our working hypothesis was based on the relationship between plasma and brain SA. We expected to find equivalent SA (i.e., a ratio of brain-to plasma SA equal to unity) if no other source but the plasma supplied free methionine to the brain tissue. This is not the case since the plasma SA was found to be approximately twice the brain SA.

TABLE 1. Regional ratios of brain to plasma methionine specific activity

Brain regions	Brain SA/plasma SA
Cerebellum	0.57 ± 0.2
Medulla	0.57 ± 0.1
Striatum	0.40 ± 0.1
Hippocampus	0.43 ± 0.1
Hypothalamus	0.49 ± 0.2
Thalamus	0.48 ± 0.1
Cortex	0.52 ± 0.1
White Matter	0.57 ± 0.1

Ratios are presented as mean ± SD of 10 rats infused for either 15, 30 or 60 min. Results after these infusion times were pooled together since no significant differences were found between times.

This result for methionine is in agreement with results reported in the literature for other amino acids (8, 12, 13, 31). Using ^{14}C-leucine, Smith et al. (12) and Hargreaves-Wall et al. (13) also measured the SA of the whole rat brain leucyl-tRNA (which is the direct presursor pool for leucine incorporation into proteins) in steady-state conditions. Similarly, Keen et al (8) made equivalent measurements for leucine after a bolus injection of ^{14}C-leucine. All of them reported direct evidence showing that an endogenous source of brain leucine (likely to be protein-breakdown) contributes to protein synthesis. This may also be the case for methionine since we found the brain SA activity of free methionine to be lower than the plasma SA. Although in the brain the SA of free methionine is not necessarily equivalent to the SA of the methionyl-tRNA, some results reported for leucine seemed to indicate that the corresponding SA for leucyl-tRNA and free leucine were rather similar under normal conditions (12, 13). In this regard, we entirely agree with these authors about the possibility of changes in the relationship between the aminoacyl-tRNA SA and the free amino acid SA under different physiological or pathological conditions.

3.3.6 *Regional contribution of plasma methionine to brain methionine*

From the results presented above we have calculated, for each brain region, the fraction of brain free methionine derived from the plasma by multiplying the regional endogenous methionine content by the regional ratio of brain-to-plasma SA (table 2).

The contribution of an endogenous source (likely to be protein breakdown) to brain free methionine can be then determined by substracting the fraction derived from plasma to the total methionine content. The methionine fraction derived from plasma did not vary significantly from region-to-region. In contrast, the amount supplied by protein breakdown was found to vary significantly between brain regions ($F = 4.5$; $df = 7, 63$; $p < 0.001$). The contribution of protein breakdown to brain free methionine was different depending on the brain region considered and, therefore, this contribution cannot be globally estimated for the brain as a whole. Moreover, even greater differences may exist locally in the brain within the relatively large regions that we have considered in this study.

TABLE 2. Estimated amounts of brain free methionine (nmol/g) derived from plasma or from protein breakdown. Values are the mean ± SD of 10 rats following infusion of ^{14}C-methionine.

Brain regions	from plasma	from protein breakdown
Cerebellum	11.6 ± 5	7.4 ± 1
Medulla	11.4 ± 4	8.5 ± 3
Striatum	13.4 ± 6	19.6 ± 4
Hippocampus	12.0 ± 5	16.6 ± 7
Hypothalamus	13.1 ± 4	15.3 ± 11
Thalamus	11.9 ± 4	14.4 ± 8
Cortex	14.1 ± 3	14.1 ± 6
White Matter	11.7 ± 3	9.4 ± 5

3.3.7 *Estimation of the regional rates of methionine incorporation into brain proteins*

Two rates of protein synthesis could be calculated for each brain region: the apparent rate (R_a), which considers the plasma source of brain methionine only; and the real rate of protein synthesis (R_i), which does also account for any source of brain methionine not directly arriving to the brain from the plasma (see Materials and Methods). This last rate can be calculated under the assumption that the SA of free brain methionine is identical to the SA of the corresponding precursor pool for incorporation into proteins (the methionyl-tRNA). Results of these calculations are presented in fig. 8.

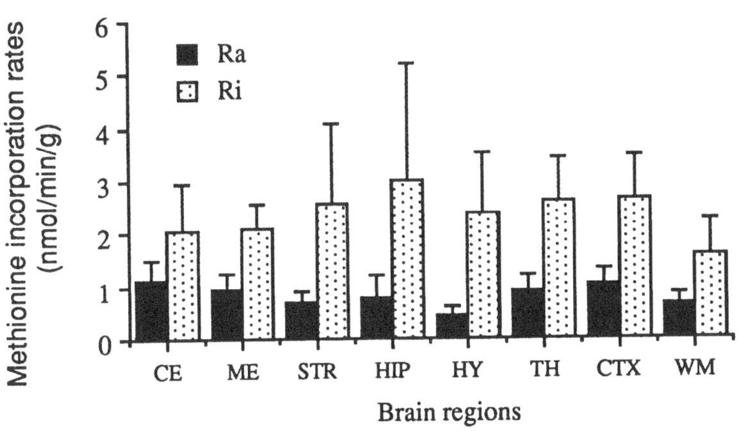

FIGURE 8. Regional rates of methionine incorporation into proteins. Rates have been estimated as described in Methods from results obtained after ^{14}C-methionine infusion to rats. Values are presented as mean ± SD (n = 10). Ra indicates apparent incorporation rates (accounting only for the plasma source) and Ri indicates real incorporation rates (accounting for the plasma source and also for the endogenous source). See fig. 6 for correspondance between names of brain regions and abbreviattions

Regarding R_a, there are significant differences between regions ($F = 17$; $df = 7, 49$; $p < 0.001$), with higher rates in the cerebellum than in the striatum and white matter ($p < 0.05$). This is in agreement with the results of the autoradiographic studies, where, for instance, the striatal area shows one of the lowest incorporation rates (see fig 1 and 2). The average R_a for the different brain regions considered was 0.87 ± 0.3 nmol/min/g of brain tissue. This value is within the range of methionine incorporation rates reported in the literature (2). The values of methionine incorporation considering the real rate R_i were higher than the R_a values (see fig. 8), with an average rate of 2.3 ± 1 nmol/min/g. R_a are the only rates that can be determined, either by PET or by autoradiography, since the local brain methionine SA cannot be measured. Therefore, these techniques would lead to underestimation of the real rates of methionine incorporation into proteins. No regional correlation exists between R_a and R_i, indicating no possibility of direct estimation of R_i (which is the rate directly related to the rate of protein synthesis) from R_a (which is the measurable rate).

3.4 HEART AND PLASMA [^{14}C-METHYL]METHIONINE STEADY-STATE

Equivalent measurements as those presented above for the brain were made in the heart tissue in rats at 15 (n = 3), 30 (n = 2) and 60 min (n = 3) after the infusion of ^{14}C-methionine (some of the rats were the same as those presented for the cerebral study). Under steady-state conditions for endogenous and for labelled methionine, the ratio of heart to plasma methionine SA was measured for each rat. Results are presented in table 3. No differences were found between times and the mean ratio of heart to plasma SA was found to be 1. This indicates that the pool of brain free methionie exchanges readily with the plasma methionine pool during the experimental time.

TABLE 3. Ratio of heart to plasma methionine specific activity

Rat	Time (min)	Ratio
1	15	1.29
2	15	0.89
3	15	0.93
4	30	0.94
5	30	1.20
6	60	0.81
7	60	1.06
8	60	1.42

Values are presented from individual rats. The time indicates the duration of the infusion of ^{14}C-methionine.

This is in agreement with results in the literature for other amino acids indicating that all heart free amino acid content may contribute to protein synthesis. For instance, for phenylalanine in vitro studies with isolated hearts (32-34) have shown that heart free phenylalanine equilibrates with the phenylalanine in the perfusate; and for leucine (35) the heart SA has been found to equilibrate with the plasma SA and both of them with leucyl-tRNA.

The apparent incorporation rate, R_a, and the real incorporation rate, R_i, were calculated for the heart tissue according to the equations in Materials and Methods. R_i can be estimated assuming that in the heart free methionine SA is equivalent to the SA of the methionyl-tRNA (if all the heart free methionine was available for protein synthesis). These rates are shown in table 4 (where method 1 indicates: R_a, and method 2 indicates: R_i). No significant differences were found between both methods suggesting that in the heart, in normal conditions, there is no need to know the endogenous methionine concentration to estimate the real rates of protein synthesis. Obviously, this would only be true if the SA of the methionyl-tRNA was equivalent to the plasma SA. In this case, the use of labelled methionine for the estimation of protein synthesis rates would be valid, since no recycling would became apparent during the experimental time. However, the use of [^{11}C-methyl]methionine in PET studies of the heart would still be subjected to the ability of modelisation taking into account methionine metabolism.

TABLE 4. Rate of methionine incorporation into heart proteins (nmol/min/g): Comparison between two estimations

Rat	Time (min)	Method 1	Method 2
1	15	1.65	1.20
2	15	1.74	1.28
3	15	2.04	2.10
4	30	2.21	2.71
5	30	2.94	2.87
6	60	2.11	2.27
7	60	2.47	3.21
8	60	2.69	2.40

Method 1 provides estimation for Ra and method 2 for Ri. No significant differences between these estimations were found. The infusion of ^{14}C-methionine lasted for different periods of time (min, in column 2).

4. Conclusions

a) From the autoradiographic and the dissection studies, we can conclude that total accumulation of label into brain regions reflects the regional incorporation of label into proteins, at least in normal conditions.

b) The brain SA is significantly diluted by an endogenous source of methionine which is likely to be protein breakdown.

c) The consequence for PET studies is that, although brain images after [^{11}C-methyl]methionine injection should reflect the accumulation of labelled methionine into brain proteins, PET can only see a part of the full process of protein synthesis (i.e., the incorporation of methionine derived from plasma). Whilst PET is blind to the incorporation of unlabelled methionine derived from protein breakdown. Hence, regional protein synthesis rates cannot be measured by PET, unless a method is found to estimate the regional endogenous contents of brain methionine. This may also be valid for other amino acids recycling in the brain.

d) Methionine in the heart tissue seems to readily exchange with plasma methionine and no evidence for methionine recycling has been observed within the one-hour experimental time. From this point of view, the heart seems to be suitable to study the rate of protein synthesis from the incorporation of labelled methionine into proteins. Even in the heart, however, the problem of methionine metabolism should to be considered and a mathematical model should to be developed to attempt accurate estimations of protein synthesis rates. In addition, the possibility that methionine recycling might became apparent in the heart under pathological conditions cannot be discarded.

5. Acknowledgements

The authors would like to thank Ms A. Jobert for assistance in the autoradiography experiment Dr A.M. Planas had support from the SHFJ, CEA, Orsay.

6. References

1. Ericson K, Blomqvist G, Bergström M, Eriksson L, Stone-Elander S. Application of a kinetic model on the methionine accumulation in intracranial tumors studied with positron emission tomography. Acta Radiol 1987; 28:505-509.

2. Lestage P, Gonon M, Lepetit P, et al. An in vivo kinetic model with l-[^{35}S]methionine incorporation into protein in the rat. J Neurochem 1987; 48:352-363.

3. Mazoyer BM, Levasseur M, Syrota A, et al. Modelization of ^{11}C-L-methionine PET kinetics in the brain of fasted and fed humans. J Cereb Blood Flow Metab 1989; 9(suppl 1):S37.

4. Phelps ME, Barrio JR, Huang S-C, Keen RE, Chugani H, Mazziotta JC. Criteria for the tracer kinetic measurement of cerebral protein synthesis in humans with positron emission tomography. Ann Neurol 1984; 15(suppl):S192-S202

5. Jones RM, Cramer S, Sargent T, Budinger TF. Brain protein synthesis rates measured in vivo using methionine and leucine. J Nucl Med 1985; 26:830.

6. Lundqvist H, Stålnacke C-G, Långström B, Jones B. Labeled metabolites in plasma after intravenous administration of [$^{11}CH_3$]-L-methionine. In: The metabolism of the human brain studied with positron emission tomography (Greitz T, Ingvar DH, Widén L eds), New York, Raven Press, 1985: 233-240.

7. Buonomo C, Mills P, Hilton J, Anderson JH, Wong DF, Dannals RF. Labelled plasma metabolites of L-methyl-hydrogen-3-methionine and L-methyl-carbon14-methionine in the dog. Am J Physiol Imag 1988; 3:178-181.

8. Keen RE, Barrio JR, Huang S-C, Hawkins RA, Phelps ME. In vivo cerebral protein synthesis rates with leucyl-transfer RNA used as a precursor pool: determination of biochemical parameters to structure tracer kinetic models for positron emission tomography. J Cereb Blood Flow Metab 1989; 9: 429-445

9. Ishiwata K, Vaalburg W, Elsinga PH, Paans AMJ, Woldring MG. Comparison of L-[1-^{11}C]methionine and L-methyl-[^{11}C]methionine for measuring in vivo protein synthesis rates with PET. J Nucl Med 1988; 29:1419-1427.

10. Ishiwata K, Hatazawa J, Kubota K, et al. Metabolic fate of L-[methyl-^{11}C]methionine in human plasma. Eur J Nucl Med 1989; 15:665-669.

11. Ishiwata K, Kameyama M, Hatazawa J, Kubota K, Ido T. Measurement of L-[methyl-^{11}C]methionine in human plasma. Appl Radiat Isot 1991; 42:77-79.

12. Smith CB, Deibler GE, Eng N, Schmidt K, Sokoloff L. Measurement of local cerebral protein synthesis in vivo: influence of recycling of amino acids derived from protein degradation. Proc Natl Acad Sci USA 1988; 85:9341-9345.

13. Hargreaves-Wall KM, Buciak JL, Pardridge WM. Measurement of free intracellular and transfer RNA amino acid specific activity and protein synthesis in rat brain in vivo. J Cereb Blood Flow Metab 1990; 10:162-169.

14. Comar D, Saudubray JM, Duthilleul A, et al. Brain uptake of ^{11}C-methionine in phenylcetonuria. Eur J Pediatr 1981; 136:13-19.

15. Bustany P, Henry JF, Soussaline F, Comar D. Brain protein synthesis in normal and demented patiens-a study by positron emission tomography with ^{11}C-L-methionine In: Functional Radionuclide Imaging of the brain (Magistretti PL, ed), New York, Raven Press, 1983: 319-326.

16. Bustany P, Henry JF, de Rotrou J, et al. Correlation between clinical state and positron emission tomography measurement of local brain protein synthesis in Alzheimer's dementia, Parkinson's disease schizophrenia, and gliomas. In: Positron emission tomography (Greitz T, Ingvar DH, Widén L, eds), New York, Raven Press, 1985: 241-249.

17. Bergström M, Ericson K, Hagenfeldt L, et al. PET study of methionine accumulation in glioma and normal brain tissue: competition with branched chain amino acids. J Comput Assist Tomogr 1987; 11:208-213.

18. O'Tuama LA, Phillips PC, Smith QR, et al. L-Methionine uptake by human cerebral cortex: maturation from infancy to old age. J Nucl Med 1991; 32:16-22.

19. Bergström M, Lundqvist H, Ericson K, et al. Comparison of the accumulation kinetics of L-[methyl^{11}C]-methionine and D-[methyl^{11}C]-methionine in brain tumors studied with positron emission tomography. Acta Radiol 1987; 28:225-229.

20. Mosskin M, von Holst H, Bergström M, et al. Positron emission tomography with ^{11}C-methionine and computed tomography of intracranial tumors compared with histopathologic examination of multiple biopsies. Acta Radiol 1987; 28:673-681.

21. Hatazawa J, Ishiwata K, Itoh M, et al. Quantitative evolution of L-[methyl-C-11]methionine uptake in tumor using positron emission tomography. J Nucl Med 1989; 30:1809-1813.

22. Mineura K, Sasajima T, Kowada M, Shishido F, Uemura K. Determination of tumor extent detected by (^{11}C-methyl)-1-methionine positron emission tomography in gliomas. J Cereb Blood Flow Metab 1991; 11(Suppl 2):S596.

23. Leskinen-Kallio S, Ruotsalainen U, Nagren K, Teräs M, Joensuu H. Uptake of carbon-11-methionine and fluorodeoxyglucose in non-Hodgkin's lymphoma: a PET study. J Nucl Med 1991; 32: 1211-1218.

24. Gati I, Bergström M, Muhr C, Langström B, Carlsson J. Application of (methyl-^{11}C) methionine in the multicellular spheroid system. J Nucl Med 1991; 32: 2258-2265.

25. Yoshimine T, Hayakawa T, Kato A, et al. Autoradiographic study of regional protein synthesis in focal ischemia with TCA wash and image subtraction techniques. J Cereb Blood Flow Metab 1987; 7:387-393.

26. Patlak CS, Pettigrew KD. A method to obtain infusion schedules for prescribed blood concentration time courses. J Appl Physiol 1976; 40:458-463.

27. Planas AM, Prenant C, Mazoyer BM, Comar D, DiGiamberardino L. Regional cerebral L-[^{14}C-methyl] methionine incorporation into proteins: evidence for methionine recycling in the rat brain. J Cereb Blood Flow Metab 1992: (in press)

28. Lowry OH, Rosebrough NJ, Farr AL, Randall RJ. Protein measurement with the folin phenol reagent. J Biol Chem 1951; 193:265-275.

29. Laemmi UK. Cleavage of structural protein during the assembly of the head of the bacteriophage T4. Nature 1970; 227: 680-685.

30. Shahbazian FM, Jacobs M, Lajtha A. Amino acid incorporation in relation to molecular weight of proteins in young and adult brain. Neurochem Res 1986; 11: 647-660.

31. Seta K, Sansur M, Lajtha A. The rate of incorporation of amino acids into brain proteins during infusion in the rat. Biochim Biophys Acta 1973; 294:472-480

32. Morgan HE, Earl DCN, Broadus A, Wolpert EB, Giger KE, Jefferson LS. Regulation of protein synthesis in heart muscle. I. Effect of amino acid levels on protein synthesis. J Biol Chem 1971; 246: 2152-2162.

33. Schreiber SS, Oratz M, Evans C, Reff F, Klein I, Rothschild MA. Cardiac protein degradation in acute overload in vitro: reutilization of amino acids. Am J Physiol 1973; 224: 338-345.

34. McKee EE, Cheung JY, Rannels DE, Morgan HE. Mesurement of the rate of protein synthesis and compartmentation of heart phenylalanine. J Biol Chem 1978; 25: 1030-1040.

35. Everett AW, Prior G, Zak R. Equilibration of leucine between the plasma compartment and leucyl-tRNA in the heart, and turnover of cardiac myosin heavy chain. Biochem J 1981; 194: 365-368.

DISCUSSION

F. Mauguière In the experiments on monocular deprivation you have shown: 1. that amino acid incorporation is depressed in the LGN layers receiving their inputs from the occluded eye and 2. that this effect can be observed even after that deoyglucose metabolism has returned to normal in the occipital visual cortex. How can you conclude from this observation that the study of protein synthesis could help to understand "regeneration processes"?

C. Smith First, glucose utilization rates in the striate cortex were not found to be normal at the time of our protein synthesis experiments. The [^{14}C]deoxyglucose autoradiograms of striate cortex (see Fig 8) show that after chronic monocular deprivation the pattern of ocular dominance columns so clearly visible in the normally developing monkey studied with one eye closed has been lost (see ref. 30). In the chronically occluded monkey the entire cortex is metabolically responsive to the open eye suggesting that the terminals in layer IV of triate cortex have reorganized. The process underlying this reorganization may be reflected in a change in protein synthesis rates in the cell bodies of origin of these terminals which are located in the cell laminae of the dLGN. An increased rate of protein synthesis in the nondeprived cell laminae might indicate that the nondeprived cells actively grow into the adjacent columns and make synaptic connections with cells normally responsive to the deprived eye. A decreased rate of protein synthesis in the deprived cell laminae might indicate that the deprived cells cease to maintain their growth rates and fail to compete effectively for synaptic connections in striate cortex. Our results suggest that the loss of ocular dominance columns in striate cortex is the consequence of depressed protein synthesis and deficient axonal growth in the developing geniculocortical pathway for the deprived eye.

F. Mauguière Have you studied protein synthesis in the visual cortex after monocular deprivation? Is the functional reorganization of the cortical circuitry, as suggested by DG studies, paralleled by protein synthesis changes?

C. Smith We have looked at protein synthesis in the visual cortex following monocular deprivation. When the deprivation takes place from birth to Day 25 we see no change in the pattern of the [^{14}C]leucine autoradiograms in the cortex. However, when the deprivation occurs later in the critical period, columns perpendicular to the cortical surface with alternating high and low rates of protein can be seen in striate cortex.

B. M. Mazoyer et al. (eds.), PET Studies on Amino Acid Metabolism and Protein Synthesis, 69–74.

L. Di Giamberardino In the oclusion studies, what part of the reduced protein synthesis rate can be attributed to neuronal loss?
C. Smith We did not carry out neuron counts in our studies but there are no reports in the literature of neuronal loss in the dLGN under the conditions of monocular deprivation used in our studies. There is evidence, however, for changes in cell size (e.g., Sloper JJ, Headon MP, Powell TPS. Dev Brain Res 1987; 31: 267-276).

L. Di Giamberardino Do you think that stimulation of axons that are not used to receive antidromic spikes, as you did in your experiments, can provide useful informations on protein synthesis response to neuronal activity?
C. Smith Yes, I think it could be useful, especially if we had found an effect. It would have indicated that acute activation of a pathway could have an effect on protein synthesis rates. The fact that we did not find an effect certainly does not rule out the possibility that other conditions of stimulation or the study of other pathways may show effects of stimulation on rates of protein synthesis.

F. Mauguière Concerning the absence of effect of nerve trunk electrical stimulation on protein synthesis in neuronal cell bodies, we should remember that electrical stimulation using 0.2 msec pulses represents only 1 msec per sec of efficient stimulation. In that respect, continuous natural stimulations are far more effective in producing a continuous flow of inputs. We observed no effects on C-14-DG metabolism in the VP4 and somatosensory cortex by electrically stimulating the median nerve of monkeys using short pulses delivered at rates of 2 to 5 Hz. It has been shown however that glucose metabolism is very sensitive to functional activation using natural somatosensory stimulation. Do you think that from your study using electrical stimulation of a nerve trunk it can be concluded that protein synthesis is not influenced by functional activation?
C. Smith In think that we can only conclude that, under the conditions we examined in our experiments, protein synthesis is not influenced by functional activation. We do know, however, that the conditions examined in our studies, i.e., continuous stimulation at 10Hz, did result in large 100% increases in rates of glucose utilization (see ref. 32). We are currently examining the effects of patterned stimulation and a delayed effect of stimulation on protein synthesis.

A. Lajtha Do changes in protein contents parallel changes in protein synthesis rate?
C. Smith In our experiments which are primarily autoradiographic we have not measured protein content.

D. Comar When you measure brain radioactivity, can you differentiate amino-acid incorporation in neurons from that in glial cells?
C. Smith No, not at the present time. Our suggestion in the regeneration studies that the effects are in the neurons is only by reference, i.e., examination of the time course of morphological changes in different cell types and the correlation between morphological changes and changes in protein synthesis. Our studies show only protein synthesis rates in a region as a whole. The resolution of ^{14}C autoradiograms is approximately 50 µm which is not good enough to differentiate among cell types. In the future we hope to increase the resolution of the method with the use of [^{3}H]leucine as a tracer. However, the metabolic disposition of ^{3}H-labeled leucine is more complex than that of [1-^{14}C]Leucine. [^{3}H]leucine may label intermediates in the TCA cycle which if quantitatively significant could render any results impossible to interpret. In order to be sure that we are looking at incorporation of label into protein in individual cells we would have to carefully examine the question of the disposition of the ^{3}H label.

D. Heiss Since a large portion of aminoacids used for protein synthesis comes from the breakdown of proteins, I wonder if you measure transport of aminoacid from plasma to brain during the stimulation experiments or if this is also an increase of protein turnover contributing to the increase of protein synthesis observed in stimulated states?
C. Smith With the experimental design of the [1-^{14}C]leucine method, i.e. a pulse injection of [1-^{14}C]leucine followed by a 60 min interval and fixation and washing of tissue sections, transport does not enter into the operational equation except in the λ. λ can be thought of as the fraction of the total flux of leucine into the precursor pool that comes from the plasma. There are only two sources of leucine in brain, plasma and protein breakdown, so:

$$\lambda = \frac{flux_{plasma}}{flux_{plasma} + flux_{protein}}$$

If the flux of leucine into the precursor pool changes due to a change in the tansport of leucine and the flux from protein degradation is unchanged then λ will increase. If, for instance, the flux of leucine from plasma increased by 50 % then λ would increase from 0.58 to 0.68. If the flux of leucine from the plasma decreased by 50 % the λ would decrease to 0.41.

As long as λ is measured under the conditions of the experiment then changes in transport would be taken into account in the operational equation.

G. Meyer In our experiments, label incorporation into proteins stays always below 80 %, which seems to be in contradiction with the data you showed.
J. Bobillier The relative rate of methionine utilization into brain proteins and into the transmethylation pathway must not be confused with the relative amount of (35S) bound to proteins and (35S) free in tissue during the time course of (35S) in brain after an intravenous bolus of L-(35s)methionine. Under these experimental conditions the amount of (35S)-bound proteins increases almost linearly, until a plateau, while the concentration of (35S)-free tissue, after an initial peak, decreases progressively. The percentage of label incorporation into proteins will then increase as a funcion of time. We have previously shown [J. Neurochem., 48, 352-363 (1987)] that this percentage was below 80 %, at 60 min, but was approximately,in relation to the examined regions, 75-90 % at 90 min after the tracer injection. I expect therefore, if you take into account the kinetic of the distribution ratios for (35S) between proteins and the free tissue pool, that the results of your experiments with L-(35S)methionine should be in close accordance with the present data.

G. Meyer What is the relationship between the influx of aminoacid in brain tissue and their incorporation into proteins?
A. Planas We found that total radioactivity accumulated in the brain is regionally correlated to the incorporation of label into proteins, after i.v. administration of labelled methionine. This may suggest that the net uptake of methionine from plasma is related to the rate of incorporation of labelled methionine into brain proteins (such a relation would not apply to the full rate of brain methionine incorporation into proteins, since it would not take into account any other source of brain free methionine different from the plasma source). The net uptake of methionine, however, results from the bidirectional transport of methionine across the blood brain barrier, from plasma into brain tissue and out of the brain to plasma. These two processes are not necessarily symmetrical. Therefore, we do not know the relationship between the influx rate of methionine from plasma to brain and the incorporation of labelled methionine into brain proteins.

A. Lajtha Would it make any difference to use longer infusion times? Also, did you notice any difference in metabolites in the 15 and 60 minutes infusion experiments?
A. Planas Assuming that the difference in free methionine SA that we found between brain and plasma were due to the contribution of protein breakdown, I would expect that longer infusion periods would make this difference to become smaller. However, infusions lasting maybe several days would be necessary, given the long half life of brain proteins.
Regarding your question about metabolites, we observed that the total amount of labelled metabolites increased from 15 to 60 min. We did not identify the different labelled products arising in plasma and brain during the experiment. However, our results indicate that at least two different labelled metabolites appear in both brain and plasma and that, at a given time, more metabolites are present in the tissue than in the plasma.

B. Sadzot You mentionned that the specific activity of free labelled methionine is lower in brain than in plasma. The proposed explanation is protein degradation and labelled methionine recycling. Is it possible that this difference in specific activity is rather due to transport of labelled methionine and accumulation of that tracer into the extracellular space?
A. Planas Studying the SA of methionine in brain and plasma in the rat after i.v. infusion of L-(14C- or 11C-methyl) methionine, we found the SA to be lower in the brain than in the plasma. We interpreted this result as evidence of the presence in the brain of an extraplasmatic source of non-labelled methionine which dilutes the SA in this tissue during the experiment. Such a source could well be protein degradation. Considering that the half life of proteins is of the order of several days, methionine liberated by protein breakdown will be essentially non-labelled. We propose that this can be a source of free unlabelled methionine in the brain which could be utilized for protein synthesis and, therefore, be recycled in the tissue. Coming back to your question, please do notice that it is the unlabelled methionine that we think would be recycled, not the labelled methionine. Your suggestion about transport of labelled methionine and accumulation of that tracer into the extracellular space could not explain the difference of SA that we found between brain an plasma. The i.v. injected labelled molecules are tracing the processes of transport from plasma and accumulation in brain that take place for the endogenous methionine normally present in the plasma. However, systematically injecting the label, we do not trace the contribution of other sources of brain free methionine different from the plasma source.

D. Comar Comparing your results in brain and heart, you explain the difference by the presence of the blood-brain barrier. In the case of a disrupted barrier in the brain, would you expect to find the same specific activity of free methionine in plasma and in brain extracellular fluid?

A. Planas In a situation of free exchange of methionine between plasma and tissue, independently of the facilitated transport system across the blood-brain barrier, we should probably expect a rapid equilibration of free methionine between plasma and extracellular fluid. Regarding pathological cases where the barrier can be broken, as it may be the case in brain tumors, a few points, however, should be considered: 1) the comparison between the SA of plasma and of tumoral extracellular fluid would only be valid under steady-state conditions, for both, labelled methionine and endogenous methionine; and 2) the heterogeneity of the tumoral tissue must be taken into account since this may contain necrosed areas and the barrier might be only locally disturbed. We have found evidence suggesting that brain gliomas may show not only a high rate of protein synthesis but also a high rate of methionine metabolism. Given the heterogeneity of this tissue, it remains to be seen whether, within the tumor, methionine derived from protein breakdown would be available for metabolism or for protein incorporation in the same way as plasma methionine, or whether the contribution of plasma methionine may be a limiting step for tissue methionine metabolism. Care must be paid to tissue heterogeneity as a factor able to restrict the exchange of methionine.

CARBON-11 AMINO ACIDS, LABELING AND METABOLITES

Willem Vaalburg, Philip H. Elsinga, Anne M.J. Paans

Abstract
The preparation methods available for ^{11}C-amino acids and the different aspects of the value of these amino acids for measuring the protein synthesis rate in vivo are discussed.

Introduction

Positron Emission Tomography (PET) using ^{11}C-labeled amino acids has prominent clinical potential in oncology, neurology and psychiatry. Many natural and a number of unnatural amino acids have been labeled [1]. Amino acids can be used for the investigation of blood brain barrier kinetics and disruptions [2, 3], however, most of the present investigations are focussed on the in vivo quantification of protein synthesis rates in tumors [4]. In tumor tissue ^{11}C-amino acids accumulate due to protein incorporation. This phenomenon is used clinically not only for the detection of tumors and metastases, but also for the investigation of the heterogenicity and necrosis of tumor tissue, for the diagnosis of recurrence and for the discrimination between tumor tissue and scar.

Carbon-11 labeled amino acids also have potential for the determination of activities of specific enzymes in tumors and other tissues. For instance ^{11}C-labeled ornithine may be useful to measure ornithine decarboxylase activity [5]. The application of carboxylic labeled DOPA is investigated for the assessment of the DOPA decarboxylase activity [6]. Carbon-11 DOPA is also applied for melanoma imaging. The amino acid is a precursor for the melanin synthesis which is enhanced in many melanoma.

An interesting and developing PET-application is the in vivo evaluation of tumor therapy by the quantification of the effect of treatment on protein synthesis rates. In a recent publication [4] the suitability is discussed of the various amino acids which are currently investigated for this purpose. For this application metabolic models must be available to quantitatively describe the fate of the tracer in the human body. Essential

B. M. Mazoyer et al. (eds.), PET Studies on Amino Acid Metabolism and Protein Synthesis, 75–80.

for a good and simple model is that the amino acid has a fast and high incorporation in tissue proteins, that it has minor non protein metabolic pathways and a fast blood clearance.

Amino acids for protein synthesis rates

The amino acids included for the quantification of protein synthesis rates are ^{11}C-carboxylic labeled leucine, tyrosine, phenylalanine, methionine and non-carboxylic ^{11}C-labeled methionine. Because of the biochemical behaviour it is apparent that the L-isomers are the most interesting isomers. Of the tracers mentioned phenylalanine has the highest brain uptake index which may be an advantage for the investigation of cerebral protein synthesis. However little information is available on the metabolic behaviour in man on the time scale of ^{11}C-investigations. It is known that phenylalanine is hydroxylated into tyrosine quite extensively [7, 8] which may make it difficult to correct for in the metabolic model. DL-[1-^{11}C]valine was used for the detection of several different tumors [9]. Animal studies with L-[1-^{14}C]valine revealed a large amount of unmetabolized tracer in tissue, so a low protein incorporation [10]. Therefore advantages are not to be expected for valine as compared to tyrosine, leucine and methionine which are the most often postulated tracers.

It is expected that carboxylic labeled amino acids are more valuable for protein synthesis than amino acids with the label in a non-carboxylic position. Although incorporation in proteins is the major metabolic route, significant decarboxylation can occur. When the label is in the carboxylic position $^{11}CO_2$ is formed. For the currently available compartment models it is assumed that the $^{11}CO_2$ formed is cleared from the tissue rapidly and expired from lung into air. In the models no tissue compartment is included for $^{11}CO_2$. Shields et al. [11] however, have found that this is a to optimistic view. They showed that the $^{11}CO_2$ clearance from the body is relatively slow and that the presence of $^{11}CO_2$ has to be taken into account. Wether a correction is necessary with carboxylic labeled amino acids, has to be verified.

L-[methyl-^{11}C]methionine is the most often used amino acid in tumor PET-studies. The main reason for this is the convenient preparation method. Methyl labeled methionine is involved in several other metabolic pathways such as transmethylation, which might lead to accumulation of a variety of labeled non protein metabolites in tissue. Such a metabolites profile requires a complicated compartment model. The only comparative metabolic study has been carried out by Ishiwata et al. [12] using tumour bearing rats. From the percentage of metabolites found at 60 minutes after injection, it was concluded that a carboxyl group label has to be preferred above a methyl group label for methionine. Ishiwata et al. [13] also investigated the input function of L-[methyl-^{11}C]methionine in man and observed that only total plasma radioactivity as an input function for the kinetic analyses is questionable. Because of considerable variations in clearance patterns they concluded that for plasma metabolites corrections should be performed. It is not expected that with respect to the clearance pattern the other amino acids have advantages.

The tracers L-[1-^{11}C]tyrosine and L-[1-^{11}C]leucine are evaluated as two outstanding amino acids for protein synthesis quantification. Both tracers have a rapid incorporation into proteins and a small intra-cellular pool of free amino acid. For tyrosine in animals detailed metabolic studies were performed [14]. In a rat model L-[1-^{11}C]tyrosine was compared with L-[methyl-^{11}C]methionine and L-[1-^{11}C]methionine. The conclusion was drawn that L-[1-^{11}C]tyrosine is a better tracer for protein synthesis rates than the ^{11}C-labeled methionines, because it has a smaller metabolites compartment and a higher protein incorporation.

Paans et al. [15] developed a four compartment kinetic model for tyrosine. Hawkins et al. [16] evaluated a kinetic model for L-[1-^{11}C]leucine in man. In rats it was observed that within 35 minutes 90% of the brain activity was protein incorporated. This figure is comparable with those for methionine and tyrosine. They concluded that, as for methionine, also for this amino acid the input function has to be measured.

Preparation methods

For the preparation of ^{11}C-amino acids several methodes are available. The synthetic approaches are reviewed by Kilbourn [17]. Radiochemical aspects, including choice of radionuclide, position of label and ease of preparation are discussed in a report of a task group organized within the framework of the EEC concerted action on PET [4]. In addition to these reviews only a few comments on the preparation methods are given here. For labeling in the carboxylic position the two most frequently applied methodes are the Bücherer-Strecker synthesis with Na^{11}CN as labeled precursor and the carboxylation of the α-lithiated isocyanides with $^{11}CO_2$. Both methodes yield DL-amino acids. After enantiomeric separation for PET studies useful amounts of the L-isomers can be obtained.

A disadvantage of the Bücherer-Strecker synthesis is that carrier has to be added to increase the radiochemical yield and to make the preparation method reliable and reproducible. Using an ^{11}C-aldehyde as precursor this method can also be applied to prepare 2-^{11}C-amino acids. The isocyanide method has as drawback that moisture sensitive α-lithio precursors have to be handled at low temperatures.

If it is desired to label an amino acid in a non carboxylic position, as alternative for the Bücherer-Strecker synthesis with a ^{11}C-aldehyde, the ^{11}C-alkylation of a glycine derivative can be used. Amino acids prepared in this way are labeled in the 3-position. The method is also applied for the asymmetric synthesis of several amino acids. When a chiral isonitrile is used as non-labeled starting material a high enantiomeric excess can be obtained. However, at present no general applicable organic synthetic method is available which yields enantiomeric pure final product. Therefore methods are needed for fast resolution of the enantiomeric mixtures. Enzymatic as well as chiral HPLC methods are used for this purpose.

Enzymatic methodes are also applied as preparation methodes. Advantages of such biosynthetic methodes are the mild reaction conditions, the stereoselectivity and the use of columns with immobilized enzymes which make automation easier.

In conclusion: several ^{11}C-labeled amino acids are potentially useful for PET, especially the quantification of protein synthesis rates in vivo is currently a topic of interest, but this application is still in its infancy.

References

1. Fowler J.S. and Wolf A.P. (1986) "Positron emitter-labeled compounds: Priorities and Problems". In: Positron emission tomography and autoradiography ch 9, Eds. Phelps M.E., Mazziotta J.C. and Schelbert H.R., Raven Press, New York.
2. Johnström P., Stone-Elander S., Ericson K., Mosskin M. and Bergström M. (1987) Carbon-11 labeled glycine synthesis and preliminary report on its use in the investigation of intracranial tumors using positron emission tomography. Appl. Radiat. Isot. 38, 729-734.
3. Wiesel F.A., Blomqvist G., Halldin C., Sjögren I., Bjerkenstedt L., Venizelos N. and Hagenfeldt L. (1991) The transport of tyrosine into the human brain as determined with L-[1-^{11}C]tyrosine and PET. J. Nucl. Med. 32, 2043-2049.
4. Vaalburg W., Coenen H.H., Crouzel C., Elsinga Ph.H., Långström B., Lemaire C. and Meyer G.J. (1992) Amino acids for the measurement of protein synthesis in vivo by PET. Nucl. Med. Biol. 19, 227-237.
5. Elsinga Ph. H., Daemen B.J.G., Ishiwata K. and Vaalburg W. (1990) Ornithine decarboxylase activity in prostate and tumor: a feasibility study for PET with L-[1-^{14}C]- and L-[5-^{14}C]-ornithine. Nucl. Med. Biol. 17, 587-600.
6. Reiffers S., Beerling H.D., Lakke J.P.W.F. (1978), Paans A.M.J., Vaalburg W. and Woldring M.G. (1978) Rapid decarboxylation of ^{11}C-labeled DL-DOPA in the brain: A potential approach for external detection of nervous structures. Brain Res. 145, 59-67.
7. Clarke J.T. and Bier D.M. (1982) The conversion of phenylalanine to tyrosine in man. Direct measurement by continuous intravenous tracer infusions of L-[^{3}H]phenylalanine and L-[1-^{13}C]tyrosine in the postabsorptive state. Metabolism 31, 999-1005.
8. Moldawer L.L., Kawamura I., Bistrian B.R. and Blackburn G.L. (1983) The contribution of phenylalanine to tyrosine metabolism in vivo. Studies in the postabsorptive and phenylalanine-loaded rat. Biochem. J. 210, 811-817.
9. Washburn L.C., Wieland B.W., Sun T.T., Hayes R.L. and Butler T.A. (1987) [1-^{11}C]DL-valine, a potential pancreas imaging agent. J. Nucl. Med. 19, 77-83.
10. Kirikae M., Diksic M. and Yamamoto Y.L. (1989) Quantitative measurement of regional glucose utilization and rate of valine incorporation into proteins by double-tracer autoradiography in the rat brain tumor model. J. Cerebr. Blood Flow Metab. 9, 87-95.
11. Shields A.F., Graham M.M., Kozawa S.M., Kozell L.B., Link J.M., Swenson E.R., Spence A.M., Bassingthwaighte J.B. and Krohn K.A. (1992) Contribution of labeled carbon dioxide to PET imaging of ^{11}C-labeled compounds. J. Nucl. Med. 33, 581-584.

12. Ishiwata K., Vaalburg W., Elsinga Ph. H., Paans A.M.J. and Woldring M.G. (1988) Comparison of L-[1-^{11}C]methionine and L-[$^{11}CH_3$]methionine for measuring in vivo protein synthesis rates with PET. J. Nucl. Med. 29, 1419-1427.
13. Ishiwata K., Kameyama M., Hatazawa J., Kubota K. and Ido T. (1991) Measurement of L-[methyl-^{11}C]methionine in human plasma. Appl. Radiat. Isot. 42, 77-79.
14. Ishiwata K., Vaalburg W., Paans A.M.J., Elsinga Ph.H., Vissering H. and Woldring M.G. (1986) A comparison of L-[1-^{11}C]methionine and L-[N-methyl-^{11}C]methionine for measuring protein synthesis rates in tumors with PET.
In: Clinincal demands in Nuclear Medicine, Proc. Europ. Nucl. Med. Congress Goslar p.37-40, Schattauer Verlag, Stuttgart.
15. Paans A.M.J., Daemen B.J.G., Elsinga Ph.H. and Vaalburg W. (1989) Determination of the in vivo protein synthesis rate using dynamic PET data from L-[1-^{11}C]tyrosine in an animal model. In: Nuclear medicine; quantitative analysis in imaging and function. Proc. Europ. Nucl. Med. Congress, Strasbourg, p. 30-33, Schattauer Verlag, Stuttgart.
16. Hawkins R.A., Huang S.C., Barrio J.R., Keen R.E., Feng D., Mazziotta J.C. and Phelps M.E. (1989) Estimation of local cerebral protein synthesis rates with L-[1-^{11}C]leucine and PET: methods, model and results in animals and humans. J. Cerebr. Blood Flow Metab. 9, 446-460.
17. Kilbourn M.R. (1985) Synthesis of carbon-11 labeled amino acids. Int. J. Nucl. Med. Biol. 12, 345-348.

QUALITY CONTROL ASPECTS IN THE PREPARATION OF [^{11}C]-METHIONINE

KJELL NÅGREN

ABSTRACT. Some quality control aspects of the preparation of [methyl-^{11}C]- and [1-^{11}C]-labelled methionine is examined. The oxidation of [^{11}C]-labelled methionine to its corresponding [^{11}C]-labelled sulphoxide and sulphone during synthesis, purification and in analytical samples is emphasized. Analytical procedures for determination of enantiomeric and radiochemical purity are also discussed.

1. Introduction

L-[methyl-^{11}C]Methionine is the most frequently used amino acid for PET oncology studies in Europe. This is also illustrated by the numerous contributions in this volume presenting its use in the study of tumors, especially in the brain but also in other parts of the human body. Synthesis methods were reported independently by the Uppsala and Orsay groups in 1976 [1,2]. Both groups have later on reported modified and improved procedures for its preparation[3,4]. Quality control aspects have also been published [5,6].

The synthesis of DL-[1-^{11}C]methionine was reported by the Groningen group in 1986 [7]. The preparation of L-[1-^{11}C]methionine by separation of the racemate on a chiral HPLC-column has also been reported [8]. So far, the clinical potency of this tracer for PET studies in humans still have to be evaluated. Some quality control aspects, like enantiomeric purity and oxidation products, can be discussed with regard to both tracers.

2. Synthesis pathways

The pathways leading to L-[methyl-^{11}C]methionine starting from L-S-benzyl-homocysteine in liquid ammonia [1] or L-homocysteine thiolactone in acetone [2] are shown in Scheme 1. The labelling reaction is in both cases the alkylation by [^{11}C]-methyl iodide on the L-homocysteine sulphide anion, however in quite different reaction media. The reaction pathway used in the preparation of DL-[1-^{11}C]methionine [7] is shown in Scheme 2. It utilizes the reaction of [^{11}C]-carbon dioxide with an in situ preparated α-lithio isocyanide, and subsequent hydrolysis of the isocyanide function.

B. M. Mazoyer et al. (eds.), PET Studies on Amino Acid Metabolism and Protein Synthesis, 81–87.

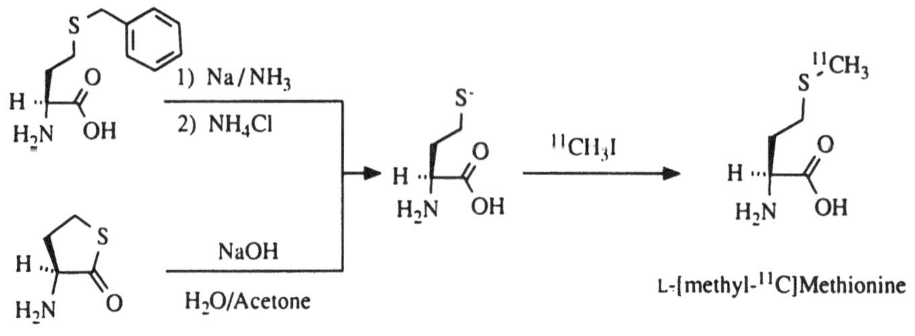

1) Na/NH_3
2) NH_4Cl
NaOH
H_2O/Acetone
$^{11}CH_3I$
L-[methyl-^{11}C]Methionine

Scheme 1.

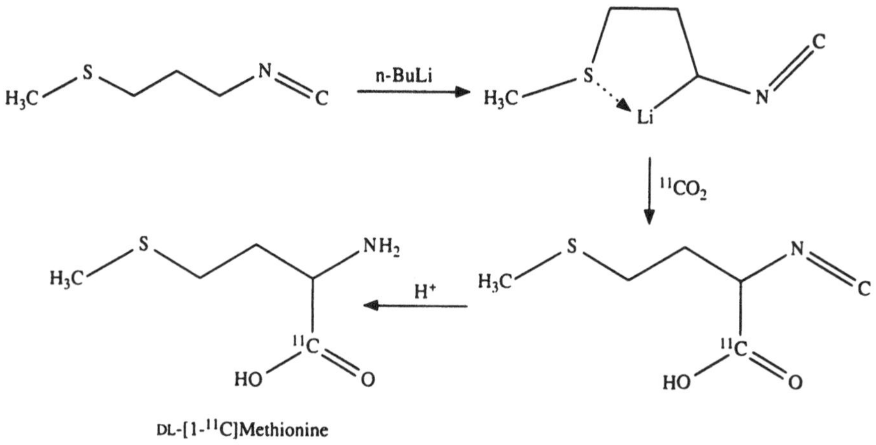

n-BuLi
$^{11}CO_2$
H^+
DL-[1-^{11}C]Methionine

Scheme 2.

3. Quality control aspects

3.1. RACEMIZATION

The separation of optical isomers can be performed by several methods using diastereomeric interactions. These methods can either be used for purification of an optical isomer from a mixture or for the analysis of the composition in a mixture of the isomers. The methods commonly used include enzymatic reaction with one of the isomers, chemical reaction with chiral compounds to form diastereomers, and HPLC separation of the isomers using either a chiral eluent or a chiral solid phase. All of these methods have been used in the analysis of the optical purity of L-S-[methyl-^{11}C]methionine.

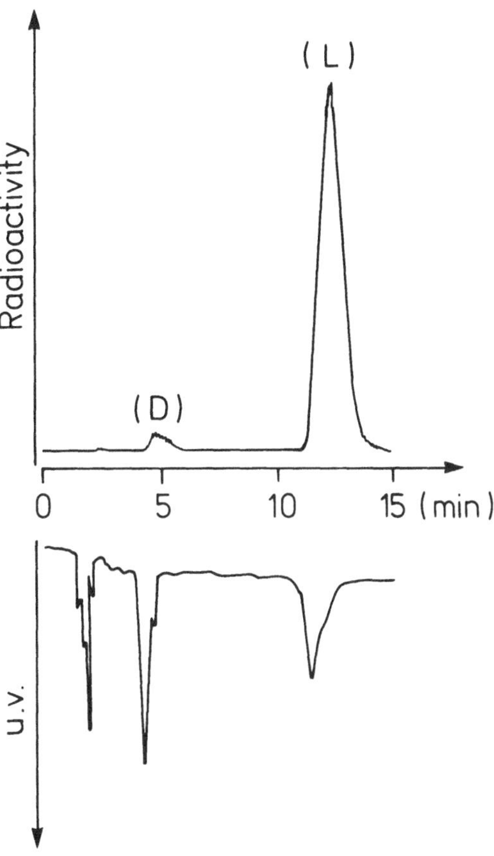

Figure 1. HPLC analysis of enantiomeric purity of [methyl-^{11}C]methionine with a Daicel Crownpak CR(+) column at 7 °C and 0.7 mL/min of 20 mM $HClO_4$. Upper chromatogram: radioactivity detector. Lower: UV-detector at 210 nm (DL-methionine co-injected).

3.1.1. *L-S-[methyl-^{11}C]Methionine.* The racemization of the thiolactone method have been investigated by two groups. Johnström et al. [9] used a chemical derivatization with (-)-camphanic acid and HPLC separation of the formed diastereomers. Their results, when using the synthesis conditions originally proposed [2], was a 95 % enantiomeric purity. Ishiwata et al. [10] have recently undertaken a study of the influence of the alkaline concentration on the racemization. They used direct separation of the reaction mixtures with HPLC and a chiral eluent. By minimizing the alkaline concentration they obtained an enantiomeric purity of 98 %.

The racemization of the S-benzyl-homocysteine method was first determined by aminoacylation of tRNA [11]. Due to limitations of this method a value of >98 % enantiomeric purity was determined [4]. The accuracy of the determination was improved by using direct separation on HPLC with a chiral column. By this method a value of >99 % enantiomeric purity was determined [4]. The Turku PET radiochemistry group has used an other type of chiral column, Daicel Crownpak CR(+), which give an excellent resolution of the L- and D-isomers of methionine as shown in Figure 1. By using this column values of 98-99.5 % enantiomeric purity for L-S-[methyl-^{11}C]-methionine, prepared by the S-benzyl-homocysteine method, have been determined (unpublished results).
Recently, the synthesis of L-Se-[methyl-^{11}C]selenomethionine using Se-benzyl-selenohomocysteine was reported [12]. The enantiomeric purity, determined by HPLC separation of the diastereomers formed on reaction with Marfey's reagent, was determined to >99 %.

3.1.2. *L-[1-^{11}C]Methionine.* The synthesis method used [7] gives a racemic product. This product was separated with HPLC using a chiral eluent, and subsequent switching of the solvent to saline using a second HPLC system [8]. No reports of the enantiomeric purity were given, but in principle it should be possible to achieve very high enantiomeric purity (>99 %) using chiral HPLC.

3.2. OXIDATION

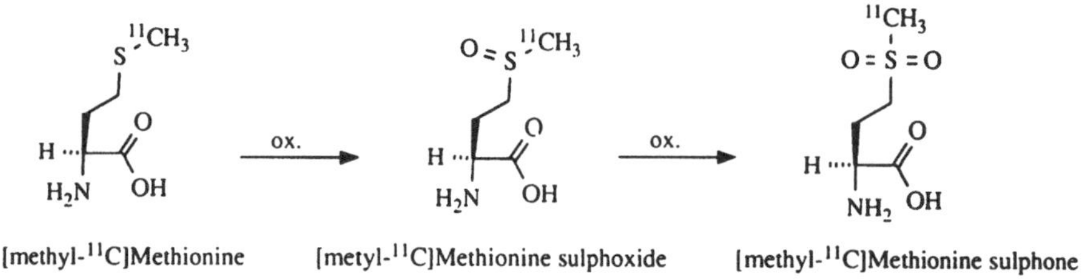

[methyl-^{11}C]Methionine —ox.→ [metyl-^{11}C]Methionine sulphoxide —ox.→ [methyl-^{11}C]Methionine sulphone

Scheme 3.

Methionine can be oxidized to the sulphoxide on prolonged exposure to air, and as the reaction is reversible the sulphoxide is often used as a protected methionine derivative in peptide synthesis [13]. The sulphoxide can in turn be oxidized to the sulphone (Scheme 3). This have implications both in the synthesis and the analysis of high specific radioactivity [^{11}C]methionine as the concentration of the product during synthesis, purification and in the final product solution is very low.
The radiochemical purity of the final products reported for [methyl-^{11}C]- and [1-^{11}C]-methionine is higher than 98 and 94 %, respectively[1-4,7,8]. Although the authors have not identified the radiochemical impurities, it is reasonable to assume that [^{11}C]methionine sulphoxide is present in minor amounts. Interestingly, the only oxidized ^{11}C-labelled methionine derivative found in the literature is a methionine sulphone containing modified ACTH fragment [14].

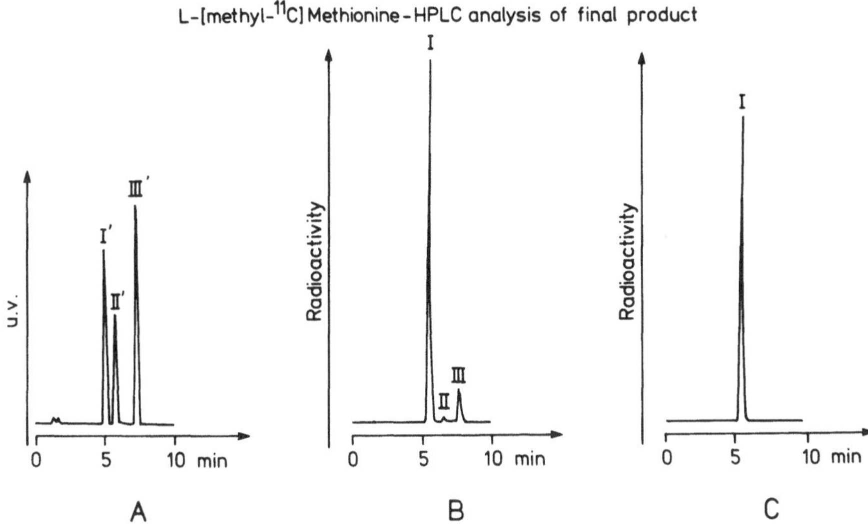

Figure 2. Analysis of a mixure of: A: I'-methionine, II'-methionine sulphone and III'-methionine sulphoxide (u.v. detector); B and C: final [methyl-^{11}C]methionine solution (radioactivity detector) from B: an old, and C: a new C-18 preparative column. HPLC conditions are descibed under 3.2.1.

3.2.1. *Analysis of oxidation products.* The Turku PET radiochemistry group has routinely analyzed [methyl-^{11}C]methionine batches with HPLC using a Waters μBondapak NH_2 column and conditions previously described [4] (system 2) except that the column has been operated at room temperature. The system was calibrated with standards for methionine and its sulphoxide and sulphone. The quality of the final product was found to be highly dependent on the quality of the HPLC column used in the purification as shown in Figure 2.
The order of elution of these three compounds are reversed on semi-preparative purification when a C-18 μBondapak column is used. When a column is new, two separate peaks are obtained for [^{11}C]methionine and [^{11}C]methionine sulphoxide (the sulphone co-elute with methionine). This separation gradually decreases with time. Addition of dilute hydrochloric acid to the crude product before purification substancially increases the lifetime of the column [Antoni, personal communication]. Anyhow, the analytical amino columns have a long life time. This ensures that degradation of the preparative column will be detected on analysis of final [^{11}C]methionine batches.

3.2.2. *How to avoid oxidation?* The amount of [^{11}C]methionine sulphoxide in the crude product of the thiolactone synthesis is 1-3 % [Johnström, personal communication]. When L-[methyl-^{11}C]methionine is prepared in Turku by the S-benzyl homocysteine method this percentage vary between 0.5 and 25 %. Although the S-benzyl homocysteine procedure is less reproducible with respect to the sulphoxide content in the crude product, the yield and purity of the final product is still very high when a high quality semi-preparative column is used.

The best way to avoid oxidation, especially after purification, is to avoid contact with oxygen (or air). In both synthesis methods the solvent (ammonia or acetone) is removed before purification. It is strongly recommended that an inert gas is used for this purpose. The most important part of the preparation, however, is to use an HPLC solvent that can be sterile filtered directly and used for injection. Evaporation of the purified product is not recommended.

3.2.3. *Analysis samples of final product solutions*. It is recommended that the analysis sample is placed into a septum-equipped vial which has been thoroughly rinsed with dry nitrogen gas. If a small sample is placed into an open vessel and analyzed later on, the composition of the sample may not be representative of the final product. It is also important in this context to remember to use freshly prepared standard solutions of methionine for reference, as these aqueous solutions are oxidized within a few days.

3.3. OTHER QUALITY CONTROL ASPECTS

Other quality control aspects like radionuclidic and chemical purity, specific radioactivity, pharmaceutical quality, stability and shelf-life have recently been covered in an other report [6].

4. Conclusions

Both the S-benzyl-homocysteine and the homocysteine thiolactone synthesis methods have proven their validity for the preparation of L-[methyl-^{11}C]methionine as a radiopharmaceutical for PET studies during the last 16 years. The choice of method is merely a matter of the laboratory routines which are employed at the site of preparation. The L-[1-^{11}C]methionine synthesis method is hampered by the need of chiral HPLC for purification. As the research and development in chiral HPLC is very intensive this will not present a severe problem in the near future. The clinical potency of L-[1-^{11}C]-methionine for PET studies in man also has to be evaluated and demonstrated.
The incentive of this report has been to demonstrate the need of a calibrated analytical procedure for the detection of ^{11}C-labelled oxidized species in [^{11}C]methionine final product solutions. This is important regardless of labelling position. In addition an example of an accurate method for determination of the enantiomeric purity has been given.

Acknowledgement
I a grateful to Dr. Pertti Lehikoinen for collaboration in the preparation and analysis of [methyl-^{11}C]methionine.

References

1. Långström B, Lundqvist H. The preparation of ^{11}C-methyl iodide and its use in the synthesis of ^{11}C-methyl-L-methionine. Int J Appl Radiat Isot 1976; 27:357-363.
2. Comar D, Cartron J-C, Maziere M, Marazano C. Labelling and metabolism of methionine-methyl-^{11}C. Eur J Nucl Med 1976; 1:11-14.
3. Berger G, Maziere M, Knipper R, Prenant C, Comar D. Automated synthesis of ^{11}C-labelled radiopharmaceuticals: imipramine, chlorpromazine, nicotine and methionine. Int J Appl Radiat Isot 1979; 30:393-399.

4. Långström B, Antoni G, Gullberg P, Halldin C, Malmborg P, Någren K, Rimland A, Svärd H. Synthesis of L- and D-[methyl-^{11}C]methionine. J Nucl Med 1987; 28:1037-40.
5. Meyer G-J, Osterholz A, Hundeshagen H. Routine quality control of ^{11}C-labelled radiopharmaceuticals by high pressure liquid chromatography. J Radioanal Chem 1983; 80:229-235.
6. Meyer G-J, Coenen HH, Waters SL, Långström B, Cantineau R, Strijckmans K, Vaalburg W, Halldin C, Crouzel C, Mazière M. Quality assurance and quality control of short-lived radiopharmaceuticals. EEC report (in press).
7. Bolster JM, Vaalburg W, Elsinga PH, Wynberg H, Woldring MG. Synthesis of DL-[1-^{11}C]methionine. Appl Radiat Isot 1986; 37:1069-1070.
8. Bolster JM, Vaalburg W, Elsinga PH, Ishiwata K, Vissering H, Woldring MG. The preparation of ^{11}C-carboxylic labelled L-methionine for measuring protein synthesis rates. J Labelled Compd Radiopharm 1986; 23:1081-1082.
9. Johnström P, Ehrin E, Stone-Elander S, Nilsson JLG. Synthesis of ^{11}C-labelled D- and L-methionine for positron emission tomography. Investigation of the enantiomeric purity of the products. Acta Pharm Suec 1984; 21:189-194.
10. Ishiwata K, Ido T, Vaalburg W. Increased amounts of D-enantiomer dependent on alkaline concentration in the synthesis of L-[methyl-^{11}C]methionine. Appl Radiat Isot 1988; 39:311-314.
11. Lundqvist H, Långström B, Malmqvist M. Determination of enantiomeric purity in biogenic ^{11}C-labelled amino acids by aminoacylation of tRNA. J Radioanal Nucl Chem 1985; 89:79-87.
12. Någren K, Långström B. Synthesis of D/L- and L-Se-[methyl-^{11}C]selenomethionine. J Labelled Compd Radiopharm 1988; 25:133-139.
13. Bodanszky M. Principles of peptide synthesis. Berlin: Springer-Verlag, 1984: 146-148.
14. Janssen WPA, van Nispen JW, Jansen JFGA, Wiegman T, Paans AMJ, de Wied D, Woldring MG, Vaalburg W. Rapid conversion of a homocysteine-containing peptide into its corresponding $^{11}CH_3$-labeled methionine sulfone analogue. PET experiments with H-[S-CH_3]Met(O_2)-Glu-His-Phe-D-Lys-Phe-OH. J Labelled Compd Radiopharm 1989; 26:248-250.

PRODUCTION OF L-[^{18}F]FLUORO AMINO ACIDS FOR PROTEIN SYNTHESIS: OVERVIEW AND RECENT DEVELOPMENTS IN NUCLEOPHILIC SYNTHESES

C. LEMAIRE

ABSTRACT. The main methods available for the preparation of [^{18}F]aromatic amino acids, such as fluorophenylalanine and fluorotyrosine, which have been suggested for the measuring of local incorporation into protein by positron emission tomography are examined. This includes old methods as the Balz-Schiemann reaction, well established methods as electrophilic fluorination, regioselective fluorodemetalation as well as recent development in nucleophilic syntheses.

1. Introduction

For the study of cerebral protein synthesis rates (PSR) in human by PET, almost exclusively α-amino acids labelled in various positions with the positron emitters ^{11}C ($T^{1/2}$= 20 min) and ^{18}F ($T^{1/2}$= 110 min) have been investigated (Vaalburg *et al.*, 1992).

Although the introduction of fluorine into organic molecules is not often easy, ^{18}F-fluorinated amino acids are, comparatively to ^{11}C-labelled amino acids, particularly interesting as radiotracers for use in positron emission tomography. This is probably due to the small sterical changes caused by the fluorination, the high bond-strength of the carbon-^{18}F bond and the decay characteristics of this tracer, especially its half life of 110 min which allows relatively long time synthesis and is compatible with the study of relatively slow cerebral PSR.

From a chemical point of view, aliphatic and aromatic fluoro amino acids are characterized by the presence of an asymmetric carbon which lead to one pair of enantiomers, having respectively D and L configuration. In the Cahn-Ingold-Prelog system, in which the substituents present on an asymmetric carbon are distributed according to a set of rules all the natural L-α-amino acids can be classified of (S) configuration and the D of (R) configuration. Glycine, the simplest amino acid, is the only one that is not optically active. Although the physical and chemical properties of two enantiomers can be very similar, the spatial distribution of the substituent around the asymmetric center may modified drastically the properties of the compound. For example, α-amino acids present dissimilarities in their taste properties, the L form can taste bitter and the corresponding D isomer be sweet (Coppola G.M. *et al.*). Moreover, most amino acids are biologically active only in one enantiomeric form and occur in proteins as the L or (S) enantiomers.

B. M. Mazoyer et al. (eds.), PET Studies on Amino Acid Metabolism and Protein Synthesis, 89–108.

Enantiomeric purity is consequently of great importance and the amino acids must, therefore, be synthesized as enantiomerically pure compounds. For PSR only the L or (S) form should be prepared.

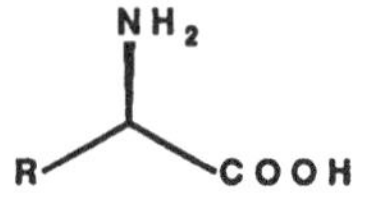

D or (R) amino acid 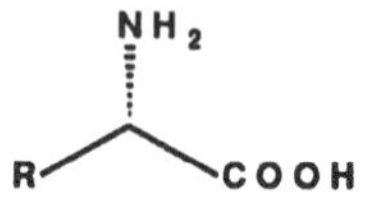 L or (S) amino acid.

The aim of this paper will not be to discuss or make recommendations on the optimal ^{18}F-fluorinated amino acids for PSR measurement, but to summarize and comment on the main methods of synthesis available for these compounds in the literature. Although synthesis of fluorine-18 labelled aliphatic amino acids such as fluoroproline are known (Van der Ley, 1983), in this paper, focus will be given on the main problems encountered in labelling of fluoro aromatic amino acids such as 4-fluorophenylalanine and 2-fluorotyrosine which have been proposed for probing protein synthesis by PET. Both electrophilic and nucleophilic forms of fluorine-18 are required to synthesize these compounds.

2. ^{18}F-Production

Among the numerous nuclear reactions capable of producing electrophilic and nucleophilic ^{18}F species which are described in the literature, only a limited number are adequate for routine fluorine-18 production (Guillaume *et al.*, 1991). The characteristics of the two main routes of fluorine-18 production which are certainly the most adequate for medical cyclotron are summarized in table 1.

Table 1. Main characteristics of electrophilic and nucleophilic ^{18}F-productions

	Electrophilic route	Nucleophilic route
Nuclear reaction	$^{20}Ne(d, \alpha)^{18}F$	$^{18}O(p, n)^{18}F$
Batch of production (E.O.B.) (1h, 10 μA)	150 mCi	600-1000 mCi
Fluorinating agent	$[^{18}F]F_2$ $CH_3COO[^{18}F]F$	$^{18}F^-$ ^{18}F-alkyl-X
Theoretical maximum radiochemical yield	50 %	100 %
Mass involved	50-200 μmol.	"n.c.a."
Specific activity	2 mCi/μmol.	> 1000 mCi/μmol.

The $^{20}Ne(d, \alpha)$ and $^{18}O(p, n)$ nuclear reactions lead to enough fluorine activities for both a moderate projectile energy and beam current (Guillaume *et al.*, 1991) and allows the production of various electrophilic and nucleophilic fluorinating agents.

On one hand, very reactive [^{18}F]fluorine gaz is produced by bombardement with deuterons of a mixture of F_2 and neon in a nickel passivated target. The activity produced by this gazeous Neon target amounts in our laboratory to 150 mCi and can only be extracted in presence of molecular fluorine as carrier (0.13 %). Consequently, the specific activity of this electrophilic fluorinating agent and others derivatives such as the less reactional acetyl hypofluorite are low (2 Ci/mmol). Moreover, in this case, half of the ^{18}F activity is lost as fluoride salt.

On the other hand, the less reactional nucleophilic ^{18}F-fluoride ion is easily produced with high specific activities by the $^{18}O(p, n)$ nuclear reaction on 18-oxygen enriched water. The available activities are high (> 600 m Ci) and the no-carrier-added fluoride ion can be recovered without the addition of any carrier (specific activity ≥ 1000 mCi/ μmol). This primary fluorinating agent ($^{18}F^-$) belong to the nucleophilic class and can be incorporated with a theoretical yield of 100 % in organic molecules. Many others n.c.a. electrophilic agents such as [^{18}F]fluoroalkylating agents, can be obtained from n.c.a. ^{18}F-fluoride (F-alkyl-X; X= tosylate, triflate, ...). Production of [^{18}F]fluorinated radiopharmaceuticals are only possible from these few primary or secondary electrophilic and nucleophilic fluorinating agents (Kilbourn, 1990).

3. Radiochemistry

3.1. THE BALZ-SCHIEMANN REACTION

Twenty-five years ago, the only method which was known for the introduction of ^{18}F fluoride into aromatic rings was the Balz-Schiemann reaction in which a diazonium tetrafluoborate precursor was labelled by exchange with no-carrier-added ^{18}F (Palmer *et al.* 1977, review). Heating of this labelled fluoroborate diazonium salt lead to the introduction of ^{18}F-fluorine into the aromatic ring of the amino acid. After hydrolysis of this intermediate fluoroester, the free fluoro amino acid was obtained. The Balz-Schiemann reaction is regioselective and all positional isomers can be synthesized.

This labelling method has been successfully applied to the preparation of the three positional isomers of [^{18}F]fluorophenylalanine (position 2, 3, 4) and of 3-[^{18}F]fluorotyrosine (Scheme 1).

If the maximum theoretical radiochemical yield obtainable from a diazonium fluoroborate salt is 25%, in practice, it is lower because by-product formation and very long time of synthesis are required. The specific activity of synthesized compounds is also low since carrier added starting precursors are needed for the synthesis. The hydrolysis stage generated at the end of the synthesis led to a mixture of D,L amino acids requiring resolutions of the enantiomers. All these reasons explain why Balz-Schiemann reaction is not in use anymore and why others methods of fluorination were developped.

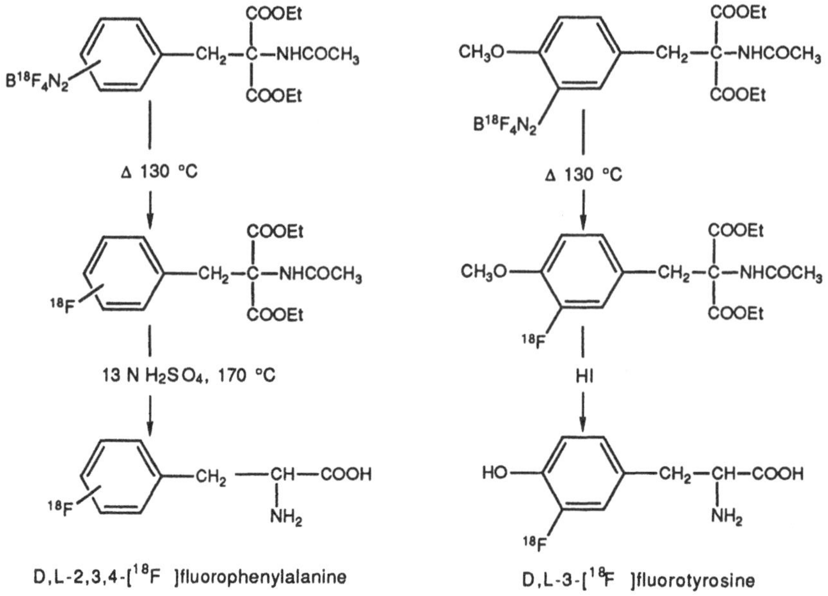

D,L-2,3,4-[^{18}F]fluorophenylalanine

D,L-3-[^{18}F]fluorotyrosine

Scheme 1.
Preparation of aromatic amino acids via the Balz-Schiemann reaction

3.2. ELECTROPHILIC FLUORINATION

Most of the others methods so far reported for the synthesis of fluoroamino acids are based on the direct electrophilic fluorination of unprotected or protected aromatic rings of amino acids. This fixedness may be explained by different reasons:

- nucleophilic substitution with [^{18}F]fluoride was considered during a long time as difficult and a challenge while enough activities of reactive electrophilic agents such as molecular fluorine or less reactive acetyl hypofluorite were currently available by the ^{20}Ne(d, α)^{18}F reaction and low energy cyclotron,
- fluoro amino acids generally exhibit low toxicity and high specific activities were not mandatory,
- the aromatic ring of amino acids was particularly suitable to react with an electrophilic agent such as molecular fluorine or acetyl hypofluorite.

These electrophilic routes imply the direct fluorination of aromatic rings of unprotected L-phenylalanine ($R_{1,2,3,4,5,6}$=H; R_7=OH; Scheme 2) and L-tyrosyne ($R_{1,2,4,5,6}$=H; $R_{3,7}$=OH) with [^{18}F]F_2 or [^{18}F]AcOF (Coenen *et al.*, 1986a, b; 1988; 1989). The influence of the reaction solvent on the radiochemical yield and positional selectivity of these two electrophilic fluorinating agent has been reported. From this recent study, it appears that trifluoroacetic acid was the solvent of choice (Coenen *et al.*, 1988). Unfortunately, due to the high reactivity of the electrophilic fluorinating agents, a mixture of all possible positional isomers is obtained, the corresponding 2 and 3 isomer of fluorophenylalanine and fluorotyrosine respectively being formed preferentially (Table 2). By this electrophilic method, the yields in 4-[^{18}F]fluorophenylalanine and 2-[^{18}F]fluorotyrosine are lower than 5 %. Similar results have been obtained by other groups (Murakami *et al.*, 1988a, b).

R_1, R_2, R_3, R_4, R_5, COR_7, NHR_6

1) [^{18}F]F_2 or [^{18}F]AcOF

2) HPLC
or hydrolysis and HPLC

COOH, NH_2, R_3, R_5

4 - [^{18}F]fluorophenylalanine
R_3= ^{18}F; R_5= H

2 - [^{18}F]fluorotyrosine
R_3= OH; R_5= ^{18}F

Scheme 2. General pathway for electrophilic syntheses of various isomers of L-fluorophenylalanine and L-fluorotyrosine

Some attempts, including more elaborated regioselective reactions such as fluorodemetallation or in some cases the use of adequate protecting group were then made to increase the radiochemical yield of these two compounds. The result of these improvements are summarized in the Table 2.

The regioselective synthesis of L-4-[^{18}F]fluorophenylalanine implies electrophilic substitution at 0°C by [^{18}F]F_2 of a 4-substituted trimethyltin derivative of L-phenylalanine in carbon tetrachloride ($R_{1,2,4,5}$=H; R_3=Sn(Me)$_3$; R_6= $COCF_3$; R_7= NHC(Me)$_3$; Scheme 2) (Coenen *et al.*, 1986). In this case, the regioselectivity in the four position is optimum and only one positional isomer was detected. The radiochemical yields are high (25%). Unfortunately, the preparation of the metallated precursor is not always easy and at this time 2-fluorotyrosine can not be prepared by such a procedure.

In the case of tyrosine, acetylation of the phenol group increases fluorination in the 2-position of the molecule ($R_{1,2,4,5,6}$=H; R_3=OAc; R_7=OH; Scheme 2) (Coenen *et al.*, 1988, 1989) . The protection of the phenol group being easily achieved with acetic anhydride in perchloric acid. This approach led to a yield of 17% in 2-[^{18}F]fluorotyrosine after fluorination with [^{18}F]acetyl hypofluorite and hydrolysis while only 3 % of the 3-isomer were formed in the same conditions (Table 2). Fluorination with acetyl hypofluorite lead generally to less by-product than molecular fluorine and facilates HPLC purification of the interesting isomer. As radiochemical yield is not very different than with F_2 (Table 2), for radioprotection problems, fluorination with AcOF is preferable as half of the radioactivity is lost on the acetate column during its preparation.

All these electrophilic methods are generally fast, require starting compounds which are often commercially available or easily synthesized and proceed without racemisation of the starting L-isomer neither during radiofluorination nor final hydrolysis. Except in the fluorination of a tin compound of phenylalanine, these syntheses are not very regioselective and a mixture of all possible positional isomers is generally obtained (Table 2). The interesting isomers are separated by HPLC with difficulty due to the incomplete separation often observed beetween the different peaks. For routine production, the peak cutting technique leads therefore to yields smaller than under analytical conditions.

Table 2. Radiochemical yields of various L-isomers of fluorophenylalanine and fluorotyrosine via electrophilic syntheses (% E.O.B.)[(x)]

Starting substrate	Fluorinating agent	Radiochemical yield (% E.O.B.) Isomers		
		2	3	4
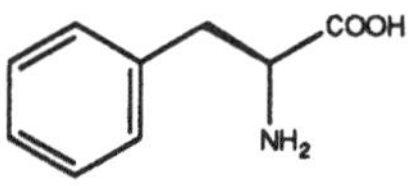 COOH, NH_2	$[^{18}F]F_2$	20.3	3.9	3.9
	$CH_3COO[^{18}F]F$	17.7	2.4	1.6
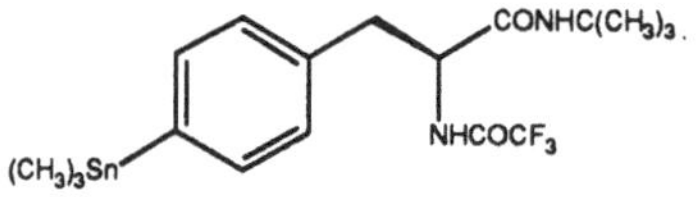 $CONHC(CH_3)_3$, $NHCOCF_3$, $(CH_3)_3Sn$	$[^{18}F]F_2$	-	-	25
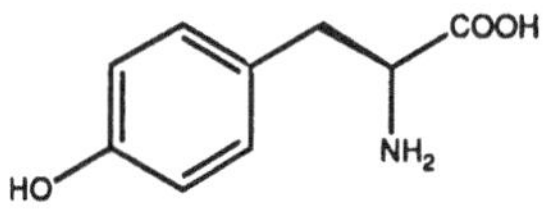 COOH, NH_2, HO	$[^{18}F]F_2$	2.3	28.5	
	$CH_3COO[^{18}F]F$	1.2	20.7	
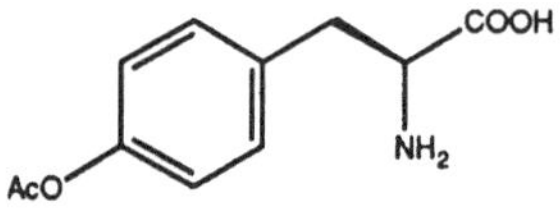 COOH, NH_2, AcO	$[^{18}F]F_2$	11.3	4.6	
	$CH_3COO[^{18}F]F$	8.2	1.7	

(x) Yields are based on activity extracted from the target

3.3. NUCLEOPHILIC SYNTHESES

Recently, a few attempts have been made to produce L-aromatic amino acids with n.c.a. $[^{18}F]$fluoride. Comparatively to the electrophilic approach, the nucleophilic route presents some potential advantages:

- the fluorinated amino acids would become available for centers which produce fluorine-18 either with thermal neutrons from a nuclear reactor or with protons-only cyclotron,
- these nuclear reactions led to high amounts of activity averaging from 0.1 to 1 Ci, the $^{18}O(p, n)$ reaction being certainly the most effective method for the production of n.c.a. ^{18}F,
- the targetry developed for the production of nucleophilic fluoride by the $^{18}O(p, n)$ ^{18}F reaction is more simple than those required for the production of electrophilic agent by the $^{20}Ne(d, \alpha)^{18}F$ reaction,
- the ^{18}F-fluoride ion has become easier to use than the electrophilic species,
- the synthesis should lead to a maximum of regioselectivity (only the desired isomer) and stereospecifity (only L form),
- the very high specific activity of the ^{18}F-fluoride ion (n.c.a.) should allow to evaluate the effect of carrier on the *in vivo* parameters of these amino acids,

- the fluoride ion can be incorporated with a theoretical yield of 100 % in a large variety of organic substrates.

3.3.1. *The Azlactone Method.* As electrophilic methods lead generally to a mixture of several isomers, a nucleophilic synthesis based on the regiospecific exchange between a good leaving group and fluorine-18 appears to be a more regioselective method. Although not feasible for all aromatic substrate, nucleophilic aromatic substitution is now considered as a general and reliable route when the phenyl ring carries an ortho or para leaving group, generally halogen, nitro or $N^+(Me)_3$, activated by an electron withdrawing functional group such as cyano, aldehyde, ketone or nitro (Angelini *et al.*, 1985; Attina *et al.*, 1983a, 1983b; Lemaire *et al.*, 1992; Shiue *et al.*, 1984).

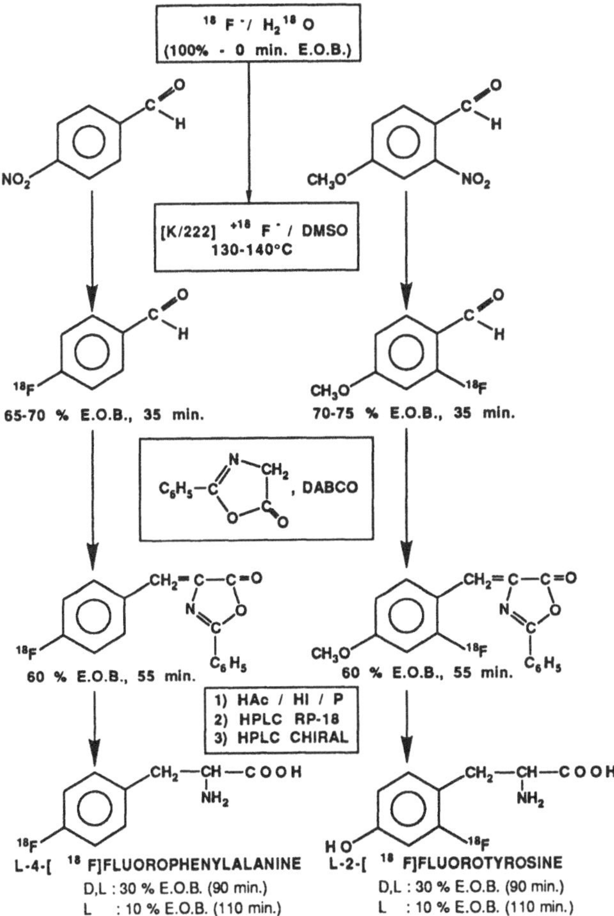

Scheme 3. NCA synthesis of L-4-[^{18}F]fluorophenylalanine and L-2-[^{18}F]fluorotyrosine

Direct nucleophilic fluorination of the inactivated 4-nitrophenylalanine and 2-nitrotyrosine being not possible, the first n.c.a. preparation of these amino acids has been carried out by the multi step synthesis outlined in Scheme 3.

This approach required first the no-carrier-added ^{18}F-nucleophilic aromatic substitution of the activated nitro groups of p-nitrobenzaldehyde (commercially available) and 4-methoxy-2-nitrobenzaldehyde (which must be prealably synthesized) (Lemaire *et al.*, 1987, 1992a). In such a synthesis, the reactivity of the non reactional ^{18}F-fluoride anion was previously increased by using K^{18}F/aminopolyether kryptofix in dipolar aprotic solvent (DMSO) (Coenen *et al.*, 1986c). The radiochemical yields of 4-[^{18}F]fluorobenzaldehyde was in the 65-70% range, and slightly higher for the 2-[^{18}F]fluoro-4-methoxybenzaldehyde (70-75%). By using a microwave oven, these labelled aldehydes are obtained with similar yields as by using classical heating (aluminium block, 140°C, 20 min). The advantage of microwave over classical heating is the short reaction time required (2 min, 300 W). These yields are comparable with those reported for isomeric substituted nitrobenzaldehydes. (Lemaire *et al.*, 1991a, 1992b; Ding *et al.*, 1990, 1991; Chakraborty and Kilbourn, 1991). Among activating groups, the aldehyde function appears as particurlarly interesting comparatively to other activating groups owing to the high radiofluorination yields and the wide scope it offers for subsequent chemical syntheses (Lemaire *et al.*, 1991a, 1992).

In the second step of the synthesis, the [^{18}F]fluorinated aldehydes were converted into their corresponding azlactones as previously described for the preparation of various ^{11}C-α-amino acids (Halldin and Langström, 1984). The following reduction and hydrolysis with hydroiodic acid and red phosphorous yielded after purification by HPLC only the 4 and 2 structural isomers of fluorophenylalanine and fluorotyrosine respectively. In order to obtain the desired L-p-[^{18}F]fluorophenylalanine and L-2-[^{18}F]fluorotyrosine, the mixture of D, L isomers (50/50) generated at the hydrolysis step is resolved by HPLC, using a chiral proline column. The decay corrected yield for the optical pure L form or D form of these two amino acids is of 10 % after 110 min of synthesis (Lemaire et al., 1987, 1991b).

If the advantage of this nucleophilic synthesis is its regioselectivity (only one positional isomer), on the other hand, the absence of stereoselectivity required at the end of the synthesis a chiral separation leading to some limitations:

- resolution of racemates is time consuming,
- as half of the activity is discarded the maximum theoretical radiochemical yield is as for the electrophilic methods of 50%,
- the final sample contains 1 to 2 mg of copper which might pose some health problems and must be removed before human studies.

3.3.2. *Asymmetric Synthesis*. From a radiochemical point of view, it would be therefore much more advantageous to not synthesize at all the unwanted isomer. An enantioselective or ideally enantiospecific synthesis should overcome such a difficulty and would give the opportunity of preparing in this last case both enantiomerically pure D or L fluoro amino acids in high radiochemical yields.

The asymmetric synthesis of natural and analog of amino acid is a major challenge for organic chemists and numerous methods have been described in the literature (Haemers *et al.*, 1989; Williams M. W., 1989). Although the most effective way of using chiral reagents in asymmetric synthesis of amino acids is their use as catalysts and that several successes in this type of asymmetric synthesis have been reported, major advance have been realized during this last year in asymmetric synthesis using stoechiometric amounts of chiral auxiliaries.

For the asymmetric construction of a α-carbon of an α-amino acid several approaches can be considered. The asymmetric center can be induced either by C_α-C_β bond formation with anionic or cationic amino acid precursor either by C_α-H bond formation or by C_α-N bond formation. As for all methods of labelling with short lived radioisotopes, the ^{18}F-labelled precursor should be introduced as late as possible in the synthesis scheme and among these various methods, the C_α-C_β bond formation is certainly the only approaches that can be used from a practical point of view in ^{18}F-fluorine radiochemistry.

The overall strategy of such an asymmetric synthesis which has been developed in our laboratory involves removal by a base at low temperature of an α-proton from a chiral protected glycine derivative to give an α-anion of glycine (Scheme 4). This last one is then alkylated diastereoselectively by an electrophile such as an aryl halide to form a new carbon α–carbon β-bond. The last step in the sequence implies the removal of the protecting group to yield after HPLC purification the desirated enriched or pure L-amino acid.

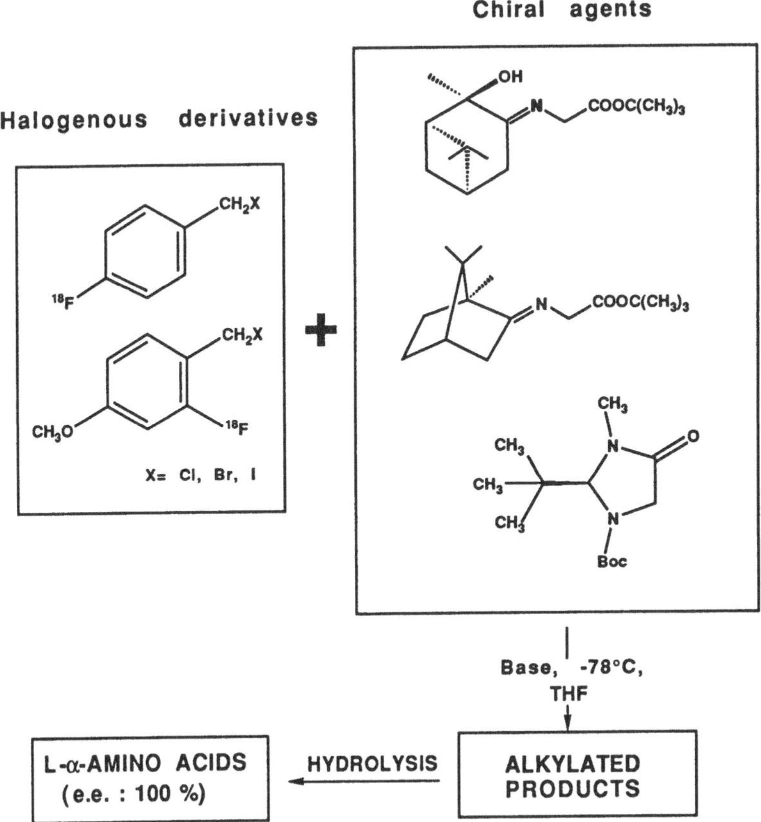

Scheme 4. Basic pathway for asymmetric syntheses of L-4-[^{18}F]fluorophenylalanine and L-2-[^{18}F]fluorotyrosine

Typically, the diastereoselective introduction of the electrophilic group in the molecule will depend of the presence of bulky steric substituent and chiral centers in the molecule. The approach being performed from the less hindered side of the molecule. Asymmetric synthesis occurs because the reagent and the substrate, at least one of which is chiral, form diastereomeric transition states which differ in energy. The magnitude of this difference in energy determine the excess of one enantiomer over the other (e.e. or enantiomeric excess). The problem in chemical asymmetric synthesis is to find reagents that maximize this values. The structure of the chiral inductor chosen is then directly related with enantiomeric excess of free amino acid observed after acid hydrolysis.

Besides a high reactivity and diastereoselectivity in C_{α}-C_{β} bond formation the chiral auxiliary chosen for the enantioselective synthesis should be readily available in both pure enantiomeric forms in order to lead to the L or D form of the radiopharmaceutical. Moreover, the cleavage of the protecting groups to free L or D amino acids must be nearly quantitative and proceed without racemisation.

In accordance with these requirements, three chiral starting compounds have been selected: the imine glycine derivative from pinanone, the (S)camphor imine of tert-butyl glycinate and the Boc-BMI. The two fluoro halogenous compounds shown on the general Scheme 4 have also been investigated.

This multi step synthesis of fluorophenylaline and fluorotyrosine involves four major steps. The first one is the preparation of the [^{18}F]fluorobenzyl bromide derivative followed by the alkylation reaction of the chiral derivative with these benzyl halide compounds, and then subsequent hydrolysis and HPLC purification.

As the first step, the preparation of the two labelled [^{18}F]fluoro benzyl halogenous compounds involves the quantitative conversion of the [^{18}F]luorobenzaldehyde previously described into the corresponding alcohol by treatment with sodium borohydride fixed on a small alumina column and then in the corresponding halogenous derivatives by thionyl bromide ($SOBr_2$) (Scheme 5) (Lemaire *et al.*, 1992a). The decay-corrected radiochemical yields for the halides averaged 40-55 % and the time required for the conversion from the aldehydes was of about 10 min. The radiochemical purity of the crude bromides was higher than 95%.

3.3.2.1. The Pinanone and Camphor Methods. The first chiral starting compound which was selected for the second step of this synthesis was the glycine tert-butyl imine of pinanone (Scheme 6)(Oguri *et al.*, 1978). This chiral agent was successfully previously used by the Upsalla group for the preparation of various ^{11}C-α-amino acids (Antoni G. and Långström B., 1986). The enolate of this Schiff base was generated by two equivalents of the lithium salt of 2,2,6,6-tetramethylpiperidine in dry THF solution at - 78°C. The [^{18}F]fluorobenzyl bromide was then added and the alkylation performed during 10 min.

As the imine group is sensitive to racemisation, hydrolysis of the protecting groups was conducted in two steps. Reaction with hydroxylamine acetate led to the *tert*-butyl ester of fluorophenylalanine and treatment with HCl yielded the free amino acid.

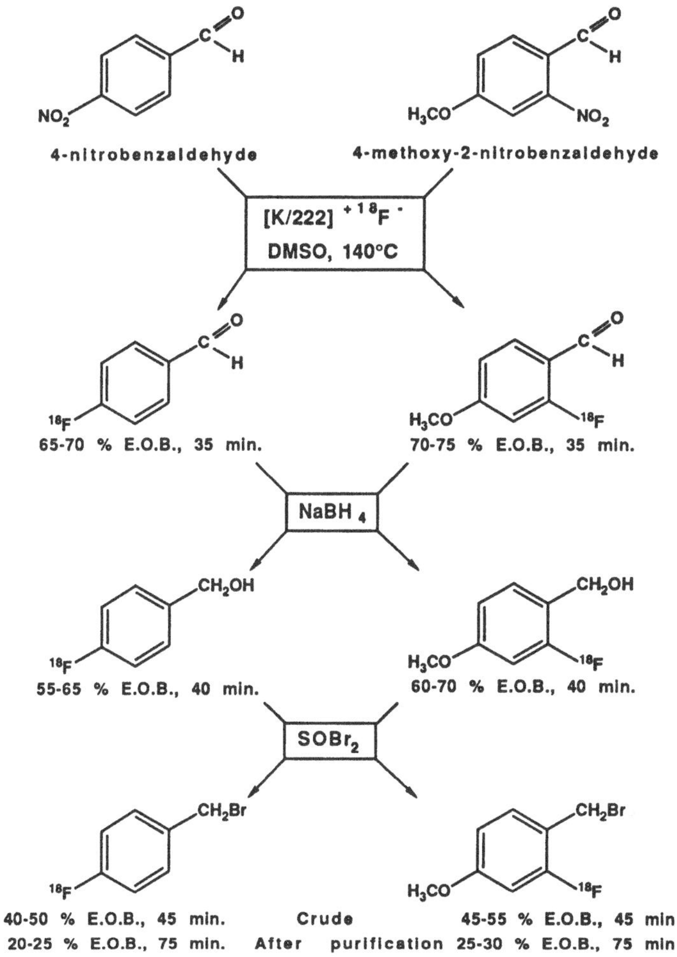

4-nitrobenzaldehyde 4-methoxy-2-nitrobenzaldehyde

[K/222] $^{+18}F^-$
DMSO, 140°C

65-70 % E.O.B., 35 min. 70-75 % E.O.B., 35 min.

$NaBH_4$

55-65 % E.O.B., 40 min. 60-70 % E.O.B., 40 min.

$SOBr_2$

40-50 % E.O.B., 45 min. Crude 45-55 % E.O.B., 45 min
20-25 % E.O.B., 75 min. After purification 25-30 % E.O.B., 75 min

Scheme 5. Radiochemical steps for the preparation of 4-[^{18}F]fluorobenzyl bromide and 2-[^{18}F]fluoro-4-methoxybenzyl bromide

In the last step, purification of the [^{18}F]fluoro amino acids was achieved by preparative HPLC on an ODS column.

The overall yield, decay corrected, averaged beetween 13 and 17% (125 min). The enantiomeric excess determined by ligand exchange chromatography (HPLC) was found of 85 % (L: 92.5 %)(Lemaire *et al.*, 1991b).

In order to increase the enantiomeric purity, an inductor of chirality derivative of camphor was also investigated (Scheme 6)(McIntosh *et al.*, 1986, 1988). The synthetic pathway was similar to those described previously, except that in this case the imine was deprotonated by one equivalent of lithium diisopropylamide (LDA).

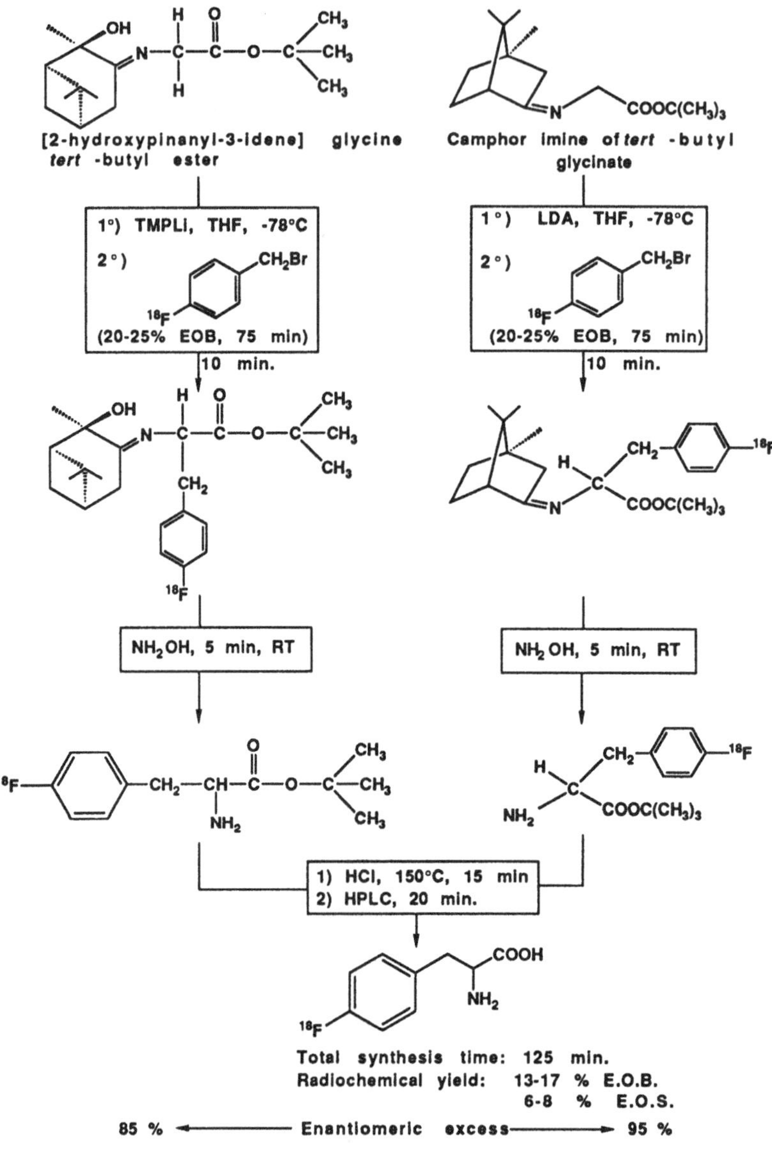

Scheme 6. Asymmetric synthesis of n.c.a. L-4-[^{18}F]fluorophenylalanine via two chiral inductors

Alkylation by the 4-[^{18}F]fluorobenzyl bromide was conducted in similar manner and the hydrolysis was also performed in two step, the imine group being sensitive to racemisation. The final radiochemical yield and enantiomeric purity were in this case of about 15 % and 95 % (L= 97.5 %) respectively. In comparison with the pinanone derivative, this camphor method lead to higher enantioselectivity.

3.3.2.2. The Boc-BMI Method. The Scheme 7 summarizes the results obtained for the synthesis of the 4-[^{18}F]fluorophenylalanine and 2-[^{18}F]fluorotyrosine with the imidazolidinone chiral inductor agent (Boc-BMI) (Seebach *et al.*, 1989).

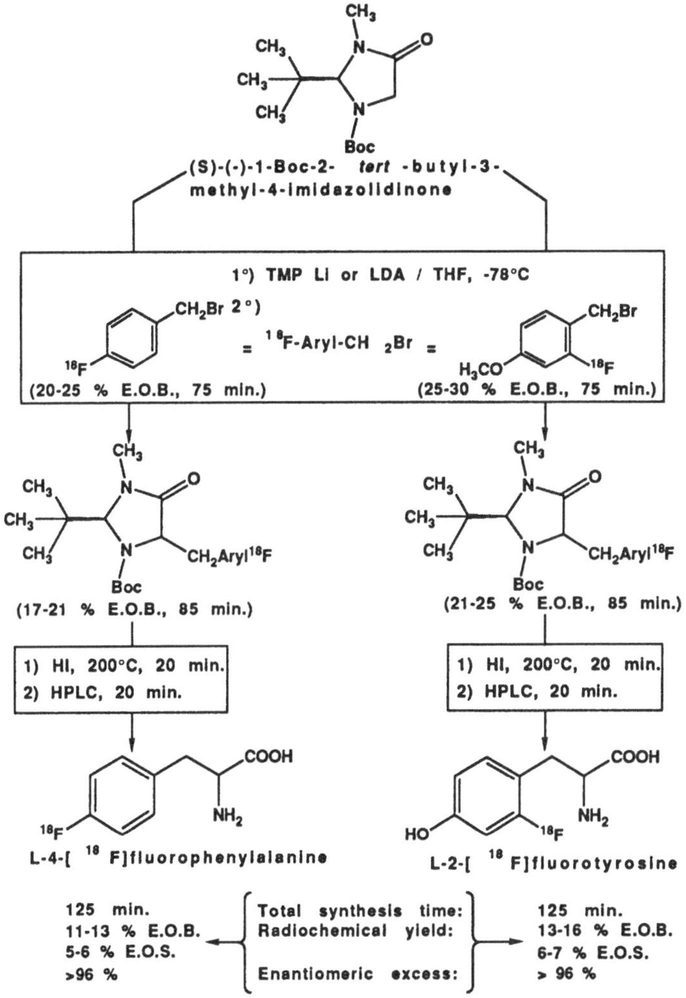

(S)-(-)-1-Boc-2- *tert* -butyl-3-methyl-4-imidazolidinone

1°) TMP Li or LDA / THF, -78°C

2°) = ^{18}F-Aryl-CH_2Br =

(20-25 % E.O.B., 75 min.) (25-30 % E.O.B., 75 min.)

(17-21 % E.O.B., 85 min.) (21-25 % E.O.B., 85 min.)

1) HI, 200°C, 20 min.
2) HPLC, 20 min.

L-4-[^{18}F]fluorophenylalanine L-2-[^{18}F]fluorotyrosine

125 min.	Total synthesis time:	125 min.
11-13 % E.O.B.	Radiochemical yield:	13-16 % E.O.B.
5-6 % E.O.S.		6-7 % E.O.S.
>96 %	Enantiomeric excess:	> 96 %

Scheme 7. Enantioselective syntheses of n.c.a. L-4-[^{18}F]fluorophenylalanine and L-2-[^{18}F]fluorotyrosine starting from Boc-BMI

Comparatively to the two other chiral agents previously investigated, the fluoroalkylated Boc-BMI derivative can be hydrolysed directly with HI without racemisation. In this case, after HPLC purification, the enantiomeric purity of the final solution was found to be higher than 96 % (L= 98%), the overall yield averanging between 5 and 7% EOS (125 min) (Lemaire C., 1991b).

In comparison with the pinanone and camphor derivatives, the Boc-BMI method is more promising since alkylation reaction leads to very high e.e.. Moreover, the imidazolidone reagent is less sensitive to racemisation during hydrolysis and is commercially available in its two enantiomeric forms. By such a procedure fluoro amino acids of R or S configuration are now available by nucleophilic syntheses without final chiral separation. Although e.e. as close as 96 % (L= 98 %) have been obtained, this procedure is difficult to apply for routine application owing to the complexity of the various step to be followed and the low yield obtained after long synthesis time due mainly to a complex purification of the benzyl bromides.

Indeed, although radiochemical purity of the alkylating agent was high (> 95 %), in order to obtain high and reproducible yield in the alkylation step of the three chiral inductors selected, high chemical purity of the electrophilic ^{18}F-agent was required. Among the various methods investigated for the purification of the crude benzyl bromide, a reproducible high yield was only obtained when purification of the fluorobenzyl derivatives was performed on a silica gel column eluted with a mixture hexane/ether (90/10). The long time required for this purification step (30 min) and the loss of activity due to unexpected secondary reaction on the support are certainly the two main limitations of this method. In order to increase the yield, a new method of preparation of the halogenous benzyl derivatives was required. Moreover, starting the synthesis from the $N^+(Me)_3$ derivative should simplify the final preparative HPLC purification.

The first optimization of this synthesis in view of routine preparation of 2-[^{18}F]fluorotyrosine has consisted in the nucleophilic substitution of the aryltrimethylammonium triflate in DMSO for 10 min (Scheme 8) (Lemaire *et al.*, 1992 b).

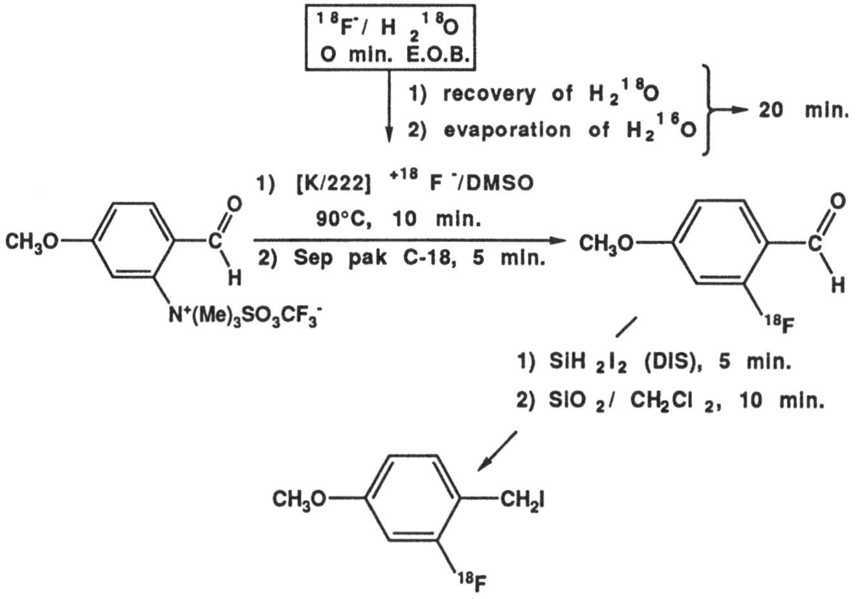

$^{18}F^-$/ $H_2{}^{18}O$
0 min. E.O.B.

1) recovery of $H_2{}^{18}O$
2) evaporation of $H_2{}^{16}O$
20 min.

CH_3O — $N^+(Me)_3SO_3CF_3^-$ — CHO

1) [K/222] $^{+18}F^-$/DMSO
90°C, 10 min.
2) Sep pak C-18, 5 min.

CH_3O — ^{18}F — CHO

1) SiH_2I_2 (DIS), 5 min.
2) SiO_2/ CH_2Cl_2, 10 min.

CH_3O — CH_2I — ^{18}F

Scheme 8. Reaction scheme for the preparation of the 2-[^{18}F]fluoro-4-methoxybenzyl iodide

The cationic nature of this precursor greatly facilitates the purification of the ^{18}F-fluorinated aldehyde which contrarily to the nitro derivative can be recovered in CH_2Cl_2 without the starting precursor after Sep pak purification (Radiochemical yield: 70 % E.O.B., 35 min.).

The second optimization has required the preparation of the corresponding labelled ^{18}F-fluorinated benzyl iodide (Scheme 8). It is easily obtained in one step by treatment of the labelled aldehyde by diiodosilane (DIS), a reagent which allows reductive iodination of aldehydes (Keinan et *al., 1990*) (Scheme 8). The halogenous compound is quickly and easily purified by column chromatography on a small silica gel column eluted with methylene chloride (Radiochemical yield: 56 % E.O.B., 50 min.).

After evaporation and solubilization in THF, the electrophilic alkylating agent (2-[^{18}F]fluoro-4-methoxybenzyl iodide) is available for reproducible alkylation of the lithium enolate of Boc-BMI previously generated by treatment with lithium diisopropylamide (Scheme 9).

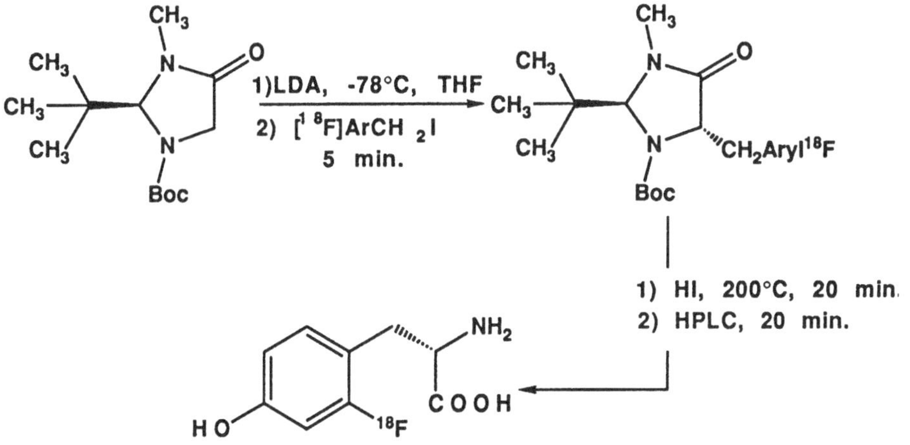

1)LDA, -78°C, THF
2) [^{18}F]ArCH$_2$I
5 min.
$CH_2Aryl^{18}F$
1) HI, 200°C, 20 min.
2) HPLC, 20 min.
NH_2
COOH
HO
^{18}F

Scheme 9. Enantioselective synthesis of 2-[^{18}F]fluorotyrosine starting from 2-[^{18}F]fluoro-4-methoxybenzyl iodide

After reaction (-78°C, 5 min.), hydrolysis and HPLC purification, radiochemically pure 2-[^{18}F]fluorotyrosine is obtained with an overall yield of 43% E.O.B. and an enantiomeric excess higher than 96% (L $\geq$ 98 %, D $\leq$ 2 %). This last high enantiomeric excess is in good agreement with those recently obtained for the preparation of 6-[^{18}F]fluorodopa (Lemaire *et al.*, 1991c) and 4-[^{18}F]fluoro-m-tyrosine (Lemaire *et al.*, 1991d) and those published by Fasth *et al.* for the asymmetric synthesis of [β-^{11}C]amino acids (Fasth *et al.*, 1991; Plenevaux *et al.*, 1992).

By such a synthesis performed in our laboratory, it has been possible starting from 100 mCi of ^{18}F-fluoride, to produce 23 mCi of radiochemically pure 2-[^{18}F]fluorotyrosine after 95 min of synthesis.

Based on these results, the potentialities of synthesis of 2-fluorotyrosine both by nucleophilic and electrophilic routes have been compared in Table 3. The first important point to be considered is the amount of electophilic or nucleophilic fluorine-18 available at the end of bombardment. The deuton, alpha reaction on neon leads in our laboratory to a maximum of 150 mCi of molecular fluorine (Table 1). The radiochemical yield of 10 % of the best electrophilic synthesis does not allow to obtain more than 10 to 15 mCi of this amino acid at the end of the synthesis (Coenen et al., 1989)(Table 3).

The larger amount of fluorine-18 which is easily produced by $^{18}O(p, n)$ reaction should give access, for a yield of 43 %, to batches of 2-[^{18}F]fluorotyrosine ranging beetween 140 mCi and 230 mCi if synthesis is performed with activity averaging beetween 600 mCi and 1000 mCi respectively (Table 3). Although these values are extrapolation of results obtained at lower level of activity (100 mCi), we think that they are realistic.

Table 3. Main characteristics of electrophilic and nucleophilic production routes of 2-[^{18}F]fluorotyrosine

	Electrophilic route	Nucleophilic route
Regioselectivity	Isomers 2, 3	Isomers 2 only
Enantiomeric excess	> 97 %	≥ 96 %
Synthesis time	60 min.	95 min.
Typical radiochemical yield	10 %	43 %
E.O.B. available activity	15 mCi	258-430 mCi
E.O.S. available activity	10 mCi	142-236 mCi
Specific activity (E.O.S.)	2 mCi//µmol.	1000 mCi/µmol.

4. Conclusion

In this paper, Balz-Schiemann reaction, electrophilic methods and nucleophilic approaches were considered for the synthesis of 4-[^{18}F]fluorophenylalanine and 2-[^{18}F]fluorotyrosine. Among these chemical syntheses, the electrophilic method is the only one which is actually in use for routine production of 2-[^{18}F]fluorotyrosine (Coenen *et al.*, 1989). The main problems in such a production results from low yield after complex HPLC purification and sterile filtration (7%) (Vaalburg *et al.*, 1992). As an alternative to this electrophilic procedure, nucleophilic asymmetric synthesis using the (S) or (R)-(-)-1-Boc-2-*tert*-Butyl-3-methyl-4-imidazolidinone [Boc-BMI] can be now considered as particularly interesting. The high enantiomeric purity observed at the end of the synthesis means that a chiral separation of the enantiomers is not necessary (L> 98 %). Moreover, the available activity (E.O.S.) is high and more than sufficient for PET studies.

Starting from differents N^+ derivatives, it is also possible to extend this general method to the preparation of other amino acids, at the condition that the $N^+(Me)_3$ compounds be activated by an electron withdrawing fonction. Actually, such a synthesis is in use in our laboratory for routine production of 6-[^{18}F]fluorodopa.

The chemical community will decide over the years to come which approach is the most useful but upon these findings, the nucleophilic synthesis appears as a very promising method for the preparation of a large variety of n.c.a. amino acids.

5. References

Angelini G., Speranza M., Wolf A.P. and Shiue C.-Y. (1985) Nucleophilic aromatic substitution of activated cationic groups by ^{18}F-labeled fluoride. A useful route to no-carrier-added (NCA) ^{18}F-labeled aryl fluorides. *J. Fluorine Chem.* 27, 177.

Antoni G. and Långström B. (1986) Asymmetric synthesis of L-[3-^{11}C]alanine. *Acta Chem. Scand.* B40, 152-156.

Attina M., Cacace F. and Wolf A.P. (1983a) Displacement of a nitrogroup by [^{18}F]fluoride ion. A new route to aryl fluorides of high specific activity. *J. Chem. Soc., Chem. Commun.* 108-109.

Attina M., Cacace F. and Wolf A. P. (1983b) Labelled aryl fluorides from the nucleophilic displacement of activated nitro groups by ^{18}F-F$^-$. *J. Labelled Compd. Radiopharm.* 20, 501-514.

Chakraborty P.K., and Kilbourn M.R. (1991) Oxidation of substituted 4-fluorobenzaldehydes: application to the no-carrier-added syntheses of 4-[^{18}F]fluoroguaiacol and 4-[^{18}F]fluorocatechol. *Appl. Radiat. Isot.* 42, 673-681.

Coenen H.H., Franken K., Metwally S. and Stöcklin G. (1986a) Electrophilic radiofluorination of aromatic compounds with [^{18}F]F_2 and [^{18}F]CH_3CO_2F and regioselective preparation of L-p-[^{18}F]fluorophenylalanine. *J. Labelled Compd. Radiopharm.* 23, 1179-1181.

Coenen H.H., Bodsch W., Takahashi K., Hossmann K.A. and Stöcklin G. (1986b) Synthesis, autoradiography and biochemistry of L-[^{18}F]fluorophenylalanine for probing protein synthesis. *Nuklearmedizin* 22, (Suppl.) 600.

Coenen H.H., Klatte B., Knöchel A., Schüller M. and Stöcklin G. (1986c) Preparation of N.C.A. [17-^{18}F]-fluoroheptadecanoic acid in high yields via aminopolyether supported, nucleophilic fluorination. *J Labelled Compd Radiopharm.* 23, 455-466.

Coenen H.H., Franken K., Kling P. and Stöcklin G. (1988) Direct electrophilic radiofluorination of phenylalanine, tyrosine and dopa.. *Appl. Radiat. Isot.* 39, 1243- 1250.

Coenen H.H., Kling P. and Stöcklin G. (1989). Cerebral metabolism of L-2-[^{18}F]fluorotyrosine, a new PET tracer of protein synthesis. *J. Nucl. Med.* 30, 1367-1372.

Coppola G.M., Schuster H.F. (1987) Asymmetric synthesis: construction of chiral molecules using amino acids. John Wiley and Sons, New York.

Ding Y. -S., Shiue C.-Y. , Fowler J. S., Wolf A. P. and Plenevaux A. (1990) No-carrier-added (NCA) aryl [^{18}F]fluorides via the nucleophilic aromatic substitution of electron-rich aromatic rings. *J. Fluorine Chem.* 48, 189-205.

Ding Y. -S., Fowler J. S., Gatley S. J., Dewey S. L. and Wolf A. P. (1991). Synthesis of high specific activity (+)- and (-)-6-[^{18}F]fluoronorepinephrine via the nucleophilic aromatic substitution reaction. *J. Med. Chem.* 34, 767-771.

Fasth K.J., Malmborg P., and Långström B. (1991) Two syntetic routes for the asymmetric synthesis of [β-^{11}C]amino acids with high enantiomeric purities. *J. Labelled Compds Radiopharm.* 30, 401-403.

Guillaume M., Luxen A., Nebeling B., Argentini M., Clark J. C. and Pike V. W. (1991) Recommendations for fluorine-18 production. (EEC Task Group Report). *Appl. Radiat. Isot.* 42, 749-762.

Haemers A, Mishra L, Van Assche and Bollaert W. (1989) Asymmetric synthesis of amino acids by enantio and diastereodifferentiating reactions. *Die Pharmazie* 44, 97-109.

Halldin C.and Langström B. (1984) Synthesis of racemic [3-^{11}C]phenylalanine and [3-^{11}C]dopa. *Int. J. Appl. Radiat. Isot.* 35, 779.

Keinan E., Perez D., Sahai M., Shvily R. (1990) Diiodosilane. 2. A multipurpose reagent for hydrolysis and reductive iodination of ketals, acetals, ketones, and aldehydes. *J. Org. Chem.* 55, 2927-2938

Kilbourn M. R. (1990) Fluorine-18 labeling of radiopharmaceuticals. National Academy Press Washington DC, USA, 149.

Lemaire C., Guillaume M., Christiaens L., Palmer A. J., Cantineau R. (1987) A new route for the synthesis of [^{18}F]fluoroaromatic substituted amino acids: No carrier added L-p-[^{18}F]fluorophenylalanine. *Appl. Radiat. Isot.* 38, 1033-1038.

Lemaire C., Guillaume M., Plenevaux A., Cantineau R., Damhaut P., Christiaens L. (1991a) ^{18}F-substituted aromatic aldehydes, key-intermediates for NCA fluorinations. *J. Labelled Compd. Radiopharm.* 30, 134-136.

Lemaire C. (1991b) Ph.D. Dissertation, Université de Liège.

Lemaire C., Guillaume M., Cantineau R., Plenevaux A., Christiaens L. (1991c) NCA asymmetric synthesis of 6-[^{18}F]fluoro-L-dopa. *J. Labelled Compd. Radiopharm.*30, 269-270.

Lemaire C., Damhaut P., Plenevaux A., Cantineau R., Christiaens L, and Guillaume M. (1991d) Asymmetric synthesis of 4-[^{18}F]fluoro-L-m-tyrosine via aromatic fluorination. *J. Nucl. Med.* 32, 935 (Abstract 112).

Lemaire C., Damhaut P., Plenevaux A., Cantineau R., Christiaens L., Guillaume M. (1992a) Synthesis of fluorine-18 substituted aromatic aldehydes and benzyl bromides, new intermediates for n.c.a. [^{18}F]fluorination. *Appl. Radiat. Isot.* 43 , 485-494.

Lemaire C., Plenevaux A., Damhaut P., Guillaume M., Christiaens L., and Comar D. (1992b). Nca asymmetric synthesis of 2-[^{18}F]fluorotyrosine. Presented at the IX International Symposium on radiopharmaceutical Chemistry. Paris, France, 1992. *J. Labelled Compd. radiopharm.* in press.

McIntosh J.M. and Mishra P. (1986) Alkylation of camphor imines of glycinates. Diastereoselectivity as a function of electronic factors in the alkylating agent. *Can. J. Chem.* 64, 726-731.

McIntosh J.M., Leavitt RK, Mishra P., Cassidy K.C., Drake J.E., and Chadha R. (1988) Diastereoselective alkylation guided by electrophile-nucleophile p-interavtions. *J. Org. Chem.* 53, 1947-1952.

Murakami M., Takahashi K., Kondo Y., Mizusawa S., Nakamichi H., Sasaki H., Hagami E., Iida H., Kanno I., Miura S. and Uemura K. (1988a) The comparative synthesis of ^{18}F-fluorophenylalanines by electrophilic substitution with ^{18}F-F_2 and ^{18}F-AcOF. *J. Labelled Compd. Radiopharm.* 25, 573-578.

Murakami M., Takahashi K., Kondo Y., Mizusawa S., Nakamichi H., Sasaki H., Hagami E., Iida H., Kanno I., Miura S. and Uemura K. (1988b) 2-^{18}F-phenylalanine and 3-^{18}F-tyrosine. Synthesis and preliminary data of tracer kinetics. *J. Labelled Compd. Radiopharm.* 25, 773-782.

Oguri T., Kawai N., Shioiri T., and Yamada S. (1978) Amino acids and peptides. XXIX. A new efficient asymmetric synthesis of α-amino acid derivatives with recycle of a chiral reagent- Asymmetric alkylation of a chiral Schiff base from glycine. *Chem Pharm Bull.* 26, 803-808.

Palmer A.J., Clarck J.C., and Goulding R.W. (1977) The preparation of fluorine-18 labelled radiopharmaceuticals. *Int. J. Appl. Radiat. Isot.* 28, 53.

Plenevaux A., Al-Darwich M.J., Lemaire C., Delfiore G., Christiaens L., Comar D. (1992) Asymmetric synthesis of n.c.a. L-[2-11C]-4-chlorophenylalanine. Presented at the IX International Symposium on radiopharmaceutical Chemistry. Paris, France, 1992. *J. Labelled Compd. Radiopharm.* in press.

Seebach D., Dziadulewicz E., Behrendt L., Cantoreggi S., and Fitzi R. (1989) Synthesis of nonproteinogenic ((R)- or (S)-amino acids analogues of phenylalanine, isotopically labelled and cyclic amino acids from tert-butyl 2(tert-butyl)-3-methyl-4-oxo-1-imidazolidinecarboxylate. *Liebigs Ann. Chem.* 12, 1215-1232.

Shiue C. Y., Watanabe M., Wolf A. P., Fowler J. and Salvadori P. (1984) Application of the nucleophilic substitution reaction to the synthesis of no-carrier-added [^{18}F]fluorobenzene and other ^{18}F-labeled aryl fluorides. *J. Labelled Compd. Radiopharm.* 21, 533-547.

Vaalburg W., Coenen H.H., Crouzel C., Elsinga Ph. H., Langström B., Lemaire C., Meyer G.J. (1992) Amino acids for the measurement of protein synthesis *In Vivo* by PET. *Nucl. Med. Biol.* 19 , 227-237.

Van der Ley M. (1983) [^{18}F]fluorine labeled aliphatic amino acids. *J. Labelled Compd. Radiopharm.* 20, 453-461.

Williams M. W. (1989) Synthesis of optically active α-amino acids. Pergamon Press, Headington Hill Hall, Oxford OX3 OBW, 1-133.

BIOCHEMISTRY AND EVALUATION OF FLUOROAMINO ACIDS

H.H. COENEN

ABSTRACT. A review is given of the main biochemical properties of ^{18}F-labelled amino acids which are substrates of enzymes for neurotransmitter and protein metabolism. Their potential as tracers for PET is evaluated with special emphasis on the brain.

1. Introduction

The important role of amino acids in metabolism induced an interest in fluorinated amino acids more than fifty years ago. A large number of mono- and polyfluorinated amino acids were prepared and examined with respect to their biological activity and interference with normal metabolism of amino acid uptake, biosynthesis and protein synthesis (for reviews see Weygand and Oettmeier, 1970; Fowden, 1972). The general role of amino acids in metabolism involve:

- Synthesis of peptides and proteins
- Formation of neurotransmitters and hormones
- Energy metabolism (degradation via citric acid cycle)
- Specific functions (as source of sulfur, for methylation, and in the anabolism of nucleobases).

So far, only the first two pathways have been made use of for the development of radiotracers labelled with fluorine-18 or other isotopes. Firstly (fluorine-18) labelled amino acids as indicators of protein synthetic activity found interest for pancreas scanning (for review see Palmer et al., 1977). A systematic study on radiohalogenated amino acids revealed the advantage of fluoroanalogues as tracers of protein synthesis (Coenen and Stöcklin, 1988). Measurement of protein synthesis with labelled amino acids in vivo by PET has recently focussed on the brain (for review see Vaalburg et al., 1992). ^{18}F-fluoroamino acids which are involved in the investigation of neurotransmission, especially of the dopaminergic system, with PET have gained great importance and potential (for review see Firnau et al., 1991). Besides dealing with [^{18}F]fluoroamino acids intended as tracers of amino acid transport and protein synthesis measurement with PET, this review will also report on ^{18}F-labelled amino acids for examination of presynaptic neurotransmission. In both cases the major focus will be on the brain. The synthetic aspects of labelling fluoroamino acids are discussed in another chapter of this book by Lemaire.

B. M. Mazoyer et al. (eds.), PET Studies on Amino Acid Metabolism and Protein Synthesis, 109–129.

The incorporation of amino acids into proteins reflects the protein synthesis rate (PSR) in tissue which in principle can be quantified in vivo by the PET technique. Principles and criteria for suitable labelled amino acids and tracer kinetic models which are prerequisites for the quantitation of PSR have been established in the literature and will not be discussed here in detail (for references see Sokoloff and Smith, 1983; Bustany and Comar, 1985; Phelps et al., 1984). Ideally the labelled amino acid should be accepted and exclusively be a substrate of the enzyme amino-acyl- t-RNA synthetase after transport into the cell. If no other metabolism or at least no labelled metabolites are formed in the tissue of interest a tracer kinetic model can be reduced to two tissue compartments. In this respect fluorinated amino acids may have advantages over their natural analogues and allow simpler modelling as will be discussed.

TABLE 1. Selected amino acids and parameters related to the uptake in the rat brain (data compiled from Oldendorf, 1971; Pardridge, 1977; Gaull,1978).

amino acid	type	transport system	BUI*	plasma level mM	influx nmol/g min	brain level μmol/g
phe	e	LN	55	0.05	6.5	0.05
leu	e	LN	54	0.10	6.2	0.05
tyr	n	LN	50	0.09	5.3	0.07
met	e	LN,A	38	0.04	1.6	0.04
val	e	A,LN	21	0.14	1.8	0.07
DOPA	n	LN	20	0.004	0.1	-

* $^{3}HOH \hat{=} 100$

The choice of amino acids labelled so far as PET-tracers of protein synthesis depended mainly on their uptake behaviour. They are listed in Table 1 together with their main features which determine their uptake in the rat brain. They are all essential amino acids which exhibit the highest brain-uptake-index (BUI) with corresponding influx constants. (Tyrosine is also "essential" for the brain which does not contain phenylalanine hydroxylase). All these amino acids are transported by the same system as for the large neutral amino acids (LN), their plasma levels differ utmost by a factor of three while their concentration in brain tissue is in the same range of 0.04 to 0.07 μmol/g. Dopa which is also a substrate of the LN transport system is added for comparison in Table 1. As a non-essential amino acid, it still exhibits a relatively high uptake index although the influx is comparatively small due to the low plasma level. It is difficult to select from these data the tracer of choice without considering the individual biochemical properties. This is especially true for fluorine analogues of these amino acids.

2. Criteria for Using [^{18}F]Fluoroamino Acids

The advantageous nuclear properties of the positron emitter fluorine-18 with a half-life of 109.8 min and a maximum β^+-energy of only 0.635 MeV, make radiopharmaceuticals labelled with this radionuclide very attractive for positron-emission-tomography (PET). While pharmaceuticals are often fluorocompounds, natural biomolecules containing fluorine are practically unknown. Introduction of a fluorine almost does not change the shape of a hydrocarbon anlogue but due to its high electronegativity it will alter the electron density and increase the ability of hydrogen bonding. The physiological properties can drastically be changed up to high toxicity although many fluoro-compounds are accepted by the cells at least in the primary metabolism like non-fluorinated analogues.

Since fluoroamino acids are non-natural analogues, no prediction can generally be made on the physiological acceptance which must be examined in each case and, for example, their incorporation into proteins must be proven. Due to the longer half life of fluorine-18, [^{18}F]fluoroamino acids are much more compatible with the protein synthesis rates than [^{11}C]amino acids. This is especially important for tissues with realtively slow PSR such as the brain. Furthermore, after degradation and loss of the fluorine-18 label this will not be reutilized; this might be a complication with ^{11}C-labelled natural amino acids.

If accepted, fluoramino acids are truly "essential" in the sense that they are not endogeously formed and their specific activity will not be diluted, either by the plasma pool of amino acids or those within the cell. This is, however, only a virtual advantage since the fluoroamino acids will compete with their natural counterparts for the active transport system in brain and for the same amino-acyl synthetases in the first step of protein synthesis. This means their apparent specific activity will be decreased leading to the same complications in quantitation as with ^{11}C-labelled amino acids. While the amino acid contents can be determined in the blood plasma, their concentration within the cells is hard to estimate in vivo due to the steady-state degradation of proteins. The degree of recycling in brain tissue, for example of leucine amounting to 0.58, must be taken into account while determining the PSR in vivo with labelled exogenous tracers (Smith at al., 1988). On the other hand, due to the slow breakdown of cerebral proteins of only 0.7% per h (Dunlop, 1983) recycling of the labelled amino acids during the time period of PET studies can be neglected. Quantitation of PSR with ^{18}F-labelled analogues would also require the knowledge of correlation factors between their enzyme rate constants and those of their natural counterparts. Without a correlation factor "metabolic rates" can only be determined as incorporation rates instead of PSR. It must be kept in mind that different amino acids are differently frequent incorporated into proteins and that various proteins are produced simultaneously.

For fluoroamino acids which act as precursors of neurotransmitters their suitable biochemistry to trace the right pathway must equally be proven. Their inclusion in this report is also motivated by the facts that nearly all fluorinated amino acid tracers used today are derivatives of α-amino-β-phenyl propionic acid and that these aromatic amino acids all use the same active transport system in passing the blood-brain-barrier. Furthermore, there is an anabolic correlation between phenylalanine, tyrosine, dopa and dopamine, and non-natural fluorine analogues might possibly be involved in both pathways. For ortho-[^{18}F]fluoro-substituted analogues this hyopothetical reaction sequence is illustrated in Fig. 1. [^{18}F]Fluoro-phenylalanine and -tyrosine can either be substrates of t-RNA synthetases or be transformed to 2-(not shown) or 6-[^{18}F]-fluoro-dopa and dopamine, respectively, besides other catabolic pathways.

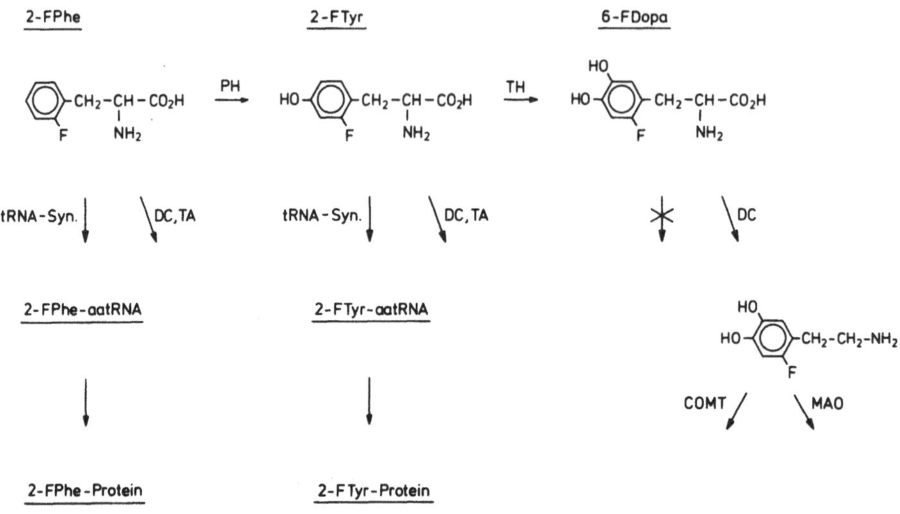

Figure 1. The anabolic pathways of L-phenylalanine, hypothetically shown for the ortho fluoroisomer.

3. Tracer for the Dopaminergic System

3.1 FALSE PRECURSORS

Fluoro-dopa is an analogue for the metabolic precursor dopa of the neurotransmitter dopamine. The 5-isomer was first labelled with fluorine-18 (Firnau et al., 1973) but the 6-isomer L-6-[^{18}F]fluoro-3,4-dihydroxyphenylalanine (6-FDOPA) proved to be most useful. It is now an established tracer to asses the dopaminergic nigrostriatal neuronal system with PET (Garnett et al., 1983; Martin et al., 1989) and kinetic modelling was recently suggested based on biochemical information (Leenders, 1991; Huang et al., 1991). The radiopharmacology and metabolism of 6-FDOPA was extensively studied by several groups (for reviews see Firnau et al., 1986, 1991; Leenders, 1991; Luxen et al., 1992).

The principal metabolites of 6-FDOPA in blood and striatal tissue are schematically summarized in Fig. 2. 6-FDOPA passes the blood-brain-barrier using the transport system for large neutral amino acids and its blocking by a dose of an amino acid cocktail in man implies standard fasting protocols for PET studies (Leenders et al., 1986). It is well documented by animal studies in rats and monkeys that 6-FDOPA is rapidly decarboxylated to 6-fluoro-dopamine (FDA) which to its larger part is stored in dopamine vesicles (Firnau at al., 1987; Cumming et al., 1987a, 1988; Melega et al., 1991a). There is some debate if FDA is taken up in a large or small dopamine pool with slow or fast turnover, respectively (Leenders, 1991).

6-FDOPA follows the metabolic pathway of L-DOPA closely as shown by comparison with ^{14}C-labelled DOPA in carbidopa pretreated rats. Only a small part of FDA is further metabolized to 3,4-dihydroxy-6-fluoro-phenylacetic acid (FDOPAC) and 6-fluoro-homovanillic acid (FHVA) by the action of monoamine oxidase (MAO) and catechol-O-methyltransferase (COMT), respectively (Melega et al., 1990a). As shown in Fig.2, COMT can also directly act on FDOPA forming intracellular 6-fluoro-3-0-methyl-L.DOPA (FOMDOPA). Corresponding to a missing catechol-metabolism, in cerebellum and cortex only FDOPA and, to a greater extent, FOMDOP were found stemming from the periphery (Melega et al., 1990a, 1991a). In MPTP treated monkeys the accumulation of 6-FDOPA decreases with the degree of striatal cell loss (Firnau et al., 1991, and references therein).The concentration of FDA in MPTM-lesioned putamen was depleted (decreased) to less than 2% of control identically to endogenous dopamine (Melega et al., 1991b).

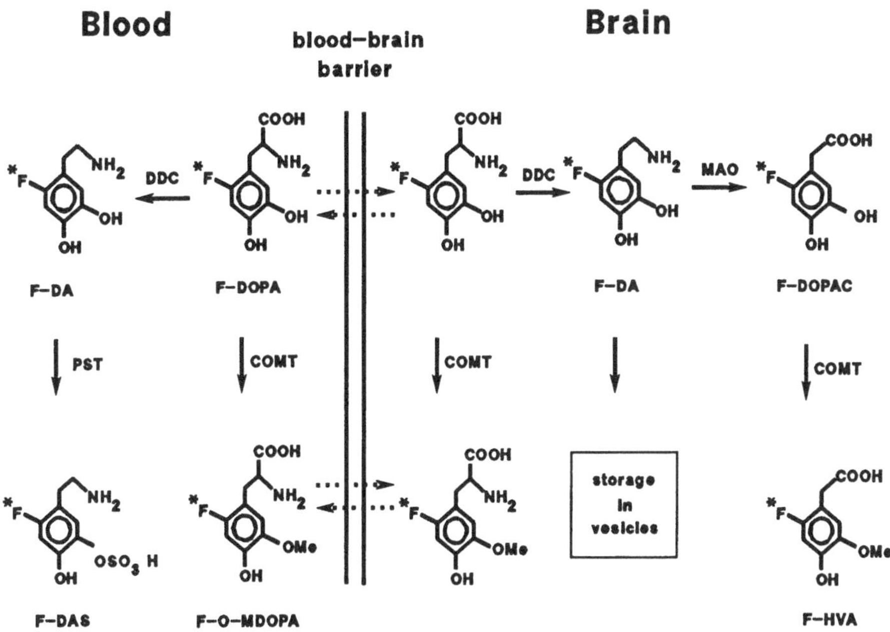

Figure 2. Principal metabolic pathways of 6-fluoro-DOPA in striatum and periphery of rodents and primates.

The problem with FDOPA especially for quantitation of striatal metabolism with PET is caused by the peripheral biochemistry. As indicated in Fig. 2, there is also decarboxylase in the periphery forming FDA which is conjugated to fluorodopamine sulfate (FDAS) by action of phenolsulfotransferase (PST). This pathway is almost completely suppressed by carbidopa increasing the plasma level of FD0PA (Boyes et al., 1986, Melega et al., 1990b).

This, however, enhances even more the major peripheral metabolism converting FDOPA into the 3-O-methylated derivative. This metabolite passes also the BBB, however equally in all regions, and contributes to half of the ^{18}F-uptake in striatum (Firnau et al., 1991). While carbidopa treatment increases the ^{18}F-uptake in striatum by almost a factor of two, it produces the same effect in non-dopaminergic tissue (Leenders, 1991); i.e. the differential signal increases while the ratio, for example of putamen to cortex, remains constant. In vitro biochemical experiments showed that O-methylation of 6-FDOPA was much less effective than of natural DOPA (Creveling and Kirk, 1985). However, after pretreatment with carbidopa the level of FOMDOPA, although it is then the only significant metabolite formed in plasma, is equal to that of FDOPA 30 min p.i. in humans (Melega et al., 1990b, 1991a). There is agreement that appreciable amounts of this second tracer cross the BBB and have significant impact on the formulation of a tracer kinetic model for quantitation with 6-FDOPA and PET (Firnau et al., 1991; Melega et al., 1991a; Leenders, 1991; Huang et al., 1991).

In summary FDOPA reflects the normal and pathological metabolism of DOPA in brain tissue and can well serve as a PET tracer for the evaluation of central dopaminergic biochemical activity in vivo in the presynaptic dopaminergic system. Quantitation of metabolic rates, however, is difficult due to the complications with FOMDOPA.

The approach of inhibiting the catechol-0-methyl-transferase in humans, as first examined in rats (Cumming et al., 1987b), by using the new inhibitor nitecapone (OR-462), led only to partially reduced levels of 3-0-methyl-FDOPA in plasma while the quality of PET images was improved (Laihinen et al., 1992). It seems however, questionable if inhibition of COMT, and possibly DDC, will still provide significant information on in vivo metabolism with PET.

5-FDOPA and 2-FDOPA are even much faster 0-methylated than DOPA (Creveling and Kirk, 1985). In addition the 2-isomer passes the BBB less readily than 6-FDOPA and it is not decarboxylated in brain (Cumming et al., 1988). Therefore both of them do not lend themselves to probing the dopaminergic system with PET.

Fluoro-meta-L-tyrosine (FMT) is another false substrate which avoids several complications of FDOPA. Based on their previous experiences with radiobrominated meta-tyrosine, De Jesus et al. (1988,1989) prepared and suggested fluorine-18 labelled FMT for use as CNS dopaminergic tracer. Meta-tyrosine is effectively decarboxylated by aromatic amino acid decaboxylase (AAAD) but it is not O-methylated by COMT like L-DOPA (see Firnau et al., 1991 and references therein). This metabolic behaviour could be verified for 4-[^{18}F]fluoro-meta-tyrosine in a careful study in rats and monkeys which also showed a clear deliniation of the striatum with PET after administration of the tracer (Melega et al.,1989).

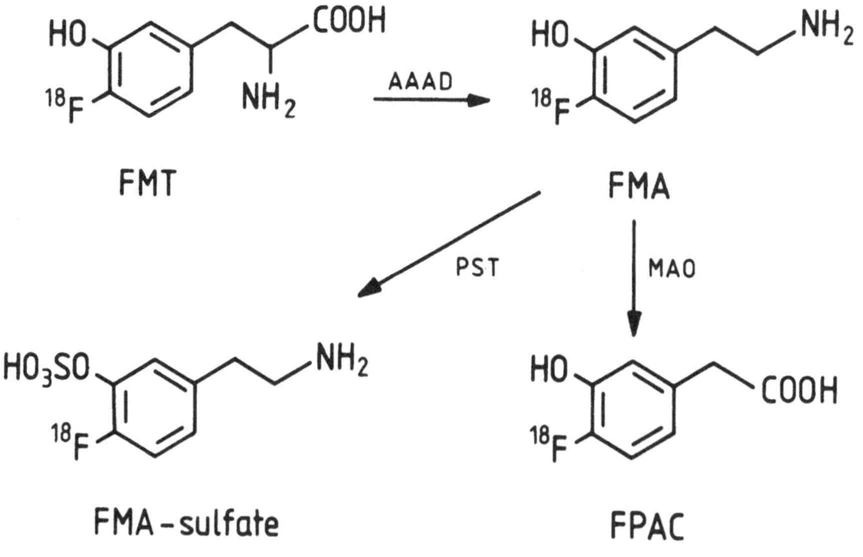

Figure 3. Principal metabolic pathways of 4-fluoro-meta-tyrosine in striatum and periphery of rodents and primates (after Melega et al.,1989).

The biochemical transformations of 4-FMT observed were qualitatively identical in striatum and periphery, and the metabolites formed are summarized in Fig. 3. After decarboxylation the 4-fluorotyramine is either oxidized by MAO to 4-fluoro-3-hydroxyphenylacetic acid (FPAC) or conjugated by phenolsulfotransferase (PST) into the 3-O-sulfate of 4-FMA. As against to FDOPA most of the tracer applied ($\approx$ 60%) was still present in plasma where FMA-sulfate was the main metabolite (35%) after 2 hours. In striatum most of the fluoroamino acid (FMT) was already decarboxylated 30 minutes p.i. to give $\approx$ 20% FMA and $\approx$ 70% FPAC and a trace of the sulfoconjugate. Since FMA, FPAC and the conjugate are rather polar compounds they do not very likely leave brain tissue. Trapping is indeed observed with still increasing ^{18}F-activity in the striatum of a monkey while cerebellar activity decreases rapidly (Melega et al., 1989).

These findings could be confirmed for the 6-[^{18}F]fluoroisomer in monkey tissue analyses and human studies with PET (Firnau et al., 1991). Again in brain tissue only the three mentioned metabolites FMA, FPAC and a conjugate were found. There seems to be no strong isomer effect like for FDOPA. Also, unlike FDOPA where 3-O-methyl-FDOPA is the major metabolite in plasma which leads to a 50% background activity in the brain (see above), ^{18}F-labelled FPAC only accumulates 20 times less in brain than 6-FMT (Firnau et al., 1991). Nigrostriatal structures are therefore much better delineated with FMT than with FDOPA and even the substantia nigra can be visualized by PET (Firnau, personal communication).

Both 6-FDOPA and FMT are transported and decarboxylated like DOPA followed by retention in brain tissue. Contrary to F-dopamine, F-meta-tyrosine is not stored in the vesicles and does not give information on dopamine metabolism. However, accumulation in both cases is only possible and determined by the action of decarboxylase, although this is fast. Both the tracers can therefore be regarded as indicators of decarboxylase activity, which may or may not be dependent on amine concentration in tissue, rather than as indicators of an existing amine pool.The rate determining step of endogenous dopamine formation is the hydroxylation of tyrosine to DOPA. Exogenous fluoro-meta-tyrosine and fluoro-DOPA enter the biochemical pathway after this step and therefore can not give information about dopamine demand and formation.

3.2 ENZYME INHIBITORS

L-α-Fluoromethyl-6-fluorodopa (AFMFD) is an amino acid analogue which is also a substrate of aromatic L-amino acid decarboxylase (AADC). It was labelled with fluorine-18 in the aromatic ring and suggested to monitor AADC-containing tissues with PET (Chirakal et al., 1989). It is known that α-methyl-substituted amino acids and amines are reversible inhibitors of decarboxylase and hydroxylase enzymes (Moore and Jerome, 1971). α-Fluoro methyl-substituted amino acids and amines are all irreversible decarboxylase inhibitors. They covalently bind to the enzyme under loss of the α-methyl fluorine after forming a Schiff base with pyridoxal phosphate (PyP) and decarboxylation. This mechanism was proposed by Kollonitsch at al. (1978) and is schematically shown in Fig. 4. Different to FDOPA and FMT, ^{18}F-AFMFD is hence covalently bound to the enzyme after decarboxylation as "suicide inhibitor". A preliminary study shows that AFMFD inhibits the activity of AADC in a dose dependent manner and that it accummulates in AADC-containing tissues in rats, while in blood the O-methylated compound and another metabolite were found (Chirakal et al., 1989).

L-α-Methyl-6-[^{18}F]fluorodopa (AMFD), the reversibly binding analogue, was reported to be unable to delineate AADC-containing tissues in vivo (Firnau et al., 1991), but no detailed data have been presented, so far.

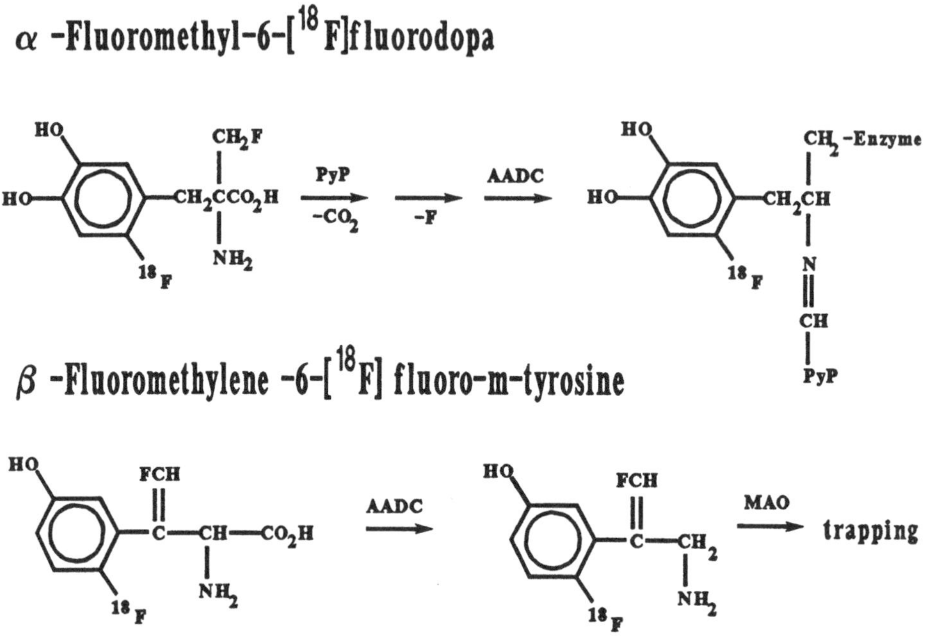

Figure 4. Reactions of suicide inhibitors of aromatic amino acid decarboxylase and monoamino oxidase (after Kollonitsch et al., 1978).

L-α-Fluoromethyl-para-tyrosine (AFMPT) was also radiofluorinated in the aromatic nucleus as a precursor of the AADC inhibitor α-fluormethyl-fluorodopa (De Jesus et al., 1992). Fluoromethyl-tyrosine is a substrate of tyrosine hydroxylase (TH), which is the rate limiting enzyme in endogeneous dopamine synthesis. α-Fluoromethyl-3-[^{18}F]fluorotyrosine is hydroxylated by TH but at a slower rate when compared to tyrosine. Further biochemical studies are needed, however, to validate this concept and usefulness of the tracer for estimation of endogenous dopamine formation (De Jesus et al., 1992).

ß-Fluoromethylene-meta-tyrosine (FMMT) is yet another fluoroamino acid analogue which after enzymatic conversion acts as an irreversible inhibitor on a second enzyme. It was also ^{18}F-labelled in the 6-position (De Jesus et al., 1990). This compound is supposed to be decarboxylated by AADC to form ß-fluoromethylene-6-[^{18}F]fluoro-meta-tyramine, the ^{18}F-analogue of a potent MAO suicide inhibitor as outlined in Fig. 4. Provided the decarboxylation is fast, it could be a potent probe for the MAO concentration and hence the dopamine catabolism in presynaptic neurons. Indeed, distribution of 6-[^{18}F]fluoro-FMMT in rats was found to be similar to 6-FDOPA and FMT with preferential striatal accumulation (De Jesus et al., 1990). However, striatum-to-cerebellum ratio obtained in mice 2 hrs post injection was only 1.7 and experiments with inhibition of various enzymes of catechol metabolism were less conclusive (De Jesus, private communication) but further biochemical evaluation is encouraged.

4. Tracer for Amino Acid Transport and Protein Synthesis

4.1 DEGRADATION AND TOXICITY

The suitability of a fluoroamino acid analogue to trace a special metabolic pathway, like protein synthesis, in vivo with PET will greatly depend on the degree to which it may enter into other biochemical reactions. Especially degradation may complicate the use of a tracer due to formation of several labelled metabolites. Indeed, the effects of fluoroamino acids on microorganisms which have been examined for a long time (for review see Richmond, 1962), were mostly attributed to degradation products. Nevertheless, there are indications in some cases that metabolic inhibition by fluoroamino acids occurs after incorporation into active sites of proteins (enzymes).

The perplexing differences in toxicity of aromatic fluoroamino acids could indeed be attributed to their degradation products. As can be seen in Table 2, ortho (and para) fluoroisomers of phenylalanine and tyrosine are well tolerated in mouse. In contrast, the meta-isomers of both compounds are highly toxic exeeding even sodium fluoroacetate with a LD_{50} value of 25.2 mg/kg (Weissman and Koe, 1967). The same authors report this discrepancy also for fluorotryptophans (not listed) where the 5-isomer is highly toxic but not the 6-isomer.

TABLE 2. Toxicity of L-fluorophenylalanines and -tyrosines (Weissman and Koe, 1967).

Compound	LD_{50} (mouse) [mg/kg]
2-fluorophenylalanine	>1000
3-fluorophenylalanine	5.9
4-fluorophenylalanine	>1000
2-fluorotyrosine	>1000 (with 30 mg no effect)[a]
3- fluorotyrosine	10

[a]Coenen et al., 1989

The large difference in toxicity could be explained by the well established pathway of degradation of tyrosine which is illustrated in Fig. 5 for the 3-fluoroisomer (Weissman and Koe, 1967). Phenylalanine follows the same sequence after hydroxylation to tyrosine by phenylalanine hydroxylase. The next step is transamination of tyrosine to p-hydroxyphenylpyruvic acid which is converted after reaction with O_2 to homogentisic acid. This is cleaved in the aromatic ring to yield 4-maleylacetoacetic acid and after isomerisation 4-fumarylacetoacetic acid which again is cleaved into fumaric acid and acetoacetic acid.

The orientation of the rearrangement in fluoro-substituted aromatic amino acids is the important step determining formation of toxic metabolites. Weissman and Koe suggested the migration of the pyruvic side chain to the ortho position to fluorine as shown in Fig. 5. This then ends up in the terminal position of acetoacetic acid which is hydrolysed to the toxic fluoroacetic acid. If the rearrangement would occur to the position labelled with a star in Fig. 5, this would lead to the relatively non-toxic fluorofumaric acid. This is the case for ortho-fluorophenylalanine and -tyrosine when fluorine is in the star position and the migration of the side chain proceeds as indicated in Fig. 5. Para-fluorophenylalanine already loses the fluorine label during hydroxylation forming free fluoride and natural tyrosine (Kaufman, 1961). The degradation of fluoroamino acids is an instructive example of the extend to which very small structural changes can influence the metabolic fate of analogues.

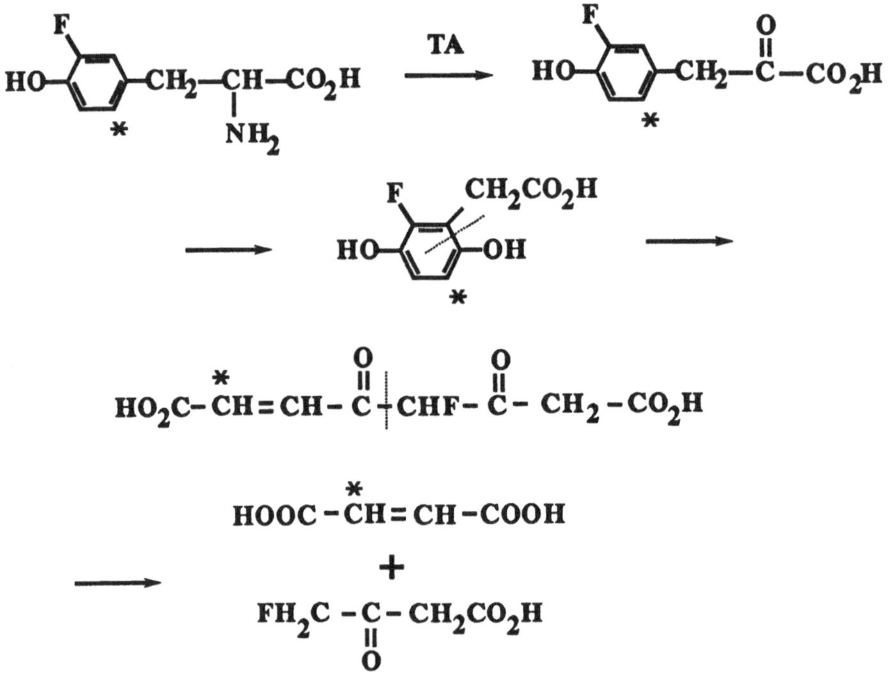

$$HO\text{-}C_6H_3(F)\text{-}CH_2\text{-}CH(NH_2)\text{-}CO_2H \xrightarrow{TA} HO\text{-}C_6H_3(F)\text{-}CH_2\text{-}C(=O)\text{-}CO_2H$$

$$\longrightarrow HO\text{-}C_6H_2(F)(CH_2CO_2H)\text{-}OH \longrightarrow$$

$$HO_2C\text{-}\overset{*}{C}H{=}CH\text{-}C(=O)\text{-}CHF\text{-}C(=O)\text{-}CH_2\text{-}CO_2H$$

$$\longrightarrow HOOC\text{-}\overset{*}{C}H{=}CH\text{-}COOH + FH_2C\text{-}C(=O)\text{-}CH_2CO_2H$$

Figure 5. Degradation of L-meta-fluorotyrosine - dotted lines indicate position of cleavage (after Weissman and Koe, 1967).

4.2 PROTEIN PRECURSORS

Para-fluorophenylalanine (PFPA) has been known to be one of the few amino acid analogues accepted for incorporation into proteins in vitro and in vivo for thirty years (Westhead and Boyer, 1961; Arnstein and Richmond, 1964). Para-fluorophenylalanine is hydroxylated to tyrosine by phenaylalanine hydroxylase which, however, does not exist in nervous system tissue (Dolan and Godin, 1966). Similar to phenylalanine hydroxylation may be mediated by tyrosine hydroxylase in the brain. All fluoroisomers of phenylalanine can be enzymatically degraded and are subtrates of L-amino acid oxidase but not of, for example, tyrosine decarboxylase (Frieden et al., 1951). There were no specific data available for the cerebral biochemistry of PFPA, but transamination to phenylpyruvate and decarboxylation to phenethylamine are unlikely to occur with phenylalanine in the brain (Axelrod and Saavedra, 1974), as must be concluded for the fluoroanalogue. From the information available it could be assumed that ^{18}F-labelled fluorophenylalanine would primarily be involved in the formation of proteins and the metabolism to tyrosine. This is schematically shown in Fig. 6.

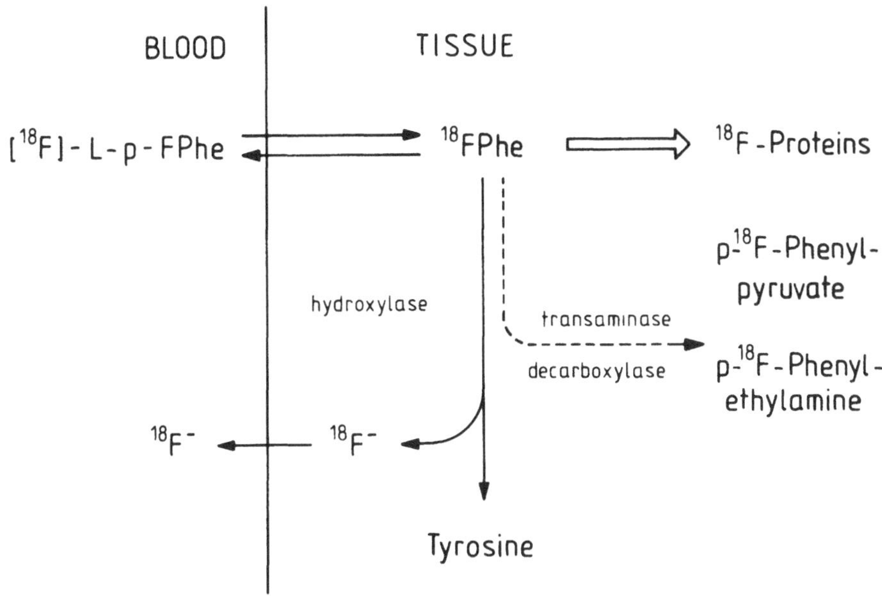

Figure 6. Possible metabolic pathways of L-para-fluorophenylalanine (after Bodsch et al., 1988).

In a theoretical evaluation para-fluorophenylalanine was considered to be unsuitable as PET-tracer (Phelps et al., 1984) due to its extensive hydroxylation (Gal, 1974). However, the conversion of para-fluorophenylalanine to tyrosine is six times slower than that of phenylalanine (Kaufman, 1961) and this process is likely to occur in the periphery (liver).The process will leave unlabelled tyrosine and ^{18}F-fluoride, and no labelled amino acid (derivative) will be formed that interferes with tracer kinetic measurements in brain. Consequently PFPA was labelled with fluorine-18 and its in vivo metabolic behaviour was examined in rodents (Coenen et al., 1986; Bodsch et al., 1988). Autoradiographic studies on gerbil brain exhibited the same regional cerebral protein pattern when compared to [^{3}H]phenylalanine. 95% of all ^{18}F-radioactivity associated with amino acyl-tRNA was identified as unchanged amino acid after hydrolysis thus proving acceptance for protein synthesis.

The incorporation rate into proteins was only half of tritiated phenylalanine (Bodsch et al., 1988). Indeed, in previous cell-free experiments p-fluorophenylalanine had shown quantitatively the same rate and amount of protein incorporation in the absence of phenylalanine. The natural analogue, however, markedly inhibited the incorporation at the stage of attachment to t-RNA (Arnstein and Richmond, 1964). Further in vivo studies with ^{18}F-PFPA in mice confirmed the slow incorporation into brain proteins with only 60% after 2 hrs p.i. as shown in Fig. 7 (supplementary data obtained in the context of the study by Bodsch et al., 1988). Formation of free fluoride could be excluded in cerebral tissue; however, about 15% of the cerebral activity could not be identified (cf. Fig. 7). In spite of its sufficient brain uptake (Coenen and Stöcklin, 1988), the low incorporation rate and considerable amounts of unidentified metabolites do not recommend the use of PFPA for quantitation of PSR and no PET-studies on humans are known.

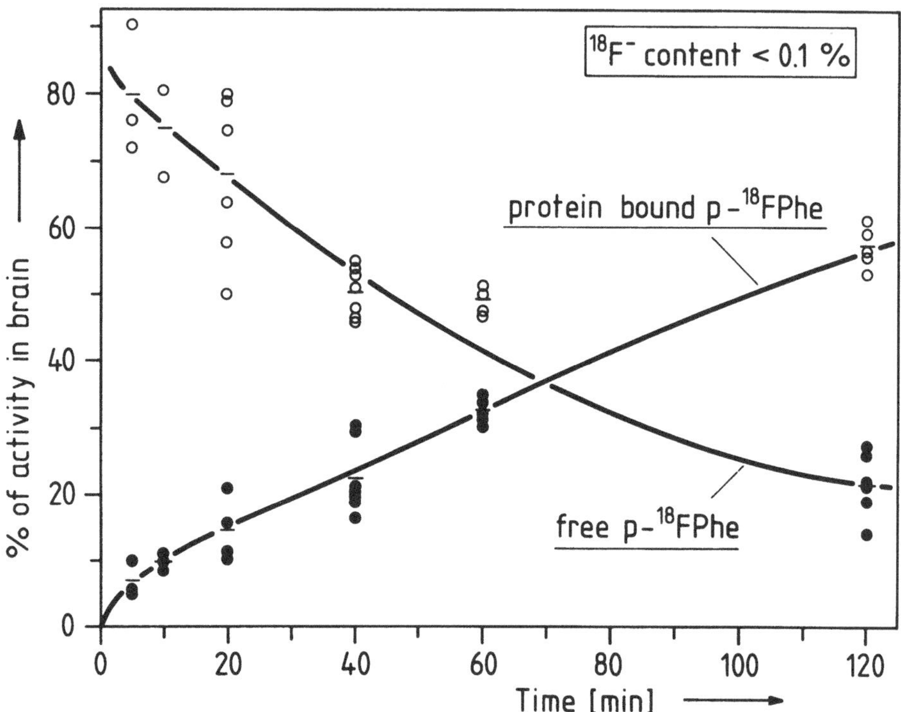

Figure 7. Time course of protein bound and free L-para-[18F]fluorophenylalanine in cerebral tissue of mice (supplementary data obtained in the context of the study of Bodsch et al.,1988).

Ortho-fluorophenylalanine (OFPA), the preferentially formed isomer in direct electrophilic fluorination reactions, exhibited an even lower protein incorporation (Coenen et al., 1986; Bodsch et al., 1988) which amounted to only a quarter of that of [^{3}H]phenylalanine. Furthermore, there was indication that incorporation into proteins was in the form of ortho-fluorotyrosine according to the possible pathway given in Fig. 1. Radiofluorinated catechols, however, were not found. These findings were confirmed in rat studies and the slow metabolism of 2-[^{18}F]OFPA left 80% of the fluroroamino acid in plasma and brain tissue unchanged (Murakami et al., 1988, 1989). This stimulated its use as a tracer of amino acid transport in PET studies on patients with gliomas (Mineura et al., 1989). It is not clear yet if the partial metabolism, which also has not been examined in tumour tissue so far, will interfere with tracer kinetic modelling.

L-Meta-fluorophenylalanine (MFPA) and *L-meta-fluorotyrosine* (MFT) do not recommend themselves for human use due to their high toxicity and the biochemical degradation involved. MFPA in its ^{18}F-labelled form has not been examined so far, while ^{18}F-MFT exhibited a low uptake and protein incorporation in mouse brain (Coenen and Stöcklin, 1988). According to the degradation scheme given in Fig. 5 significant amounts of ^{18}F-radioactivity existed as organic acids in the brain of rats (Murakami et al., 1988).

L-Ortho-fluorotyrosine (OFT) had not been examined for its chemical and biochemical properties until it was labelled with fluorine-18. As mentioned, there was an indication of formation and incorporation of ^{18}F-OFT into the aminoacylated t-RNA pool after i.v. application (Bodsch et al., 1988). Preliminary screening experiments then demonstrated that it is even more readily incorporated into cerebral proteins of mice than ^{18}F-PFPA (Coenen and Stöcklin, 1988). This prompted detailed biochemical studies of this new [^{18}F]fluoroamino acid analogue in the brain and periphery of mice (Coenen et al., 1989, 1990). The same analytical procedures were applied as elaborated for ^{18}F-PFPA (Bodsch et al., 1988) following standard methods. Aminoacylated t-RNA was determined by phenol extraction and identity of aminoacylated acids proven by HPLC after alkaline hydrolysis. Since nothing was known about protein incorporation of OFT and the aa-tRNA fraction was very small with $<2\%$ of tissue activity, TCA-precipitated proteins were further submitted to SDS gelelectrophoresis (Coenen et al., 1989). This proved a random distribution of ^{18}F-OFT in proteins of all molecular weights and thus a general acceptance for protein synthesis.

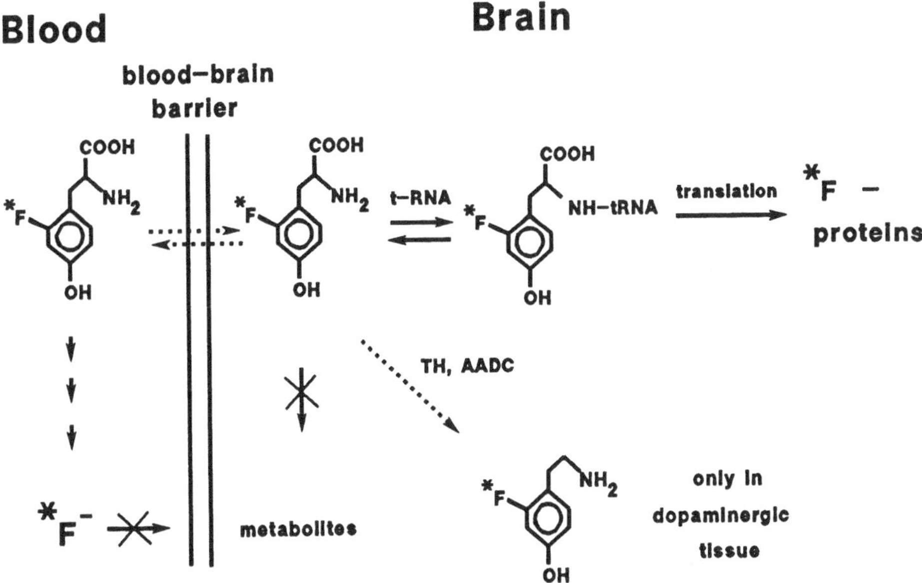

Figure 8. Metabolic pathways of L-ortho-fluorotyrosine in brain and periphery of rodents.

In contrast to ^{18}F-PFPA, the uptake and cerebral protein incorporation of ortho-fluorotyrosine is not much influenced by endogenous amino acids. This can be concluded from the identical rates of incorporation of ^{18}F-OFT and ^{14}C-labelled tyrosine with >80% at 60 min after administration (cf. Ishiwata et al., 1988). Furthermore no other metabolites were found in brain tissue besides striatum and the sum of free aminoacylated and protein bound ^{18}F-OFT amounted to 97 ± 4% of total tissue activity at all times examined up to 2 hrs (Coenen et al., 1989, 1990). Thus, considering the aminoacylated (< 2%) and free fraction together, ^{18}F-OFT fulfils the requirement for a two tissue compartmental model. This is illustrated in Fig. 8. The same unequivocal biochemistry was also found in the pancreas of mice (Coenen et al., 1990). Considering the irreversible binding in proteins during the time span of a PET study, ortho-[^{18}F]fluorotyrosine exhibits ideal prerequisites for quantitation of PSR, i.e. protein incorporation measurements.

In blood plasma as well no radiolabelled organic metabolites were found besides free ^{18}F-OFT and ^{18}F-labelled proteins. However, considerable amounts of free [^{18}F]fluoride were formed with a peak activity of 25% 20 min p.i. in mice (Coenen et al., 1990). ^{18}F is subsequently accumulated in bones. Detailed analyses suggested that fluoride is cleaved in the liver where a peak activity occurs 10 min p.i.. The extensive liberation of fluoride suggests that fluoride is cleared in an early step of tyrosine degradation. In analogy to the considerations of Weisman and Koe (1967) on the degradation of meta-fluorotyrosine (cf. Fig. 5) it is concluded that fluoride is already lost in the step of homogentisic acid formation (Coenen at al., 1990).

TABLE 3. Metabolites of [U-^{14}C]tyrosine in cortex and striatum of mice 60 min p.i. (Coenen et al., 1990).

		protein bound	free Tyr	dopamine	polar metabolite
Cortex	^{18}F	82 ± 2	11 ± 2	n.d.	n.d.
	^{14}C	70 ± 3	3 ± 1	n.d.	22 ± 3
Striatum	^{18}F	76 ± 3	14 ± 2	5 ± 1	n.d.
	^{14}C	62 ± 3	3 ± 1	6 ± 1	23 ± 2

n.d. = not detectable

Since the comparison of ^{18}F-OFT and [^{14}C]tyrosine indicated very similar biochemical behaviour it was also tested if the fluoroanalogue enters the catechol metabolic pathway (cf. Fig. 1). A double tracer experiment with ^{18}F-OFT and [U-^{14}C]tyrosine in cortex and striatum of mice indicated some differences in metabolism. (Coenen et al., 1990). As can be seen in Table 3, the protein bound fraction in both tissues is higher by 10% for the fluoroanalogue. This can partially be explained by decarboxylation and formation of a >20% fraction of polar metabolites of the ubiquitous labelled natural tyrosine, since these were not found with [1-^{14}C] tyrosine (Ishiwata et al., 1988). Correspondingly the fraction of free amino acid is also much lower for the ^{14}C-labelled tyrosine than for the ^{18}F-labelled analogue. While no dopamine was found in cortex, this was formed in similar amounts with both analogues (cf Table 3). It must, however, be taken into account that the non-protein metabolism of ^{18}F-OFT is much slower considering the still high fraction of free amino acid and the missing polar metabolites. Correction for the formation of ^{18}F-labelled dopamine, if necessary at all, must only be done with striatal tissue.

The unequivocal biochemical pathway of L-ortho-[^{18}F]fluorotyrosine into proteins allowing an irreversible three compartment model could be clearly confirmed in first PET studies on monkeys. These were performed as a clinical pretest in the context of first studies with tumour patients (Wienhard et al., 1991). As demonstrated in Fig. 10, a perfect linear correlation between measured data and a fitted curve was obtained after first in vivo distribution when the uptake kinetics were graphically presented in the form of a Gjedde-Patlak plot. This could be confirmed for uptake kinetics in normal and tumour cerebral tissue of patients. The uptake was much higher in tumours and testing of the BBB-integrity demonstrated active membrane transport. In agreement with other ^{11}C-labelled amino acids, transport was the dominating factor for uptake in tumours while the protein incorporation rate, as indicated by the rate constant k_3, was rather low. In addition a four compartment model described the uptake kinetics better for tumour tissue (Wienhard et al., 1991). This demands further studies on the biochemistry of ^{18}F-OFT in tumour tissue which is presently not known. Modelling of ^{18}F-OFT tracer kinetic uptake is discussed in another chapter of this book by Wienhard. The wider use of ^{18}F-OFT in PET studies, however, is encouraged also for other areas of protein synthesis than in tumours.

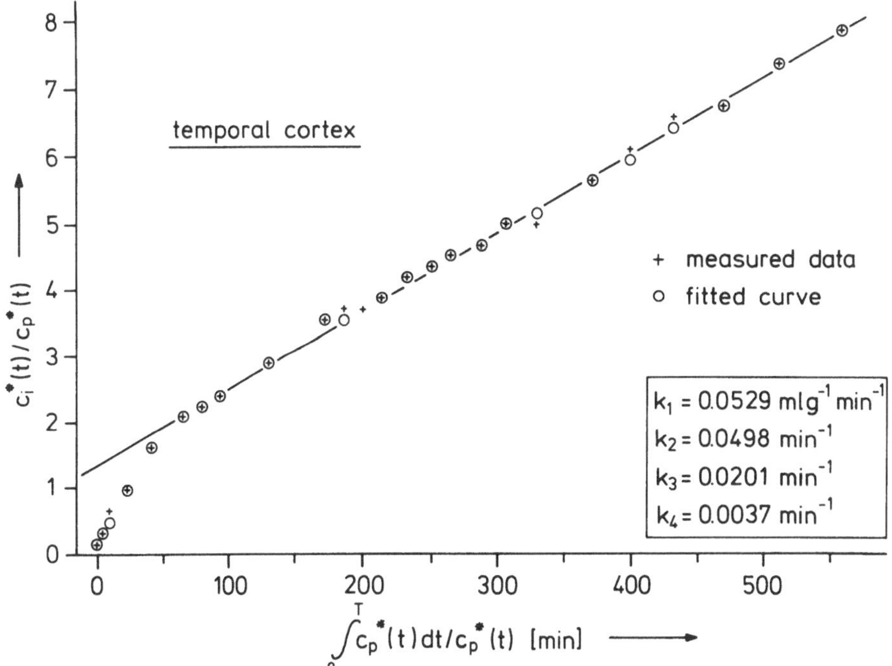

Figure 9. Uptake kinetics of L-ortho-[^{18}F]fluorotyrosine in baboon temporal cortex in form of a Gjedde-Patlak plot. Data agree with a unidirectional compartment model (supplementary data obtained in the context of the study by Wienhard et al., 1991).

5. Other Fluoroamino Acids

Besides the fluoroamino acids discussed above which have been biochemically evaluated and of which some have already found use in PET application, there are a few for which only the radiofluorination is known.

5- and 6-Fluorotryptophan were radiolabelled for pancreas scintigraphy but there are only reports on their organ distribution and not on metabolism (for review see Palmer et al., 1977). The 5-hydroxy analogue was radiofluorinated in 4- and 6-position with the goal to prepare a precursor for the neurotransmitter serotonin (Chirakal et al., 1988). However, biodistribution and metabolic studies are not known so far.

Aliphatic fluoroamino acids have found almost no interest which may be due to expected metabolic instability. *3-Fluoro-D-alanine* was however shown to be extensively metabolized in several mammalian species by action of amino acid oxidase to fluoropyruvate which is in equilibrium with fluorolactic acid (Darland et al., 1986). Its interference with this important metabolism might be of interest, but first attempts on labelling failed (van der Ley, 1983).

4-Fluoroproline has also been labelled with fluorine-18 (van der Ley, 1983) and recently in enantiomeric pure cis- and trans-form (Hamacher and Stöcklin, unpublished results). Provided in vivo stability, it is possibly a precursor for special proteins, e.g. collagens (Weygand and Oettmeier, 1970) and is presently being evaluated in vivo.

6. Conclusion

The collection of biochemical data on fluoroamino acids shows that several metabolites are formed from most analogues. No prediction can generally be made on the physiological acceptance of fluorinated derivatives of amino acids. A detailed study of the metabolism is therefore mandatory for each new tracer intended especially for in vivo quantitation of biochemical rates with positron-emission-tomography.

From the fluoroamino acids examined as precursors of proteins, only L-ortho-[^{18}F]fluorotyrosine exhibits a metabolism straight enough to allow a simple tracer model for quantitation. This is at least true for normal brain and probably pancreas tissue. If and how much free fluoride formed in the periphery will interfer has still to be studied, like validation of calculated protein incorporation rates in primates. ^{18}F-OFT has the advantage of a longer half-life which is compatible with slow protein synthesis when compared with ^{11}C-labelled amino acids. Its biochemical rate constants of the various steps of protein incorporation are not known. Identically with [^{11}C]amino acids,however, quantitative determination of PSR with PET appears presently anyhow not achievable due to the unknown intracellular pool of competing amino acids.

The other approach using fluoroamino acids as tracers of neurotransmitter and hormone formation has focused on the dopaminergic system. ^{18}F-DOPA was shown to be useful in clinical studies in spite of its complex metabolism making its quantitation difficult. [^{18}F]fluoro-meta-tyrosines appear superior for simpler modelling, although there are no clinical experiences

While fluoro-DOPA and fluoro-meta-tyrosine reflect primarily the activity of AADC, provided effective retention of their cerebral metabolites occurs, the potential PET-tracers which are suicide inhibitors indicate the concentration of active enzymes. Thus, specific steps in dopamine metabolism are traced, a concept to reduce and simplify modelling for quantitation. This approach is also very promising for other enzyme systems. Wether it is achievable and will become clinically useful awaits further studies. With respect to their potential toxicity no-carrier-added syntheses appear necessary prior to application of the irreversible enzyme blockers for PET studies in humans.

7. References

Arnstein HRV, Richmond MH (1964) Utilization of p-fluorophenylalanine for protein synthesis by the phenylalanine-incorporation system from rabbit reticulocytes. Biochem J 91:340-346.

Axelrod J, Saavedra JM (1974) Octopamine, phenylethanolamine, phenylethylamine and tryptamine in the brain. Ciba Found Symp 22:51-59.

Bodsch W, Coenen HH, Stöcklin G, Takahashi K, and Hossmann KA (1988) Biochemical and autoradiographic study of cerebral protein synthesis with ^{18}F- and ^{14}C-fluorophenylalanine. J Neurochem 5O:979-983.

Boyes BE, Cumming P, Martin WRW, McGeer EG (1986) Determination of plasma [^{18}F]-6-fluorodopa during positron emission tomography: elimination and metabolism in carbidopa treated subjects. Life Sci 39:2243-2252.

Bustany P, Comar D (1985) Protein synthesis evaluation in brain and other organs in human by PET. In: Reivich M, Alavi A (eds.) Positron emission tomography. Alan R.Liss:183-201

Chirakal R, Sayer BG, Firnau G, Garnett ES (1988) Synthesis of F-18 labelled fluoro-melatonins and 5-hydroxy-fluoro-tryptophanes. J Label Compds Radiopharm 25:62-71.

Chirakal R, Firnau G, Garnett ES (1989) A positron emitting inhibitor of aromatic L-amino acid decarboxylase: α-fluoromethyl-6-[^{18}F]fluoro-L-dopa. J Label Compds Radiopharm 26:228-229.

Coenen HH, Bodsch W, Takahashi K, Hossmann KA, Stöcklin G (1986) Synthesis, autoradiography and biochemistry of L-[^{18}F]fluorophenylalanines for probing protein synthesis. Nuklearmedizin 22:(Suppl)600-602.

Coenen HH, Stöcklin G (1988) Evaluation of radiohalogenated amino acid analogues as potential tracers for PET and SPECT studies or protein synthesis. Radioisot Klinik Forschung 18:402-440.

Coenen HH, Kling P, Stöcklin G (1989) Cerebral metabolism of L-[2-^{18}F]fluorotyrosine, a new PET tracer of protein synthesis. J Nucl Med 30:1367-1372.

Coenen HH, Wutz W, Stöcklin G, DeGrado T, Kling P (1990) Pharmacokinetics and metabolism of L-[2-^{18}F]fluorotyrosine in the brain and periphery of mice. J Nucl Med 31:716.

Creveling CR, Kirk KL (1985) The effect of ring fluorination on the rate of O-methylation of dopa by catechol-O-methyltransferase: significance in the development of ^{18}F-PET agents. Biochem Biophys Res Commun 30:1123-1131.

Cumming P, Boyes B, Martin WRW et al (1987a) The metabolism of [^{18}F]6-fluoro-L-3,4-dihydroxy-phenylalanine in the hooded rat. J Neurochem 48:601-608.

Cumming P, Boyes BE, Martin WRW, Adam M, Ruth TJ, McGeer EG (1987b) Altered metabolism of [^{18}F]-6-fluorodopa in the hooded rat following inhibition of catechol-O-methyltransferase with U-0521. Biochem Pharmacol 36:2527-2531.

Cumming P, Häusser M, Martin WRW et al (1988) Kinetics of in vitro decarboxylation and the in vivo metabolism of 2-^{18}F- and 6-^{18}F-fluordopa in the hooded rat. Biochem Pharmacol 37:247-250.

Darland G, Kropp H, Kahan FM, Hajdu R, Walker R, VandenHeuvel WJA (1985) The metabolism of 2-deutero-3-fluoro-D-alanine (DFA). In: Muccino RR (ed.) Synthesis and applications of isotopically labelled compounds. Amsterdam: Elsevier Science Publishers :277-281

DeJesus OT, Mukherjee L (1988) Radiobrominated m-tyrosine analog as potential CNS L-dopa PET tracer. Biochem Biophys Res Comm 150:1027-1031.

DeJesus OT, Mukherjee J, Appelman EH (1989) Synthesis of o- and m-tyrosine analogs as potential tracers for CNS dopamine. J Label Compds Radiopharm 26:133-134.

DeJesus OT, Murali D, Sunderland JJ, Chen CA, Weiler M, Nickels RJ (1990) [^{18}F]Fluoro-MDL 72394, a potentially trappable tracer for presynaptic dopamine neurons. J Nucl Med 30:902.

DeJesus OT, Murali D, Oakes TR, Holden JE, Nickles RJ (1992) Synthesis of ^{18}F-labelled α-fluoromethyl-p-tyrosine, a tyrosine hydroxylase-activated decarboxylase suicide inhibitor with potential as imaging agend for dopamine nerve terminals. J Label Compds Radiopharm in press.

Dolan G, Godin C (1966) In vivo formation of tyrosine from p-fluorophenylalanine. Biochemistry 5:922-925.

Dunlop DS (1983) Protein turnover in brain; synthesis and degradation. In: Lajtha A (ed) Handbook of neurochemistry. New York: Plenum Press :25-63

Firnau G, Nahmias C, Garnett ES (1973) The preparation of [^{18}F]5-fluoro-dopa with reactor produced fluorine-18. Int J Appl Radiat Isot 24:182-183.

Firnau G, Garnett ES, Chirakal R, Sood S, Nahmias C, Schrobilgen G (1986) [^{18}F]fluoro-L-dopa for the in vivo study of intracerebral domapine. Appl Radiat Isot 37:669-657.

Firnau G, Sood S, Chirakal R et al (1987) Cerebral metabolism of 6-[^{18}F]fluoro-L-3,4-dihydroxylphenylanaline in the primate. J Neurochem 48:1077-1082.

Firnau G, Sood S, Chirakal R, Nahmias C, Garnett ES (1988) Metabolites of 6-[^{18}F]fluoro-L-dopa in human blood. J Nucl Med 29:363-369.

Firnau G, Chirakal R, Nahmias C, Garnett ES (1991) Tracers for the investigation of cerebral presynaptic dopaminergic function with positron emission tomography. In: Kuhl D (ed.) Frontiers in nuclear medicine: In vivo imaging of neurotransmitter function in brain, heart and tumors. Washington DC: Amer Coll Nuc Phys :67-92

Fowden L (1972) Fluoroamino acids and protein synthesis. In Ciba Foundation Symposium, Carbon-Fluorine Compounds, North Holland, Amsterdam :141-159

Frieden E, Hsu LT, Dittmer K (1951) Enzymatic degradation of amino acid antagonists. J biol Chem 192:425-433.

Gal EM, Millard SA (1971) The mechanism of inhibition of hydroxylases in vivo by p-chlorophenylalanine: the effect of cycloheximide. Biochem Biopphys Acta 227:32-41.

Gal EM (1974) Synthetic p-halogenophenylalanine and protein synthesis in the brain. Ciba Found Symp 22:343-359.

Garnett ES, Firnau G, Nahmias C et al (1983) Dopamine visualized in the basal ganglia of living man. Nature 305:137-138.

Gaull GE (1978) Biology of brain dysfunction, Vol. 3. New York: Plenum Press

Huang SC, Yu DC, Barrio JR et al (1991) Kinetics and modeling of L-6-[^{18}F]fluoro-DOPA in human positron emission tomographic studies. J Cereb Blood Flow Metab 11:898-913.

Ishiwata K, Vaalburg W, Elsinga PH, Paans AMJ, Woldring MG (1988) Metabolic studies with L-[1-^{14}C]tyrosine for the investigation of a kinetic model to measure protein synthesis rates with PET. J Nucl Med 29:524-529.

Kaufmann S (1961) The enzymatic conversion of 4-fluorophenylalanine to tyrosine. Biochim Biophys Acta 61:619-621.

Kollonitsch J, Patchett AA, Marburg S, et al (1978) Selective inhibitors of biosynthesis of aminergic neurotransmitters. Nature 274:906-908.

Laihinen A, Rinne JO, Rinne UK, et al (1992) [^{18}F]-6-Fluorodopa PET scanning in Parkinson's disease after selective COMT inhibition with nitecapone (OR-462). Neurology 42:199-203.

Leenders KL, Poewe WH, Palmer AJ, Brenton DP, Frackowiak RSJ (1986) Inhibition of L-[18F]fluorodopa uptake into human brain by amino acids demonstrated by positron emission tomography. Ann Neurol 20:258-262.

Leenders KL (1991) 6-[^{18}F]Fluorodopa uptake in brain. In: Baron JC, et al. (eds) Brain dopaminergic systems: imaging with positron emission tomography. Dordrecht: Kluwer Academic Publishers :97-110

Luxen A, Guillaume M, Melega WP, Pike VW, Solin O, Wagner R (1992) Production of 6-[^{18}F]fluoro-L-DOPA and its metabolism in vivo - a critical review. Nucl Med Biol 19:149-158.

Martin WRW, Palmer MR, Patlak CS, Calne DB (1989) Nigrostriatal function in humans studied with positron emission tomograpy. Ann Neurol 26:535-537.

Melega WP, Perlmutter MM, Luxen A et al (1989) 4-[^{18}F]fluoro-L-m-tyrosine: An L-3,4-dihydroxyphenylalanine analog for probing presynaptic dopaminergic function with positron emission tomography. J Neurochem 53:311-314.

Melega WP, Luxen A, Perlmutter MM, Nissenson CHK, Phelps ME, Barrio JR (1990a) Comparative in vivo metalosim of [^{18}F]fluoro-L-dopa and [^{3}H]L-dopa in rats. Biochem Pharmacol 39:1853-1860.

Melega WP, Hoffmann JM, Luxen A, Nissenson CHK, Phelps ME, Barrio JR (1990b) The effects of carbidopa on the metabolism of 6-[^{18}F]fluoro-L-dopa inb rats, monkeys and humans. Life Sciences 47:149-157.

Melega WP, Grafton ST, Huang SC, Satyamurthy N, Phelps ME, Barrio JR (1991a) L-6-[^{18}F]Fluoro-DOPA metabolism in monkeys and humans: biochemical parameters for the formulation of tracer kinetic models with positron emission tomography. J Cereb Blood Flow Metab 11:890-897.

Melega WP, Hoffman JM, Schneider JS, Phelps ME, Barrio JR (1991b) 6-[^{18}F]Fluoro-L-DOPA metabolism in MPTP-treated monkeys: assessment of tracer methodologies for positron emission tomography. Brain Res 543:271-267.

Mineura K, Kowada M, Shishido G (1989) Brain tumor imaging with synthesized ^{18}F-fluorophenylalanine and positron emission tomography. Surg Neurol 31:468-469.

Moore KE, Jerome AD (1971) Tyrosine hydroxylase inhibitors. Fed Proc 30:859-870.

Murakami M, Takahashi K, Kondo Y et al (1988) 2-^{18}F-Phenylalanine and 3-^{18}F-tyrosine - synthesis and preliminary data of tracer kinetics. J Label Compds Radiopharm 25:773-782.

Murakami M, Takahashi K, Kondo Y et al (1989) The slow metabolism of L-[2-^{18}F]fluorophenylalanine in rat. J Labelled Cmpd Radiopharm 27:245-255.

Oldendorf WH (1971) Brain uptake of radiolabelled amino acids, amines and hexoses after arterial injection. Am J Physiol 221:1629-1639.

Palmer AJ, Clark JC, Goulding RW (1977) The preparation of fluorine-18 labelled radiopharmaceuticals. Int J Appl Radiat Isot 28:53-65.

Pardridge WM (1977) Kinetics of competitive inhibition of natural amino acid transport across the blood-brain barrier. J Neurochem 28:103-108.

Phelps ME, Barrio JR, Huang SC, Keen RE, Chugani H and Maziotta JC (1984) Criteria for the tracer kinetic measurement of cerebral protein synthesis in humans with positron emission tomography. Ann Neurol 15: (Suppl)192-202.

Richmond MH (1962) The effect of amino acids analogues on growth and protein synthesis in microorganisms. Bacteriol Rev 26:398-420.

Smith CB, Deibler GE, Eng N,Schmidt K, Sokoloff L (1988) Measurement of local cerebral protein synthesis in vivo: influence of recycling of amino acids derived from protein degradation. Proc Natl Acad Sci 85:9341-9345.

Sokoloff L, Smith C (1983) Biochemical principles for the measurement of metabolic rates in vivo. In: Heiss WD, Phelps ME (eds.) Positron Emission Tomography of the Brain New York: Springer :2-18

Vaalburg W, Coenen HH, Crouzel C, Elsinga PH, Langström B, Lemaire C, Meyer GJ (1992) Amino acids for the measurement of protein synthesis in vivo by PET. Nucl Med Biol 19:227-237.

Van der Ley (1983) [^{18}F]Fluorine labeled aliphatic amino acids. J Label Compds Radiopharm 20:453-461.

Weissman A, Koe BK (1967) m-Fluorotyrosine convulsions and mortality: relationship to catecholamine and citrate metabolism. J Pharmacol exp Ther 155:135-144.

Wienhard K, Herholz K, Coenen HH, Rudolf J, Kling P, Stöcklin G, Heiss WD (1991) Increased amino acid transport into brain tumors measured by PET of L-[2-^{18}F]fluorotyrosine. J Nucl Med 32:1338-1346.

Westhead EW, Boyer PD (1961) The incorporation of p-fluorophenylalanine into some rabbit enzymes and other proteins. Biochem Biophys Acta 54:145-156.

Weygand F, Oettmeier W (1970) Fluorine-containing amino acids. Russian Chem Rev 39:290-300.

DISCUSSION

N. Leenders By wich MAO is the meta-tyrosine analogue trapped and are we sure that we are measuring MAO activity?

H. Coenen It is not the β-fluoro-methylene-meta-thyrosine itself but the corresponding tyramine wich is formed in DA neurons by decarboxylation and which irreversibly binds to mitochondrial MAO. Depending on the relative concentrations of AAADC and MAO we will either measure the activity of decarboxylase if its concentration is smaller, or the concentration (not activity) of MAO if decarboxylase is in excess.

H. Coenen Is there any experience in adding ascorbic acid in order to prevent oxidation of methionine? This was helpful to avoid oxidation of fluoro-dopa.

K. Nagren There is no experience with respect to ^{11}C-labelled methionine in this matter. However, I dont think that it is necessary to add anti-oxidants. The oxidation products can be removed by the final HPLC step, and the final product solution is stable at least for one hour.

B. M. Mazoyer et al. (eds.), PET Studies on Amino Acid Metabolism and Protein Synthesis, 131.

KINETIC MODELING OF FLUOROTYROSINE UPTAKE

K. WIENHARD, K. HERHOLZ, H.H. COENEN, J. RUDOLF,
P. KLING, G. STÖCKLIN, W.-D. HEISS

ABSTRACT. L-(2-F-18)fluorotyrosine (F-Tyr) shown recently to be a tracer of cerebral protein synthesis in the mouse brain has several promising features as a PET tracer in humans: it penetrates easily into the brain, while almost no metabolites are found in brain tissue; the rather long half-life of fluor-18 permits to follow the uptake until accumulation becomes linear. Therefore, F-Tyr accumulation was studied for the first time in 15 patients with various brain tumors. Curve fitting to 2-hour F-Tyr accumulation data indicated a three tissue compartment (5 kinetic constants) model to be better suited for tumor tissue than a 3 rate constant approach. Increased uptake of F-Tyr in tumors was due to two times higher transport rates while the rate constants describing irreversible incorporation were on average decreased in tumors.

INTRODUCTION

The individual steps in developing a PET-tracer model like: tracer selection, going from a comprehensive model to a workable model, model validation and model application impose several requirements for the measurement of protein synthesis rates. Among these are: high tracer uptake in brain tissue; no labeled metabolites in brain tissue; no uptake of labeled plasma metabolites; irreversible tissue kinetics in the time window of the PET study, which is mainly defined by the half-life of the tracer; long half-life of the tracer to follow tissue kinetics until they become linear in a Patlak plot; the model should be as simple as possible; an autoradiographic model might be of advantage; a labeled analog molecule has the advantage that there is no disturbance from the recycling of unlabeled amino acids in the precursor pool; the model should be sensitive to pathologies and the whole procedure should be suitable for clinical application.

The estimation of local cerebral protein synthesis rates in man by positron emission tomography (PET) of the accumulation of labeled amino acids is limited since incorporation rates into proteins and metabolic pathways are different for various amino acids and metabolites as well as competing substrates for t-RNA

B. M. Mazoyer et al. (eds.), PET Studies on Amino Acid Metabolism and Protein Synthesis, 133–147.

cannot be quantified in the usual clinical setting. Therefore, variables and constants defining applicable kinetic models cannot be described with sufficient accuracy to provide reliable measurements of in vivo protein synthesis.

Most amino acids used for PET applications so far have been labeled with ^{11}C [1 - 6] limiting their use in the assessment of the rather slow incorporation into proteins by the short half-life of this positron emitter. The recently synthetized L(-^{18}F)fluorotyrosine (F-Tyr) is a promising tracer of protein synthesis [7] since as evaluated in mice this compound penetrates easily into the brain, while its metabolites do not cross the blood-brain barrier and no further metabolites are found in the brain tissue. These advantages are combined with the rather long half-life of the label which permits to follow the uptake into the tissue until accumulation becomes linear. It has also been shown that F-Tyr unlike other halogenated amino acids is almost quantitatively incorporated into proteins [7]. Recent biochemical studies with (1-^{14}C)-tyrosine [8] showed almost identical uptake and metabolic behavior as F-Tyr. Despite the up to now undefined affinities of F-Tyr to carrier enzymes and t-RNA in relation to natural tyrosine it seemed justified to evaluate F-Tyr in patients with brain tumors of different malignancy. For that purpose uptake kinetics for F-Tyr in a two and three tissue compartment model were compared to kinetic constants and metabolic rates of glucose as measured by 2(^{18}F)fluoro-2-deoxy-D-glucose (FDG) and to tracer transport in relation to damage of the blood brain barrier assessed by ^{68}Ga-EDTA.

MATERIALS AND METHODS

Patients population

We studied 15 patients, 12 men and 3 women, with mean age 50.6 ± 9.8 (SD) years. Five patients had low-grade gliomas, which were classified histologically as astrocytoma of grade 2 in 3 cases, whereas the other two were diagnosed clinically and had focal hypodense lesions without contrast enhancement in CT-scan. Two patients had histologically proven anaplastic oligoastrocytoma (WHO-grade III). Six patients suffered from glioblastomas, of which were 5 histologically proven. Two additional patients did not fit into those groups: a 67-year old man had a cerebral lymphoma, and a 33-year old man had diffuse gliomatosis of the frontal lobe. In both cases diagnosis was based on brain biopsy.

All patients were studied with FDG and F-Tyr, in 11 cases also with ^{68}Ga-EDTA, usually within three consecutive days. In no case was there any substantial therapeutic intervention, such as surgery, radiotherapy, or chemotherapy between scans. Five patients had not received any treatment prior to scanning, apart form corticosteroids or antiepileptic drugs, and 4 patients had stereotactic tumor biopsy only. One patient with an astrocytoma underwent incomplete tumor resection 3 months before the PET studies, with no other treatment. The remaining 5 patients suffered from recurring tumors and were examined after previous surgery, radio-therapy or chemotherapy. In all cases CT-scans showed clear evidence of a solid tumor mass.

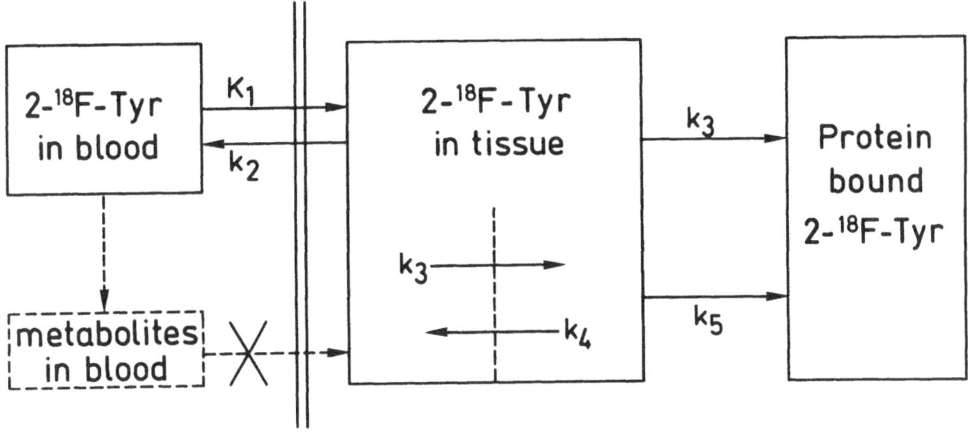

Fig. 1 Compartment model for 2-^{18}F-Tyrosine uptake with three or five rate constants. K_1 and k_2 refer to forward and reverse transport of F-Tyr across the blood brain barrier, respectively. In the upper part k_3 is the rate constant for incorporation into proteins. In the lower part the three rate constant model is extended to a five parameter model by adding a further serial tissue compartment with inward and outward transport rate constants k_3 and k_4, respectively; the incorporation into proteins is then described by rate constant k_5. It is assumed that metabolites in blood do not cross the blood brain barrier and that no further metabolites exist in tissue.

Mathematical model and model analysis

COENEN et al. [7] demonstrated in mouse experiments that F-Tyr is distributed almost exclusively in two compartments of the brain, namely as free F-Tyr in tissue and as a protein bound fraction. From this simple metabolism a two tissue compartment model as shown in Fig. 1 was proposed. This model is very similar to the deoxyglucose-(DG)-model of SOKOLOFF et al. [9]: free F-Tyr is transported from the cerebral blood pool into brain tissue where it is either incorporated into protein or transported back to blood. The mathematical form of the model equation is identical to the DG-model equations [9] with the time course of tissue activity given by

$$C_t^*(t) = K_1 \left[\frac{k_3}{k_2+k_3} \int_o^t C_p^*(t')dt' + \frac{k_2}{k_2+k_3} \exp[-(k_2+k_3)t] \int_o^t C_p^*(t') \exp[(k_2+k_3)t']dt' \right] \quad (1)$$

The accumulation rate, K_{MR}, is then expressed by $\frac{K_1 k_3}{k_2 + k_3}$.

The mice data cannot give any information whether it is necessary to divide the tissue precursor pool into an extracellular and an intracellular space as shown in Fig. 1. However, a significant improvement in the goodness of fit of the model equations to the PET time activity data may be an indication for the more refined model. From the differential equations for the three tissue compartment - five rate constant-model in Fig. 1 the time dependence of the total tissue activity as measured by PET can be derived as:

$$C_i^*(t) = \frac{K_1}{\alpha_1 - \alpha_2}\left[\left(\alpha_1 + k_3 + k_4 + k_5 + \frac{k_3 + k_5}{\alpha_1}\right)e^{\alpha_1 t} - \left(\alpha_2 + k_3 + k_4 + k_5 + \frac{k_3\ k_5}{\alpha_2}\right)\cdot e^{\alpha_2 t}\right] \otimes C_p^*(t) + \frac{K_1 k_3 k_5}{k_2 k_4 + k_2 k_5 + k_3\ k_5}\int_o^t C_p^*(t')dt' \qquad (2)$$

where $C_i^*(t)$ is the total ^{18}F-Tyr activity in tissue,

$\otimes$ denotes the operation of convolution and

$$\alpha_1 = -\frac{k_2 + k_3 + k_4 + k_5}{2} - \frac{1}{2}\cdot\sqrt{(k_2 + k_3 + k_4 + k_5)^2 - 4(k_2 k_4 + k_2 k_5 + k_3 k_5)}$$

$$\alpha_2 = -\frac{k_2 + k_3 + k_4 + k_5}{2} + \frac{1}{2}\sqrt{(k_2 + k_3 + k_4 + k_5)^2 - 4(k_2 k_4 + k_2 k_5 + k_3 k_5)}$$

The accumulation rate in this case is given by

$$K_{MR} = \frac{K_1 k_3 k_5}{k_2\ k_4 + k_2\ k_5 + k_3\ k_5}$$

$C_p^*(t)$ is the plasma activity corrected for metabolites, thus representing the concentration of F-Tyr in plasma. In the fits of the model equations to the PET data, the vascular part of the tissue activity was taken into account as an additional fit parameter by setting

$$\mathrm{PET_i(t)} = (1-\mathrm{CBV_i})\ C_i^*(t)\ + \mathrm{CBV_i}\ \ C_p^*{}_{,\mathrm{total}}(t) \tag{3}$$

where $PET_i(t)$ are the time activity data in region i, CBV_i is the vascular fraction of the tissue activity in that region and $C_p^*{}_{,total}$ is the time course of total ^{18}F-activity in blood plasma. The FDG time activity data were fitted to the Sokoloff DG-model with three rate constants taking also the local blood volume into account as in equ 3. For calculation of metabolic rate a lumped constant of 0.42 was used. The ^{68}Ga-EDTA data were fitted with a one tissue compartment model with two rate constants (setting k_3 and k_4 equal to zero in equ 1) and the blood volume contribution as a third parameter.

In addition to regional fits of tissue time activity data with the model equations, parametric images of local blood volume, accumulation rate, transport and incorporation rate constants were generated by weighted nonlinear least-squares fits on a pixel by pixel basis [10].

Tracer preparation

L-(2-^{18}F)fluorotyrosine was prepared by electrophilic radiofluorination of O-acetyltyrosine as previously described in detail by COENEN et al. [11]. The product was chemically and radiochemically pure ($\geq$ 97 %) with a specific activity of about 20 BGq/mmol.

FDG was synthetized according to EHRENKAUFER et al. [12]. ^{68}Ga was applied as ^{68}Ga-EDTA, eluated from a commercially available ^{68}Ge-^{68}Ga generator. Between 150 - 200 MBq of the tracers were administered in 3 - 5 ml pyrogen free saline solution as an i.v. bolus.

PET measurements

The PET measurements with F-Tyr, FDG and Ga-EDTA were performed on consecutive days. For that purpose, the patients were carefully repositioned using individual markers and laser beams, and the standard procedure established in our laboratory for several years [13] was applied. Data recording was started with injection of the tracer. Seven equally spaced and parallel planes, centered from the canthomeatal line to 81 mm above were simultaneously scanned for 120 min with F-Tyr, for 40 min with FDG and for 30 min with Ga-EDTA. The consecutive scan-time-intervals were gradually increasing from 1 to 5 min. The four-ring positron camera (Scanditronix PC 384-7B) with a spatial resolution of approx. 8 mm width at half maximum in 11 mm slices [14] gave with this procedure dynamic information about tracer accumulation in virtually all major structures of the brain. About twenty-five arterialized venous blood samples were obtained during the whole scan time, at intervals gradually increasing from 15 sec to 20 min. They were centrifuged, the plasma activity concentration was measured in a cross calibrated well counter, and for F-Tyr the percentage of unmetabolized ligand in plasma was determined by HPLC (RP-18, column, 250 x 4 mm; 2 % acidic acid as eluant). Data from the tomographic device and from the sample changer used for plasma counting as well as plasma glucose values determined in duplicate by a standard enzymatic method were stored in the memory of a VAX 11/780 (Digital) computer for later processing. On the reconstructed tomographic images, regions of interest (ROI) were outlined individually with respect to corresponding X-ray CT scans or MRI images, representing different tumor areas and contralateral tissue. The time course of tissue activity in these ROI's was sampled and fitted to the model equations (see mathematical model and model analysis) using an advanced fitting routine [15].

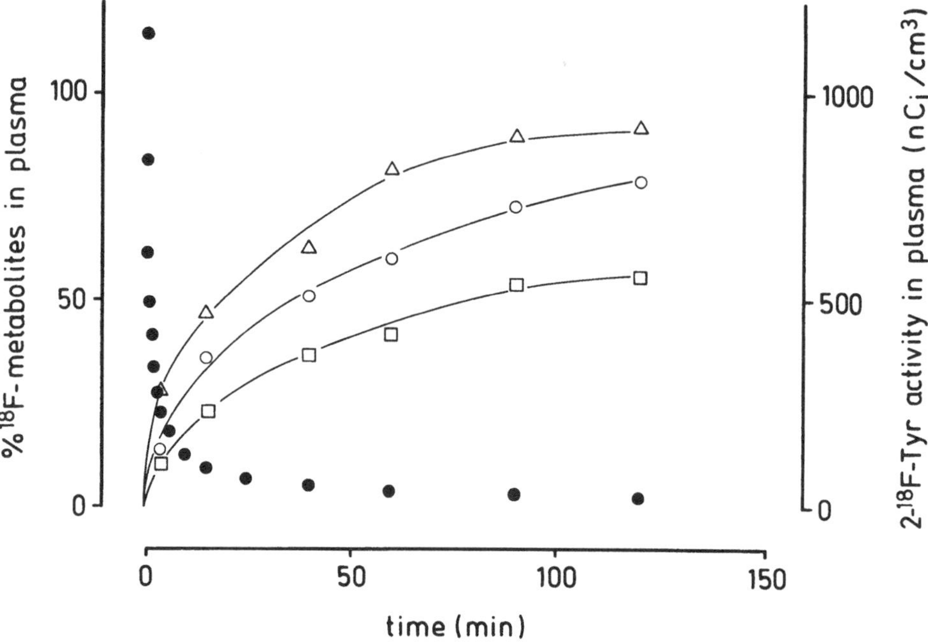

Fig. 2 Time course of unmetabolized F-Tyr activity and labeled metabolites in blood plasma. Open symbols represent the increase of labeled F-Tyr metabolites in blood plasma as percentage of total plasma activity in three different patient studies. Full dots show a typical time course of HPLC-measured unmetabolized F-Tyr activity in plasma representing the input function for the model analysis.

RESULTS

F-TYR accumulation

Fig. 2 shows the typical time course of unmetabolized F-Tyr activity in blood plasma after i.v. injection and the progressive increase in F-Tyr metabolites, presented as percent of the total blood plasma activity in three typical patient studies. It is evident from this figure that F-Tyr activity in blood plasma decreased rapidly and that F-Tyr is immediately metabolized in the body. After one hour about 50 % of the blood activity represents labeled metabolites, at two hours after injection metabolite portions over 90 % were observed in some cases. The major part of the metabolites was F-Tyr incorporated into proteins as determined by TCA precipitation. Because of the large variations no attempt was made to extract an average distribution curve form the individual data. The time course of the activity of unmetabolized F-Tyr in plasma represents the input function for the model analysis of the tissue data.

The typical time course of tissue uptake data of F-Tyr in a malignant tumor and in normal cortical tissue contralateral to the tumor is shown in Fig. 3. There is a rapid uptake during the first 5 - 10 min followed by a continuous efflux over the study interval of 2 hours. The difference between pathological and normal tissue is much bigger in the early part of the uptake curve with a factor of four between the maxima of tissue activity compared to twofold higher values at the end of the study after two hours. These differences are typical for malignant tumors and the detailed compartmental model analysis will show that there is a striking distinction in the transport from blood into tissue for tumors.

Fig. 3 also illustrates model fits for the time course of F-Tyr tissue activity. A three rate constant, two tissue compartment model, similar to the Sokoloff FDG model with k_3 describing the irreversible incorporation of the tracer, gives generally good fits to the data. However, in several tumor regions, especially those with high fit values for K_1, the three rate constant model gave only poor fits to the data especially for the first 20 min of the time activity curves. The addition of a third tissue compartment resulting in a five rate constant model improved the unsatisfactory fits considerably. For normal tissue the five rate constant model gave only insignificantly better fits than the three rate constant model.

Fitted rate constants with the two models are given in Table 1. Direct comparison is only possible for K_1 and K_{MR}. K_1 is on average 25 % lower in the 3-rate-constant model than in the 5-rate-constant model, while K_{MR} is 20 % higher. Despite these differences the general results remain stable: In particular a close correlation (Fig. 4) between K_1 ($r = 0.98$) and K_{MR} ($r = 0.98$) respectively, as calculated from the two models was obtained indicating the robustness of the approach, irrespective of details of the model configuration. Both models and the accumulation curves demonstrate that the main difference between normal and tumorous tissue is the transport rate constant of F-Tyr from plasma to tissue, K_1, which is significantly increased in tumors, and clearly separates normal from pathological tissue. The rate constant describing the irreversible step of F-Tyr incorporation, k_3 or k_5, respectively, is in the normal range or even slightly reduced in tumors. Due to the high K_1 values the accumulation rate K_{MR} is increased in tumors with a relationship to the malignancy separating low grade from high grade gliomas (Fig. 4).

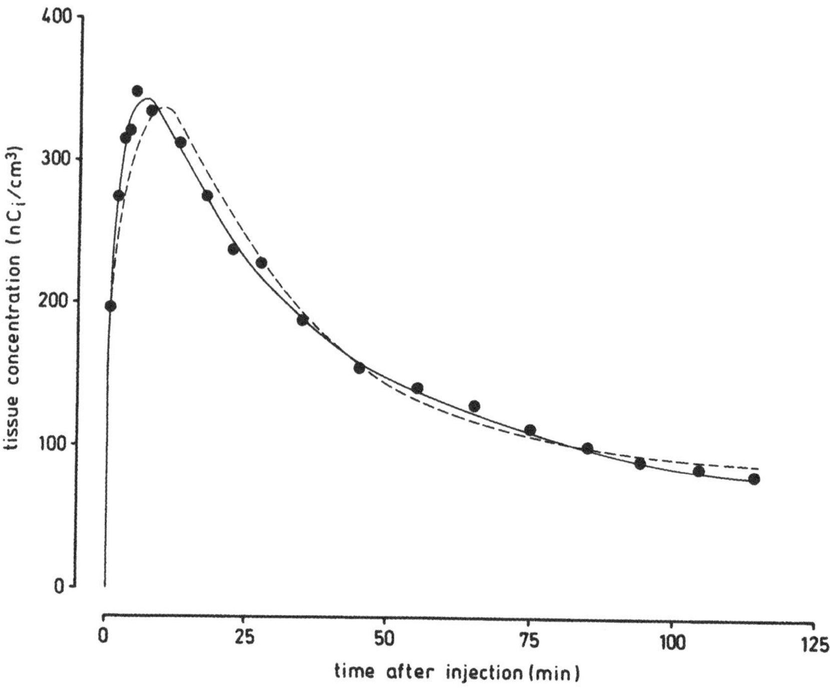

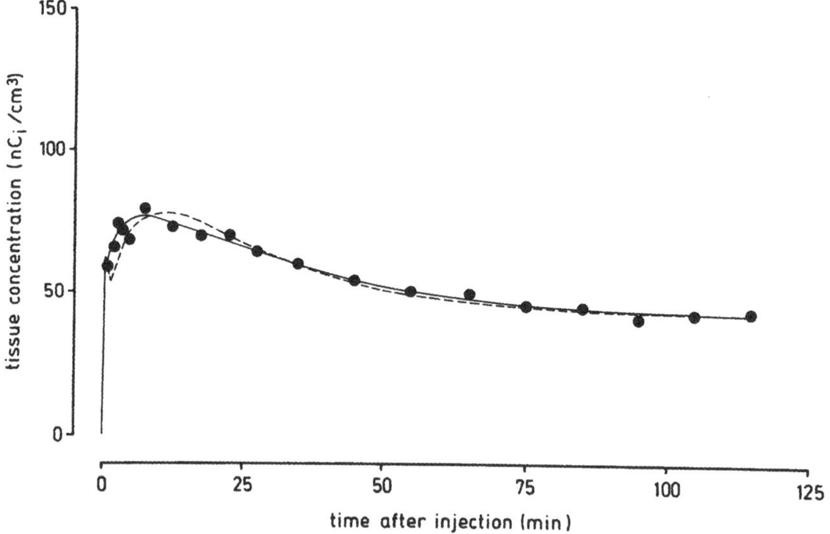

Fig. 3 F-Tyr tissue time activity data in an astrocytoma grade II (upper part) and in normal tissue contralateral to the tumor (lower part). Dashed lines are fits with a three rate constant model. Solid lines are fits with five rate constants.

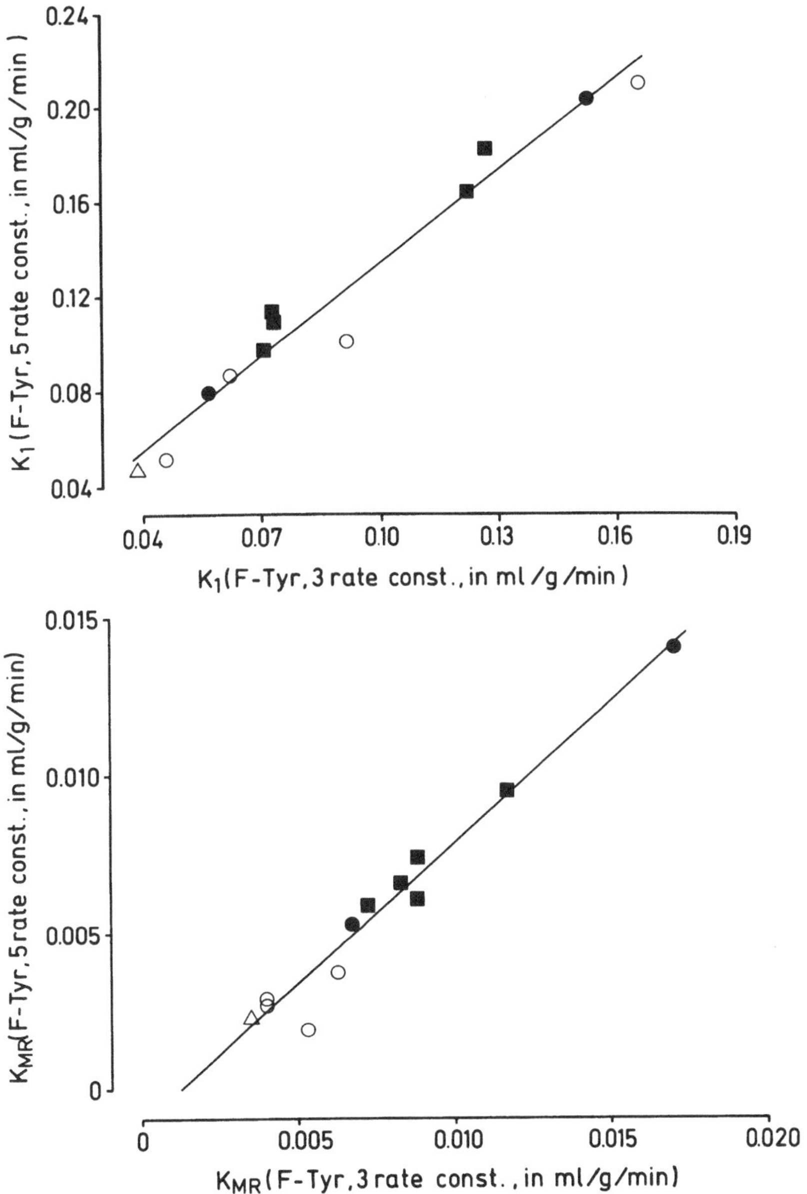

Fig. 4 Correlations between three and five rate constant models in tumor tissue upper part: transport rate constant K_1, lower part: accumulation rate K_{MR}. The symbols represent: open circles: low grade gliomas, filled circles: anaplastic oligoastrocytomas (WHO-grade III), filled squares: glioblastomas, open triangle: gliomatosis.

Table 1: Comparison of F-Tyr rate constants obtained with three and five rate constant model

	Tumor		Contral. Brain	
Parameter	3 r. c.	5 r. c.	3 r. c.	5 r. c.
K_1	0.090±0.040	0.117±0.057	0.045±0.016	0.057±0.020
k_2	0.069±0.014	0.138±0.039	0.065±0.020	0.161±0.079
k_3	0.006±0.002	0.052±0.025	0.009±0.004	0.092±0.056
k_4		0.047±0.012		0.070±0.057
k_5		0.008±0.003		0.014±0.008
K_{MR}	0.0064±0.0037	0.0056±0.0035	0.0050±0.0018	0.0044±0.0016

Comparison of F-Tyr with FDG and ^{68}Ga-EDTA

Fig. 5 shows matched PET images from a ^{68}Ga-EDTA, a F-Tyr and a FDG study of a patient with a glioblastoma. As in most instances the extension of the tumor was best demarcated at the F-Tyr image, while the FDG scan separates the solid tumor rim and the necrotic core with low metabolism. Increased ^{68}Ga-EDTA uptake was preferentially seen in tumor areas with contrast enhancement on X-ray CT, such as the rim around the central necrosis of the glioblastoma, but also appeared in tumor core regions probably housing necrosis from CT appearance and low metabolic activity. In contrast, increased F-Tyr and FDG accumulation was usually located in solid, probably actively proliferating tumor parts; their uptake was below normal in necrotic tissue.

Patlak plots [16] of the three tracers in normal and tumorous tissue of the patient presented in Fig. 5 are shown in Fig. 6. In tumors, F-Tyr and ^{68}Ga-EDTA activity show a fast initial rise related to increased blood-tissue penetration; the accumulation rate reaches zero for ^{68}Ga-EDTA after this initial slope. For FDG, tracer uptake in tumor tissue continues to rise after the initial extraction indicating increased irreversible metabolic turnover. The Patlak plots characterize the differences of tumor to normal tissue:, ^{68}Ga-EDTA indicates a parallel shift of the activity after the increased initial diffusion; for F-Tyr the curves are separated in the initial slopes indicating increased transport in tumors, later on the curves have similar slopes as consequence of comparable incorporation rates (k_3 or k_5); FDG curves diverge progressively due to the different phosphorylation rates in tumor and normal tissue.

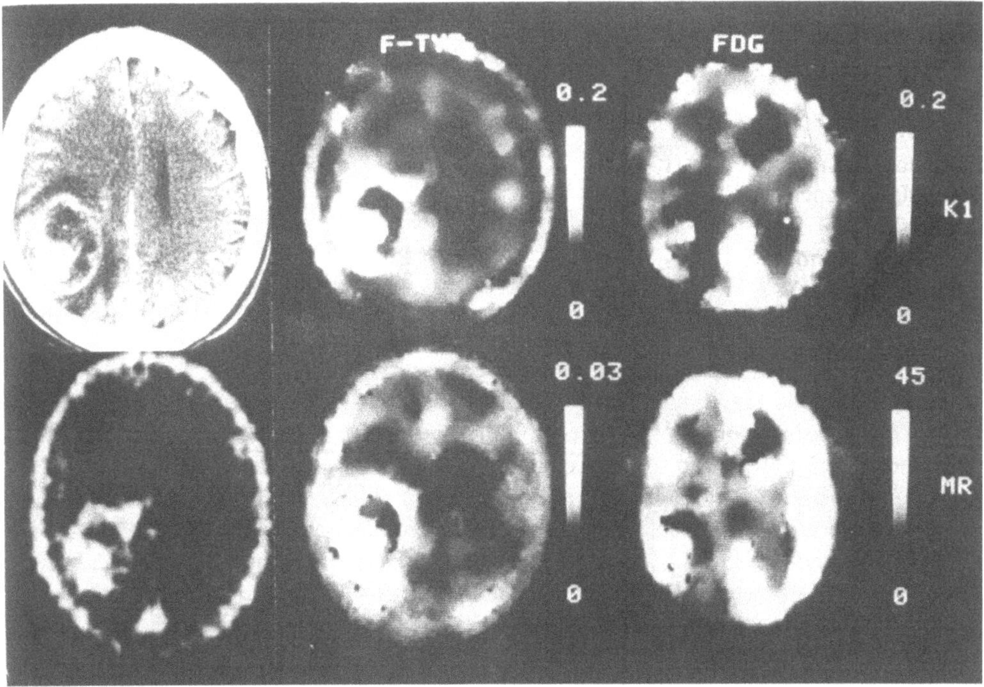

Fig. 5 Comparison between CT, ^{68}Ga-EDTA accumulation and functional images of the transport rate constant K_1 and of the accumulation rate K_{MR} for F-Tyr and FDG of a patient with a glioblastoma.

DISCUSSION

Tissue time activity data as measured by PET generally show a rather smooth curve which can easily be characterized by a small number of parameters. For practical purpose the model used to describe complex biochemical processes therefore can be rather simple. The validity of the model and of its assumptions must be tested by its capability to describe the data and by additional information about various metabolic pathways which often can only be determined in animal experiments. In the studies of F-Tyr in the brain of the mouse by COENEN et al. [7] it was shown that the incorporation of this amino acid into cerebral proteins is fast and that ^{18}F-activity in cerebral tissue is distributed almost exclusively into two pools which are the protein formed fraction and free F-Tyr. Based on these data a two tissue compartment model was used which described adequately the F-Tyr accumulation in normal tissue, but only unsatisfactorily fitted the data in tumorous tissue. In order to improve the fit of the calculated curves to the actual activities

Fig. 6 Patlak plots of F-Tyr (top), FDG (middle), and ^{68}Ga-EDTA (bottom) accumulation in a glioblastoma (open triangles) and contralateral normal tissue (full circles). The solid curves are model fits to the data.

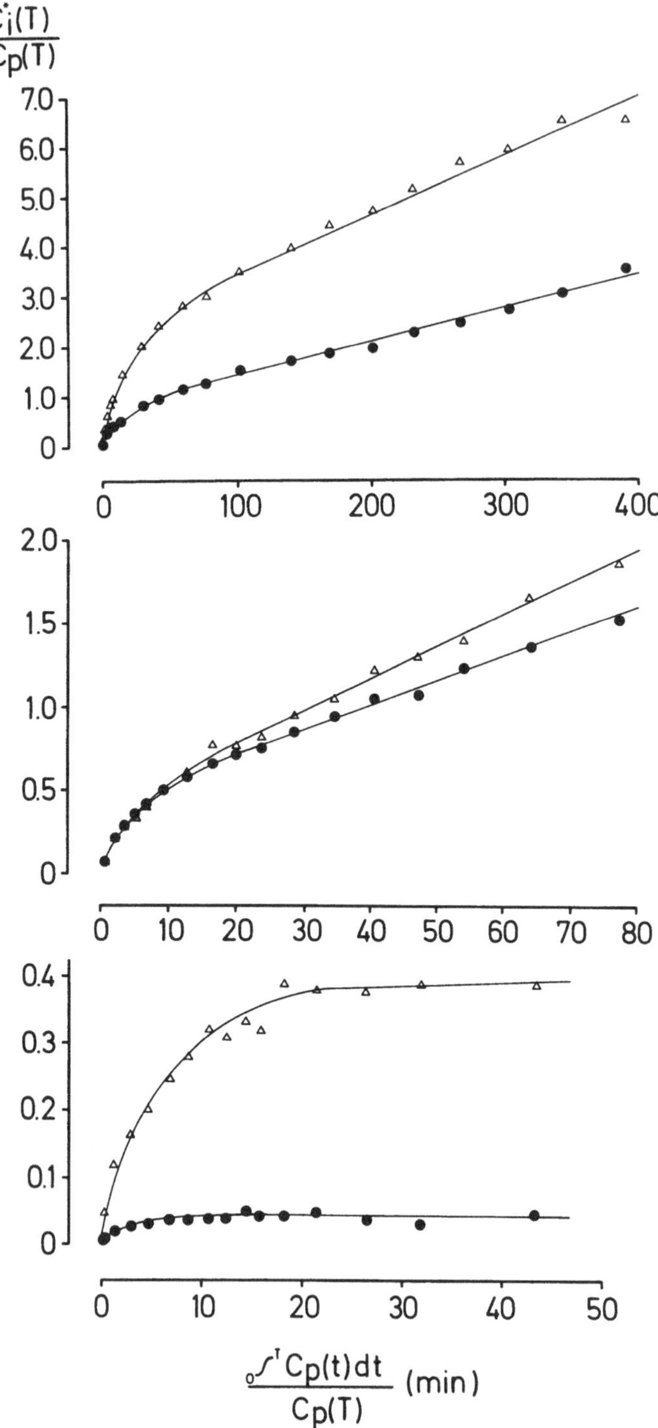

an additional third tissue compartment was introduced into the model. The physiological meaning of this additional compartment remains unclear, since the improvement of the mathematical fit to the data is no proof of the physical presence of such a compartment. The large heterogeneity of tissue compartments in tumors characteristic for high grade gliomas [17] could cause inadequate fits in a three rate constant model which will always be improved by addition of further parameters. However, in tumors also other processes could represent additional metabolic compartments, as reversible binding of the tracer to special enzymes, abnormal proteins or pathways different from normal tissue. These questions cannot be answered by the mathematical analysis of PET data but require extensive in vitro investigations of tumor metabolism.

As demonstrated by simulation procedures [18] any two or three tissue compartment model with a final irreversible step tends to overestimate metabolism in inhomogeneous tissue, since all slow processes without significant backflux during the observation period will be covered within the unidirectional metabolic step. Therefore, the absence of an increase in that step - in our data of F-Tyr incorporation into proteins - is a more reliable finding than the opposite would have been.

F-Tyr permits prolonged recordings sufficient to determine both the transport into tissue and the incorporation into proteins accurately. Quantification of absolute rates of protein synthesis with F-Tyr is currently precluded by insufficient knowledge about the intracellular level of competing amino acids, which may originate from plasma, from other metabolic pathways and from protein degradation [19, 20]. In addition, the lumped constant accounting for differences between natural tyrosine and F-Tyr in the relative affinities to carrier enzymes and to t-RNA is not known. However, comparison with (1-^{14}C)-tyrosine indicates a similar behavior [8]. Despite the limitations inherent in all approaches to quantitation of protein synthesis in man today F-Tyr has several properties suggesting its application as a PET tracer for protein synthesis in brain. It can be reproducibly synthetized; it has a high BBB permeability; no labeled metabolites cross the BBB; no metabolites are formed in brain tissue except tracer amounts of (6-^{18}F)dopamine in the striatal area [21]; its unidirectional uptake into brain tissue leads to accumulation characteristics approaching linearity asymptotically; the long half-life of the ^{18}F label permits to follow the uptake for sufficient time periods. When as done in our study plasma activity is determined and corrected for labeled metabolites, the quantification of accumulation kinetics - transport and irreversible incorporation - can be expected to be more accurate than in previous studies employing ^{11}C-methionine or its methyl or S-adenosyl-derivatives [22] since the metabolic pathways of F-Tyr are less complex than those of methionine [7]. For methionine and its derivatives the significant portion of metabolic pathways other than incorporation into proteins limited its use for the quantitative analysis of protein synthesis. The less complex metabolism of F-Tyr suggestive of unidirectional kinetics indicate the potential of this compound as a tracer for protein synthesis. However, before a quantitative model similar to that for determination of glucose metabolism with FDG can be established, further experiments are needed to determine all the interrelated variables of the kinetic model.

REFERENCES

1. Bustany, P., Chatel, M., Derlon, J.M. et al. (1986) Brain tumor protein synthesis and histological grades: A study by positron emission tomography (PET) with C-11-L-methionine. J. Neuro-Oncology 3, 397-404.
2. Schober, O., Meyer, G.J., Gaab, M.R. et al. (1986) Grading of brain tumors by C-11-L-methionine PET. J. Nucl. Med. 27, 890-891.
3. Bolster, J.M., Vaalburg, W., Paans, A.M.J. et al. (1986) Carbon-11-labelled tyrosine to study tumor metabolism by positron emission tomography (PET). Eur. J. Nucl. Med. 12, 321-324.
4. LaFrance, N.D., O'Tuama, L., Villemagne, V. et al. (1987) Quantitative imaging and follow-up experience of C-11-L-methionine accumulation in brain tumors with positron emission tomography. J. Nucl. Med. 28, 645.
5. Ericson, K., Blomqvist, G., Bergström, M. et al. (1987) Application of a kinetic model in the methionine accumulation in intracranial tumors studied with positron emission tomography. Acta Radiol. 28, 505-509.
6. Hawkins, R.A., Huang, S.C., Barrio, J.C. et al. (1989) Estimation of local cerebral protein synthesis rates with L-(1-^{11}C)Leucine and PET: methods, model, and results in animals and humans. J. Cereb. Blood Flow Metab. 9, 446-460.
7. Coenen, H.H., Kling, P. and Stöcklin, G. (1989) Cerebral metabolism of L-(2-^{18}F)fluorotyrosine, a new PET tracer of protein synthesis. J. Nucl. Med. 30, 1367-1372.
8. Ishiwata, K., Vaalburg, W., Elsinga, P.H. et al. (1988) Metabolic studies with L(1-^{14}C)tyrosine for the investigation of a kinetic model to measure protein synthesis rates with PET. J. Nucl. Med. 29, 524-529.
9. Sokoloff, L., Reivich, M., Kennedy, C. et al. (1977) The (14C)deoxyglucose method for the measurement of local cerebral glucose utilization: theory, procedure, and normal values in the conscious and anesthetized albino rat. J. Neurochem. 28, 897-916.
10. Herholz, K. (1988) Non-stationary spatial filtering and accelerated curve fitting for parametric imaging with dynamic PET. Eur. J. Nucl. Med. 14, 477-484.
11. Coenen, H.H., Franken, K., Kling, P. et al. (1988) Direct electrophilic radiofluorination of phenylalanine, tyrosine and dopa. Appl. Radiat. Isot. 39, 1243-1250.
12. Ehrenkaufer, R.E., Potocki, J.E. and Jewett, D.M. (1984) Simple synthesis of F-18-labeled-2-fluoro-2-deoxy-D-glucose: concise communication. J. Nucl. Med. 25, 333-337.
13. Heiss, W.-D., Pawlik, G., Herholz, K. et al. (1984) Regional kinetic constants and CMRGlu in normal human volunteers determined by dynamic positron emission tomography of (^{18}F)-2-fluoro-2-deoxy-D-glucose. J. Cereb. Blood Flow Metab. 4, 212-223.

14. Litton, J., Bergström, M., Eriksson, L. et al. (1984) Performance study of the PC-384 positron camera system for the brain. J. Comput. Assist. Tomogr. 8, 74-87.
15. James, F. and Roos, M. (1976) MINUIT - a system for function minimization and analysis of the parameter errors and correlations. Comput. Ph. 10, 343-376.
16. Patlak, C.S., Blasberg, R.G. and Fenstermacher, J.D. (1983) Graphical evaluation of blood-to-brain transfer constants from multi-time uptake data. J. Cereb. Blood Flow Metab. 3, 1-7.
17. Heiss, W.-D., Heindel, W., Herholz, K. et al. (1990) Positron emission tomography of fluorine-18-deoxyglucose and image-guided phosphorus-31 magnetic resonance spectroscopy in brain tumors. J. Nucl. Med. 31, 302-310.
18. Herholz, K. and Patlak, C.S. (1987) The influence of tissue heterogeneity on results of fitting nonlinear model equations to regional tracer uptake curves: with an application to compartmental models used in positron emission tomography. J. Cereb. Blood Flow Metab. 7, 214-229.
19. Phelps, M.E., Barrio, J.R., Huang, S.C. et al. (1984) Criteria for the tracer kinetic measurement of cerebral protein synthesis in humans with positron emission tomography. Ann. Neurol. 15 (Suppl.), S192-S202.
20. Smith, C.B., Deibler, G.E., Eng, N. et al. (1988) Measurement of local cerebral protein synthesis in-vivo: influence of recycling of amino acids derived from protein degradation. Proc. Natl. Acad. Sci. USA 85, 9341-9345.
21. Coenen, H.H., Wutz, W., Stöcklin, G. et al. (1990) Pharmacokinetics and metabolism of L-(2-^{18}F)fluorotyrosine in the brain and periphery of mice. J. Nucl. Med. 31, 786.
22. Ishiwata, K., Ido, T., Abe, Y. et al. (1988) Tumor uptake studies of S-adenosyl-l-(methyl-C-11)methionine and L-(methyl-C-11)methionine. Nucl. Med. and Biol. 15, 123-126.

KINETIC MODELLING OF CARBON-11 LABELLED AMINO-ACIDS

G. BLOMQVIST

Introduction

This contribution treats some aspects of the tracer-kinetic analysis of positron emission tomographic (PET) measurements of the brain function using the labelled amino-acids L-[methyl-^{11}C]methionine, L-[1-^{11}C]tyrosine, and L-[1-^{11}C]glycine. The biological and chemical properties of these tracers are identical to the corresponding unlabeled compounds. In one aspect this is an advantage, because there is a large accumulated knowledge about the behavior of these compound in the living tissue. On the other hand the amino acids enter other metabolic pathways in addition to protein incorporation and a considerably amount of labelled metabolites are retained in the tissue, which makes the interpretation of the data difficult. There may also be endogenously produced amino acids (from protein degradation) and other competing amino acids in the tissue. The influence of these processes are difficult to estimate.

This report concerns mainly the measurement of the rate of influx from blood to tissue and the "rate of tracer accumulation" in the tissue. The second quantity is a combination of the rate of protein incorporation and the rate of other metabolic pathways. No attempt to separate these two quantities is made here. Such a separation has been performed in analysis of L-[1-^{11}C]leucine data [1]. This method will be briefly commented, but otherwise two kinetic methods to analyze the tracer data will be discussed : first, the so called "Patlak analysis" which utilizes the asymptotic properties of the tracer uptake and, second, a more complete analysis using a simple compartmental model. All treated examples are taken from studies of the human brain.

In summary, a simple model with one reversible and one irreversible tissue compartment gives a satisfactory description of the tissue uptake of L-[methyl-^{11}C]methionine and L-[1-^{11}C]tyrosine. However, a good fit is no guarantee for the validity of the model assumptions. More complex models that better mirrors the biochemical fate of the tracer in the tissue are often useless in analyzing PET data because of too many unknown parameters. Models that requires estimates of unmeasured losses of labelled metabolites should be avoided and should in any case be validated by measurements of the arterio-venous (AVD) difference of these labelled metabolites. The tracers treated are suitable for determination of the unidirectional influx of the amino acids across the blood-brain barrier (BBB) and the net rate of tracer accumulation in the tissue. The estimate of these two parameters are comparatively insensitive to the exact compartmental configuration.

The applied compartment model and the Patlak analysis

To describe the kinetics of a labelled amino acid with all possible pools taken into account will in general result in a very complicated compartment model. With PET, only the *sum* activity as a function of time in each volume element of the tissue is detected, and no information about the different pools and compounds that contribute to this activity is provided directly. The unknown rate constants must be estimated by comparison of the theoretical expression with the measured time course, which normally is quite featureless. Due

B. M. Mazoyer et al. (eds.), PET Studies on Amino Acid Metabolism and Protein Synthesis, 149–160.

to the limited information provided by the PET technique, only few constants can be estimated with good accuracy and precision. Inclusion of too many unknown parameters makes them difficult to resolve. Fortunately, there is often little need to analyze data with as many compartments as demanded by the a priori knowledge of the tracer kinetics, because the parameters usually requested, the rate of influx and the accumulation rate, are rather insensitive to the exact compartmental configuration.

A simple kinetic model including only one reversible and one irreversible tissue compartment has been applied to the ^{11}C- labelled amino acids considered here. The model contains three rate constants : K_1 and k_2 for in- and outflux of tracer across the BBB, respectively, and k_3 for the transfer of tracer from the reversible into the irreversible compartment. The reversible compartment corresponds to the pool of unmetabolized amino acid in the tissue and the irreversible corresponds to labelled metabolites and tracer bound to protein. Loss of metabolites into the blood stream and degradation of labelled protein back to the pool of amino acids during the measurement time are thus neglected. With this model the total radioactivity concentration, $C^*_{tot}(T)$, is expressed as :

$$C^*_{tot}(T) = rCBV \cdot C^*_{bl}(T) + K_1 \int_0^T e^{[-(k_2+k_3)(T-t)]} C^*_{pl}(t)dt + \frac{K_1 k_3}{k_2+k_3} \int_0^T \{1 - e^{[-(k_2+k_3)(T-t)]}\} C^*_{pl}(t)dt. \quad (1)$$

Here $C^*_{pl}(T)$ is the tracer concentration in the plasma ("the input function"), $C^*_{bl}(T)$ is the radioactivity concentration in the blood and rCBV is the regional blood volume.

In this model the rate of tracer accumulation is expressed as $C_{pl} \cdot K_1 k_3/(k_2+k_3)$, where C_{pl} is the concentration of unlabeled amino acid in the plasma. It is clear from Eq. 1 that the sensitivity function [2] for the macro parameter $K_1 k_3/(k_2+k_3)$, the "accumulation rate constant", increases with time. Therefore, this parameter is best determined by measurements a long time after the tracer injection when the third term dominates the left side of eq. 1. On the other hand, during the time period following a bolus injection of the tracer, the first two terms dominate in Eq. 1 and the second term is well approximated by K_1 times the time integral of the input function. Accordingly, the influx rate constant K_1 is mainly determined by the initial measurements following a bolus injection.

If the tracer is trapped in a pool which is effectively irreversible during the measurement time, the time activity curve will tend to a simple shape independently of the number of other pools that the tracer may enter. This limiting behavior is fully utilized in the "Gjedde-Patlak" analysis [3], in which only the measurements at late times, when these asymptotic conditions are considered to be valid, are used to estimate the rate of tracer accumulation. In the Patlak plots the quantities on the x- and y-axes are, respectively :

$$\int_0^T C^*_{pl}(t)dt/C^*_{pl}(T) \quad \text{("effective time") and}$$

$$C^*_{tiss}(T)/C^*_{pl}(T), \quad (2)$$

where $C^*_{tiss}(T)$ is the radioactivity concentration in tissue. Based on general tracer kinetic principles it is shown in [3] that a tracer with an irreversible pool approaches an asymptotic time-activity distribution which is a straight line in the above x- and y-variables. Only two parameters are fitted to the data, the tracer accumulation rate constant (the slope of the line) and the "apparent volume of distribution" (the y-intercept). The method simplifies the kinetic analysis considerably and is fast and robust. As a comparison, besides the accumulation rate constant, $K_1 \cdot k_3/(k_2+k_3)$, the three parameter model gives also the influx rate constant (K_1) and volume of distribution of the free amino acid in the tissue, $K_1/(k_2+k_3)$.

Influence of measurement errors

Compared to other tracers such as labelled glucose and flow tracers, labelled amino acids accumulate slowly in the tissue. Therefore, a long time after administration of the tracer, during the period when the accumulation rate can be accurately determined, the loss of labelled metabolites from the tissue can influence the estimate of this rate seriously. Without correction for this loss, only a "net rate" of accumulation can be determined. Also, a long time after tracer administration the radioactive decay may have weakened the signal from carbon-11 labelled amino acids so much that the estimate of this parameter becomes very imprecise due to noise.

The input function, $C^*_{pl}(t)$, to be used in the kinetic analysis is the concentration of the labelled amino acid in the arterial plasma with correction for the labelled metabolites. Except for the time interval immediately following a bolus injection, these labelled metabolites contribute substantially to the total radioactivity concentration in the plasma for the tracers considered here. This contribution can also vary considerably from subject to subject. Without proper correction for this radioactivity component, large systematic errors in the estimate of the accumulation rate can be introduced.

Eq. 1 shows that the estimate of the influx rate constant, K_1, is very sensitive (a) to the measurement of the total amount of tracer remaining in the vascular volume, C^*_{bl}, (b) to the determination of rCBV, and also (c) to the synchronization between the tracer time-activity curve in the plasma and the time-activity curve in the tissue. On the other hand, if only the measurements in a short time interval following a bolus injection are utilized to determine the influx rate, the fraction of metabolites in the blood is of minor importance.

The influx rate from blood to brain of the treated amino acids across the BBB is not fast (net extraction fraction well below 0.1). Therefore, the radioactivity remaining in the blood can often be distinguished as an initial peak in the time-activity curve, provided the sampling intervals in the camera are small enough in the beginning of the scan (cf. figs. 5 and 7). The blood peak allows an estimate of rCBV directly from the uptake curve without any separate rCBV measurement using [^{11}C]CO.

In order to achieve good temporal resolution of the input function, automated blood sampling should be performed. The examples treated in this contribution are all based on automated blood sampling second by second. The used device only measures the radioactivity in full blood and must therefore be supplemented by manual measurements of radioactivity in full blood and plasma in a well counter. No measurements of possible loss of labelled metabolites from the brain tissue have been performed. For L-[methyl-^{11}C]methionine an eight ring positron camera, Scanditronix PC2048-15B, was utilized, whereas for the other two tracers a four ring positron camera, Scanditronix PC384-7B, was utilized.

Examples of Patlak analysis

Figs. 1 and 2 show Patlak plots for L-[methyl-^{11}C]methionine and L-[1-^{11}C]tyrosine, respectively. Data have been taken from normal volunteers. Clearly, the limiting distributions of these tracers are well described by straight lines. No correction for losses of labelled metabolites from the tissue have been performed. The question is: can one infer from the shown Patlak curves that there are no losses of labelled metabolites for these tracers?

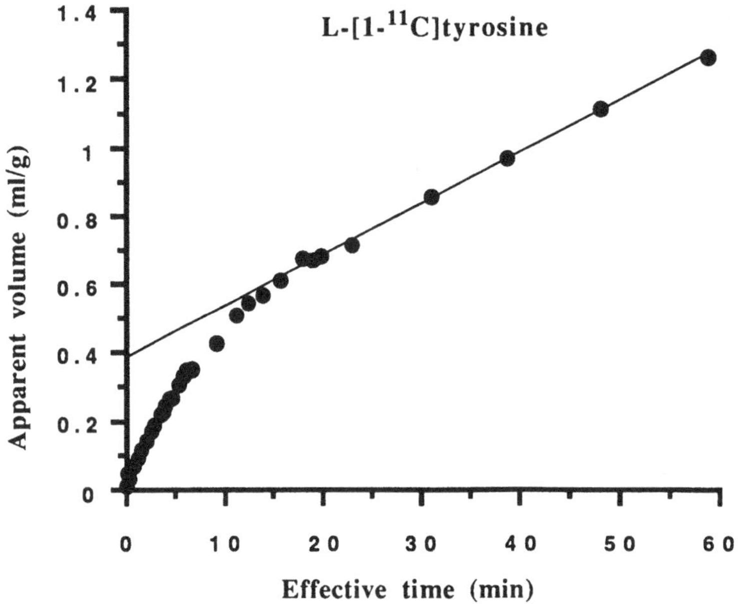

Fig. 1. Patlak plot for L-[1-^{11}C]tyrosine in the brain (healthy volunteer). The x- and y- variables are explained in the text (Eq. 2). A straight line has been fitted to the data above the effective time equal to 20 min.

To answer this question, Patlak plots of another tracer, D-[1-^{11}C]-glucose, shown in Fig. 3, is illustrative. Three curves are displayed in the plot. Due to losses of labelled metabolites the time course of the uncorrected data (lower curve) has a concave shape. Direct (AVD) measurements show that, after injection of D-[1-^{11}C]-glucose there are two components in the loss of labelled metabolites from the tissue [4]. The loss of [^{11}C]CO_2 increases slowly and monotonously, whereas the loss of other labelled, acidic metabolites increases more rapidly and reaches a maximum about 15 min after injection. After correction for loss of all labelled metabolites, the Patlak plot is well described by a straight line (upper curve), but the Patlak plot based on data corrected for the loss of [^{11}C]CO_2 only is also straight (middle curve). The reason for this behavior is the different time dependencies of the two components in the loss.

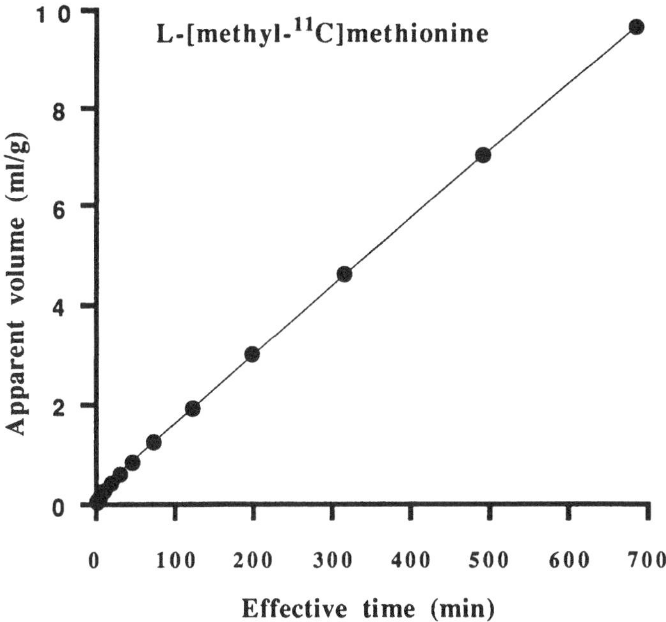

Fig 2. Patlak plot for L-[methyl-^{11}C]methionine in the brain (healthy volunteer). The x- and y-variables are explained in the text (Eq. 2). A straight line has been fitted to the data above the effective time equal to 50 min

The given example demonstrates that a Patlak plot can show a straight line behavior despite a substantial loss of tracer from the metabolic pool. At the end of the measuring interval (25 min) the accumulated loss of [^{11}C]CO_2 is, on the average 8%. It is therefore impossible to judge from the shape of the Patlak plots for L-[methyl-^{11}C]methionine and L-[1-^{11}C]tyrosine if there are any losses or not in these cases. The question can only be answered by measurements av the arterio-venous difference (AVD measurements) of labelled metabolites over the brain. In general, details in the kinetics of a tracer cannot be determined from fit of a model to the uptake curve only, but must be supplemented by further measurements and a priori knowledge about the biochemistry of the tracer.

In contrast to the plots for labelled tyrosine, methionine, and glucose, the Patlak plots of L-[1-^{11}C]glycine shown in Fig. 4 do not have any straight lines as limiting distributions. There are at least three possible explanations for this. First, the asymptotic conditions may not have been reached during the measuring time (48 min) due to the comparatively slow influx and accumulation rates of glycine (K_1 = 0.005 min^{-1}, and $K_1k_3/(k_2+k_3)$ = 0.003 min^{-1} for normal tissue in the present example), second, there can be a substantial loss of labelled metabolites that accelerates at late times and, third, there can be an increasing fraction of labelled metabolites in the blood, which is not corrected for. Again, no conclusions can be drawn from the Patlak plot alone but additional information is necessary.

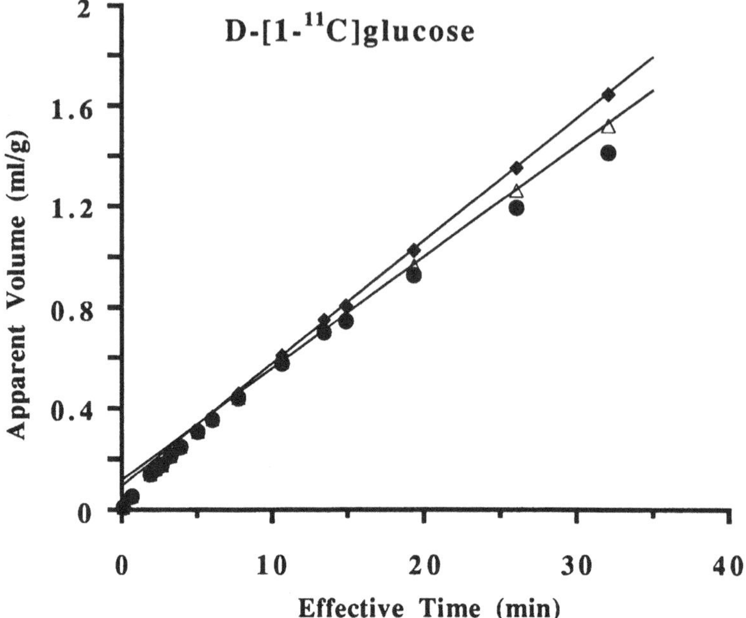

Fig. 3. Patlak plots for uptake of D-[1-^{11}C]-glucose in the brain (healthy volunteer). The x- and y variables are explained in the text (Eq. 2)The lowest points (•) are uptake data without correction for any losses, the middle points (Δ) are uptake data corrected for loss of [^{11}C]CO_2, and the upper points (♦) are uptake data corrected for loss of all labelled metabolites. In the middle and upper curves straight lines have been fitted to the data above the effective time 10 min .

Examples of compartmental analysis

L-[1-^{11}C]TYROSINE

Compared to most other amino acids, tyrosine has a high uptake rate in the brain. The main metabolic pathway is protein incorporation. In rats almost 80% of the label was found to be protein bound 60 min after injection of L-[1-^{14}C]tyrosine, and [^{14}C]CO_2 produced in decarboxylation was rapidly removed from the tissue [5]. Therefore the accumulation rate of L-[1-^{11}C]tyrosine should be a quite good measure of the incorporation rate of tyrosine into proteins.

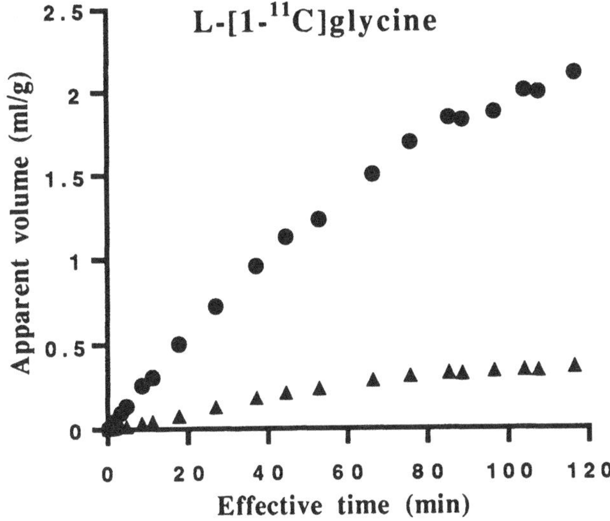

Fig. 4. Patalk plot for L-[1-^{11}C]glycine. The x- and y-variables are explained in the text. The upper curve is obtained for tumor tissue in the brain, the lower curve for normal brain tissue in a tumor patient

Fig. 5 shows an uptake curve of L-[1-^{11}C]tyrosine in the brain fitted by the model with one reversible and one irreversible tissue compartment. Evidently the time course is well described by the model. No measurements of labelled metabolites in the plasma and no measurements of the loss of labelled metabolites from the tissue were performed in the study. Therefore, the model fit only gives a measure of the net rate of accumulation. However, the main purpose of the study was to compare the rate of influx across the BBB of tyrosine between schizophrenic patients and normal controls [6]. The estimate of the influx rate constant was confined to the first 3.5 minutes following the bolus injection of the tracer, and during this initial time period no appreciable amounts of labelled metabolites are expected to be present in the blood and no appreciable loss of labelled metabolites from the tissue to occur.

To achieve good time resolution, the time activity was measured in 10-sec intervals during this initial period and the radioactivity in the blood was detected every second. The measurement with L-[1-^{11}C]tyrosine were combined with a separate measurement of rCBV using [^{11}C]CO. The bolus injection and the comparatively slow influx of tyrosine into the brain ($K_1 \approx 0.05$ min^{-1}) implies that a blood peak can be distinguished in the uptake curve (cf. fig. 5) and enables a good estimate of rCBV directly from the uptake data.

Fig 6. shows a comparison of K_1 determined in two ways for a number of regions. On the x-axis is the K_1-value obtained when rCBV was determined from the separate [^{11}C]CO experiment, and on the y-axis is the K_1-value obtained when rCBV was fitted simultaneously from the uptake curve. Evidently the difference is not large and the auxiliary [^{11}C]CO measurement could have been deleted without jeopardizing the result.

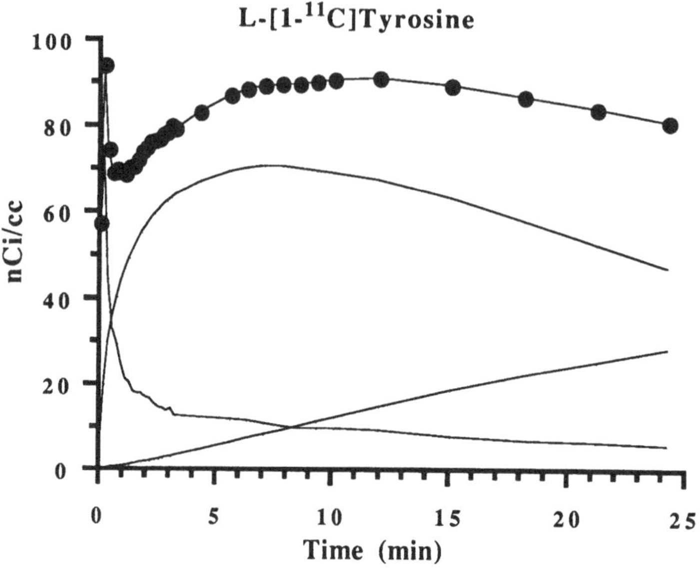

Fig. 5. Time-activity curve for L-[1-^{11}C]tyrosine in the brain (healthy volunteer). Measurements were made on a healthy volunteer. The standard compartment model of eq.1 with four parameters (rCBV, K_1, k_2, and k_3) was applied to the data.

Others [7] have used a fluorine-18 labelled analog, L-[2-^{18}F]tyrosine, to investigate the uptake and utilization of tyrosine. The labelling with fluorine-18 admits a longer measurement time to determine the accumulation rate more precisely. In rat studies it was found that the metabolites from L-[2-^{18}F]tyrosine do not cross the blood-brain barrier, and only small amounts of labelled metabolites can be detected in brain tissue.

The drawback with L-[2-^{18}F]tyrosine is that the kinetics differs from that of natural tyrosine. To measure the tyrosine accumulation rate using L-[2-^{18}F]tyrosine, it is necessary to determine a "lumped constant", i.e. the ratio between the accumulation rates for the analog and the natural compound. From published data [6-7] the lumped constant can be estimated to be of the order of 0.3, but the lack of correction for losses of metabolites from the tissue and for metabolites in the arterial plasma makes this value very approximate. The (global) lumped constant for L-[2-^{18}F]tyrosine can be determined by AVD measurements, but the problem with competing amino acids cannot be circumvented with analogs. However, if the aim of the study is confined to measure the rate of unidirectional influx across the BBB of tyrosine only, there is no need to use any analog.

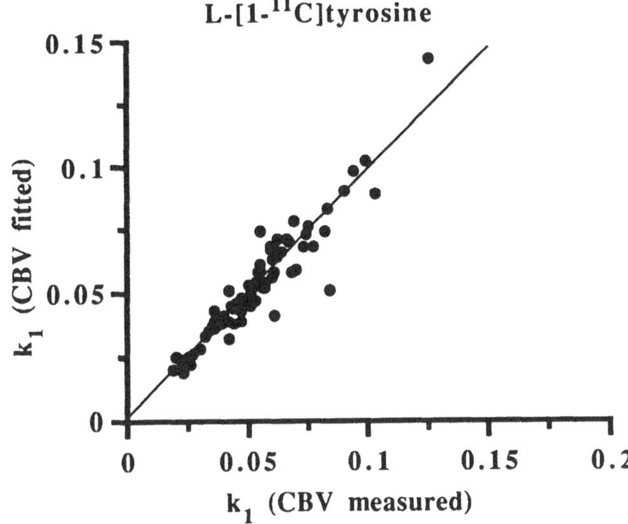

Fig. 6. Data from measurements with L-[1-^{11}C]tyrosine. Scatter plot of K_1 estimated simultaneously with rCBV, versus K_1 estimated with rCBV predetermined in a separate [^{11}C]CO experiment. Each point corresponds to the pair of values obtained in a brain region of a healthy volunteer.

L-[METHYL-^{11}C]METHIONINE

Previous studies indicate that the uptake rate of L- [1-^{11}C]methionine mainly reflects the rate of protein incorporation [8-9], but like the case for other amino acids there are a number of minor metabolic pathways that contribute substantially to the accumulation rate of this tracer.

Using previous measurements [10], the plasma radioactivity for this tracer was corrected for the contribution from labelled metabolites. However, Ishiwata et al. [11] have shown that the fraction of labelled metabolites in the plasma can vary considerable between subjects and, therefore, a correction based on the average loss can introduce errors that makes it difficult to compare the accumulation rate between subjects. Fig. 7 shows that the simple model with two tissue compartments can describe the uptake data well [12], but so far no AVD-measurements of possible losses of labelled metabolites from the tissue have been reported. As already discussed, the asymptotic straight line behavior of the Patlak plot (Fig. 2) is no guarantee that such losses do not occur.

Methionine has also been labelled in the carboxyl group [9]. With L- [1-^{11}C]methionine the label entering other metabolic pathways besides protein synthesis is then to a larger extent transferred to [^{11}C]CO_2 which leaves the tissue rapidly. Thus, the uptake of L- [1-^{11}C]methionine should reflect protein synthesis somewhat better than the uptake of L-[methyl-^{11}C]methionine does. However,with both tracers it is found that besides the tracer bound to protein there is a considerable fraction of labelled metabolites in the tissue. Therefore, with both labellings the uptake reflects a mixture of protein incorporation and other metabolic pathways.

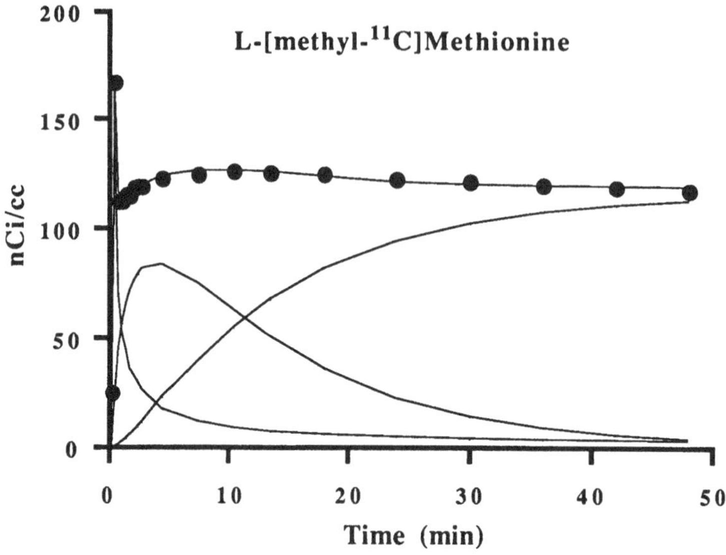

Fig. 7. Time-activity curve for L-methyl-^{11}C]methionine in the brain (healthy volunteer).Measurements were made on a healthy volunteer. The standard compartment model of eq.1 with four parameters (rCBV, K_1, k_2, and k_3) was applied to the data.

Figs. 5 and 6 show that there are two quite distinct parts in the uptake curves of labelled tyrosine and methionine. First there is (after correction for the vascular contribution) a comparatively steep increase where the unidirectional influx of the tracer dominates, and thereafter the uptake curve varies more slowly, reflecting that the rate of accumulation is much smaller than the unidirectional influx rate. This clear difference enables a quite accurate determination of a third quantity : the volume of distribution of the labelled compounds in the reversible pools, expressed as $K_1/(k_2+k_3)$ in the simple compartment model applied here.

More complex models

If the tracer is administered as a bolus and is effectively trapped in some irreversible pool, the rate of tracer accumulation is determined by the late measurements, when the tracer concentrations in the reversible pools and the plasma have reached approximate steady state and the time course of the total tracer concentration is mainly governed by the change of tracer concentration in the irreversible pool. If such asymptotic conditions are reached within the measuring time, the time course of the tracer becomes insensitive for the number of pools entered. Accordingly, the estimate of the net rate of accumulation is insensitive to the exact compartmental configuration used in the model. Like the simple two tissue compartment model considered here, the resulting expression for the uptake curve in a more complex model will also contain a term which is the product of the accumulation rate constant and the integral of the input function (cf. eq. 1), and this term will become more and more dominant with time.

In the time interval immediately following a bolus injection the time course of the tracer in the tissue is dominated by two components (a) the tracer remaining in the vascular compartment and (b) the unidirectional influx of tracer across the BBB. The exact number of pools that the tracer may enter has again little influence on the initial time course. Clearly, the input rate constant K_1 is best determined by the measurements in a short time interval following a bolus injection, irrespectively of the compartmental configuration applied.

Measurements have shown that $[^{11}C]CO_2$ and other labelled metabolites are produced by minor metabolic pathways in the tissue after injection of L-[1-^{11}C]tyrosine [5] and L-[methyl-^{11}C]methionine [9]. Produced $[^{11}C]CO_2$ rapidly leaves the tissue, whereas the labelled compounds to a large extent are retained in the tissue. It is therefore tempting to take the production and loss of these metabolites into account by a special compartment. It will then be necessary to include at least two extra rate constants in the model, one for the metabolism of the amino acid and another for the loss of metabolites to the blood. Such a compartmental model has been applied to L-[1-^{11}C]leucine and fits of K_1-k_5 have been obtained [1] .

To get a good measure of the protein incorporation rate with this 5-parameter model, two requirements must be fulfilled. First, the fit must give a good estimate of the loss of labelled metabolites from the tissue. In other words, based on the observed time course of the radioactivity remaining in the tissue, it must be possible to estimate how the theoretical time-activity curve should have been without any losses. Second, the fit must be able to discriminate between the contributions from protein incorporation and other metabolic pathways. To achieve this, very accurate and precise measurements are required. The main problem is the lack of information about the loss of labelled metabolites from the tissue. The estimate of the loss is necessarily very model dependent, and also sensitive to small changes in the uptake curve, due to systematic and random errors.

It must be noted that the loss of label is a necessary condition for the method to work. Without loss the model should contain two irreversible compartments (protein bound label and labelled metabolites) and the rate constants for transfer to these compartments is necessarily unidentifiable, because their sensitivity functions are strictly proportional from time zero to infinity. Any model of this kind should be validated by comparison of the loss predicted by the model fit with the loss $C^*_L(t)$ determined by direct AVD measurements of the labelled metabolites combined with rCBF measurements :

$$\frac{d}{dt}C^*_L(t) = AVD \cdot rCBF. \tag{3}$$

As a comparison, in studies with L-[1-^{11}C]-glucose [4] it was found that the loss of labelled metabolites actually measured could not be predicted by incorporation of an extra rate constant in the model expression.

References

1. Hawkins RA, Huang S-C, Barrio JR, et al. Estimation of the cerebral protein synthesis rates with L-[1-^{11}C]leucine and PET: Methods,model, and results in animals and humans. J Cereb Blood Flow Metab (1989) 9:446-460

2. Beck JV and Arnold KJ. Parameter estimation in engineering and science. John Wiley & Sons, New York (1977)

3. Patlak CS, Blasberg RG, Fenstermacher JD. Graphical evaluation of blood-to-brain transfer constants from multiple time uptake data. J Cereb Blood Flow Metab (1991) 3:1-7

4. Blomqvist G, Stone-Elander S, Halldin C, et al. Positron emission tomographic measurement of cerebral glucose utilization using [1-^{11}C]-D-glucose. J Cereb Blood Flow Metab (1990) 10:467-483.

5. Ishiwata K, Vaalburg W, Elsinga PH, Paans AMJ, Woldring MG. Metabolic studies with L-[1-^{14}C]tyrosine for investigation of a kinetic model to measure protein synthesis rates with PET. J Nucl Med (1988) 29:524-529

6. Wiesel F, Blomqvist G, Halldin C, et al. The transport of tyrosine into the human brain as determined with L-[1-^{11}C]tyrosine and PET. J Nucl Med (1991) 32: 2043-2049.

7. Wienhard K, Herholz K, Coenen H H, et al. Increased amino acid transport into brain tumors measured by PET of L-[2-^{18}F]tyrosine. J Nucl Med (1991) 31:1338-1346.

8. Bustany P, Henry JF, Soussaline F, Comar D. Brain protein synthesis in normal and demented patients. A study by positron emission tomography with 11C-L-methionine. In: Magistretti PL (ed) Functional radionuclide imaging of the brain. Raven Press, New York (1983) pp 319-326

9. Ishiwata K, Vaalburg W, Elsinga PH, Paans AMJ, Woldring MG. Comparison of L-[1-^{11}C]methionine with L-[methyl-^{11}C]methionine for measuring in vivo protein synthesis rates with PET. J Nucl Med (1988) 29:1419-1427

10. Lundqvist H, Stålnacke CG, Långström B, Jones B. Labeled metabolites in plasma after intravenous administration of [11CH3]methionine. In : Greitz T, Ingvar DH, Widén L (eds.)The metabolism of the human brain studied with positron emission tomography. New York. Raven Press (1991) pp. 233-240.

11. Ishiwata K, Hatazawa J, Kubota K, Kameyama M, Itoh M, Matsuzawa T, Takahashi T, Iwata R, Ido T, Metabolic fate of L-[methyl-11C]methionine in human plasma. Eur J Nucl Med (1989) 15:665-669

12. Ericson K, Blomqvist G, Bergström M, Eriksson L, Stone-Elander S : Application of a kinetic model on the methionine accumulation in intracranial tumors studied with positron emission tomography. Acta Radiol (1987) 28: 505-509.

CARBON-11 LABELED TYROSINE AS A PROBE FOR MODELLING THE PROTEIN SYNTHESIS RATE

Anne M.J. Paans, Philip H. Elsinga and Willem Vaalburg

Abstract

L-[1-^{11}C]tyrosine was used to measure the protein synthesis rate in two tumor models in rats. Dynamic PET-data were acquired using a double headed scintillation camera system operated in a coincidence mode. From blood samples the plasma concentration of L-[1-^{11}C]tyrosine was measured as a function of time. A four compartment model was used to analyze the data. The unique solutions for the two different tumor models showed a significant difference in protein synthesis rate.

Introduction

Due to their elevated growth rate most tumors show an increased demand for amino acids. The application of ^{11}C-labeled amino acids in combination with Positron Emission Tomography (PET) may yield a method for measuring the protein synthesis rate (PSR) in-vivo [1]. An increased uptake of ^{11}C-labeled amino acids in tumors has been observed and provided e.g. in the case of brain tumors accurate stereotactic coordinates for surgery [2]. The effect of therapeutic interventions has been evaluated in terms of alterations in amino acid uptake by Dunzendorfer et al. [3] and by Daemen et al. [4, 5]. Schober et al. [6] used the amount of accumulation of ^{11}C-labeled amino acids as a parameter for the determination of the tumor grade.

B. M. Mazoyer et al. (eds.), PET Studies on Amino Acid Metabolism and Protein Synthesis, 161–174.

The properties required for an amino acid as a measure for the PSR by a tracer kinetic method are [7]:

- Rapid clearance of the labeled amino acid from the blood
- Fast turnover of the amino acid precursor pool in tissue. The time scale of the turnover process should be faster than the half life of the radioactivity.
- The turnover of the protein pool should be long compared to the half life of the radioactivity.
- The number of metabolites should be minimal. The choice of amino acid and the position of the label should be evaluated very critically.

The metabolism of an amino acid selected on the above mentioned criteria probably can then be described by a tracer kinetic model which allows for an operational equation yielding the PSR.

A number of ^{11}C-labeled amino acids have been described in the literature. L-[methyl-^{11}C]methionine, an amino acid that can be synthetized in a short time with a high radiochemical yield, has been used for the detection of tumors [2, 6, 8, 9, 10]. As a measure of the PSR methyl labeled methionine is less appropriate since methionine is the main biochemical source of the methyl group [11, 12]. For the measurement of the PSR carboxylic labeled amino acids have a greater potential if both the metabolite profile and the amount of metabolites are known [13]. Metabolic studies with L-[1-^{11}C]leucine, L-[1-^{11}C]methionine and L-[1-^{11}C]tyrosine showed that all three amino acids provide a prospect for the measurement of the PSR because of their high incorporation into proteins and their low amount of labeled metabolites [12, 14, 15]. In this paper the measurement of the PSR in two tumor models is presented using L-[1-^{11}C]tyrosine (^{11}C-tyr). L-tyrosine is an essential amino acid with a small free pool in tissue and plasma and with a high turnover rate. The ^{11}C-label of carboxylic labeled amino acids is expected to be incorporated into proteins because this is the main metabolic pathway. The $^{11}CO_2$ formed in the minor metabolic pathway is diluted in the [CO_2]-bicarbonate pool and removed rapidly from the tissue as was shown in metabolic studies using the ^{14}C-analogue [12, 14].

Material and Methods

Radiopharmaceutical

$^{11}CO_2$ is produced in a pressurized target containing high purity nitrogen gas by the $^{14}N(p,\alpha)^{11}C$ reaction. L-[1-^{11}C]tyrosine was synthetized according to Bolster et al. [16]. In brief, after lithiating p-methoxyphenylethylisocyanide carboxylation was carried with $^{11}CO_2$. After acid hydrolysis the racemic mixture was purified and separated enantiomerically by HPLC on a chiral column. 185 MBq (5 mCi) injectable ^{11}C-tyrosine were obtained after a synthesis time of 1 hour starting with an initial amount of 7.4 GBq (200 mCi) of $^{11}CO_2$.

Tumor models

Two tumor models with different growth rates were used. As the slower growing tumor model the rhabdomyosarcoma was used. Female Wag/Rij rats weighing 140 g were inoculated subcutanously in the left flank with 100 mg rhabdomyosarcoma tissue. Vital homogeneous tumors with an average volume of 4 ml were found after 18 days. As the faster growing tumor model the Walker 256 carcinosarcoma was used. Male Wistar rats were injected with 10^6 Walker 256 carcinosarcoma cells in the left hind leg. After 7 days palpable tumors were found.

Positron camera

As imaging device a dual-headed uncollimated scintillation camera system, operated in a coincidence mode, which is very useful for small animal PET-studies, was used [17]. This longitudinal tomographic system allows for dynamic studies in one selectable tomographic plane with a spatial resolution of 5.5 mm FWHM and a sensitivity of 5 cps/kBq for a source in the geometrical center at a detector separation of 50 cm. Since all coincident events are back-projected in hardware onto the selected plane, overlying tissue structures will be measured by their integral value. In thin animals this so called "blurring effect" is minimized and can be compared to a partial volume effect. A zoom mode data acquisition was used. In this mode the field of view of 20x20 cm^2 is stored in a matrix of 64x64 pixels.

Data acquisition procedure

The rats were kept under light anaesthesia during the experiment and the body temperature was prevented to drop below 35 °C by external heating. ^{11}C-tyr was administered in amounts up to 7 MBq (200 μCi) through an iv tail catheter while the rat was already positioned in the camera. The injected volume of 0.4 ml at maximum, average 0.2 ml, is that small that disturbances in blood flow, blood pressure and uresis are avoided. The catheter is flushed with 0.05 ml saline to achieve complete administration of the radioactivity into the general circulation. Dynamic acquisition of PET data was started at the moment of injection with a frame rate of one image per minute during the first 10 minutes followed by 10 frames of 5 minutes. The total study lasted for 1 hour. During data acquisition blood samples were drawn. In order to obtain a good insight into the input function blood samples, volume 0.05 ml, were drawn at 0.25, 0.5, 1, 2, 3, 4, 5, 10, 15, 30 and 60 minutes after injection. The blood samples were treated as described by Ishiwata [12, 14] and the free plasma concentration was measured in a cross calibrated well counter.

Data processing

The dual-headed positron camera has a non-uniform response due to the operation of two large position sensitive detectors in a coincidence mode. The maximum sensitivity is reached in the geometrical center of the system and the sensitivity is decreasing linear with increasing radius in the selected tomographic plane [17]. Since this decrease in sensitivity is determined by the solid angle available for coincident events it is possible to correct for this decrease by calculating the solid angle for coincident events in each pixel element of the selected tomographic plane. In practice it is easier to acquire data from homogeneous plane phantom and to use this image for the correction for the radially decreasing sensitivity. This procedure is than carried out in such a way that the sensitivity at the center of the plane is normative for the whole plane. Using a point source at the center of the plane will than yield an absolute calibration factor for the images which are corrected for the non-uniform response of the system.

Next a decay correction was applied on all images. By correcting for the physical decay of carbon-11 (20 minutes) the biological behaviour of the

radiopharmaceutical used can be displayed, c.q. analyzed directly.
The images were not corrected for attenuation of the radiation within the body of the animal. Tests with rats revealed that the attenuation correction for these animals in this longitudinal tomographic system is marginal (<5%).
The center of the selected tomographic plane is cross calibrated against a 3"x3" NaI well counter. Since the difference in sensitivity between the positron camera and the well counter is about a factor of 1000, a weak ^{22}Na source was used in order to avoid dead time problems in the well counter. This procedure in combination with the correction for the non-uniform response of the system provides a calibrated system which yields information in the selected tomographic plane in terms of Bq/cm^2.

Image analysis

The response and decay corrected dynamic studies were analyzed by delineating the tissues of interest. Since the rhabdomyosarcoma were located in the flank of the animal and since the animal was carefully positioned for an optimal view the tumor region was clearly visualized and no surrounding tissue had to be accounted for. In the case of the Walker 256 carcinosarcoma the tumor was in the left hind leg. Since all radioactivity is back-projected onto the selected tomographic plane an integral value of tumor and muscle radioactivity is obtained. The tumor region was clearly visualized and no problems in delineating of the tumor were encountered. To obtain a measure of the amount of radioactivity in the tumor an identical area outlined in the right hind leg, giving the muscle radioactivity, was subtracted from the integral value of the left hind leg. In this way the amount of radioactivity in terms of Bq/cm^2 was obtained for both tumor models. These data were then converted into a concentration (Bq/cm^3) by an external measurement of the tumor thickness with a vernier calliper.

Plasma concentration

The blood samples, taken according to the already mentioned time schedule, were treated as described by Ishiwata [12, 14] in order to measure the non-protein bound pool in plasma with help of the well counter. Due to the absolute calibration of the system these data are available in terms of Bq/cm^3.

Mathematical model

The model to describe the tyrosine metabolism consists of four compartments according to Smith et al. [18]: the plasma pool (C_p), the precursor pool in tissue (C_e), the metabolite pool in tissue (C_m) and the protein bound pool in tissue (C_{pr}). The four compartments are related with each other as shown in fig. 1. Based on the time scale of the experiments an irreversible way of incorporation of the ^{11}C-tyr into the protein is assumed. The influx into the metabolite pool is irreversible due to the time scale of the experiments. The fate of $^{11}CO_2$, being the main metabolite, needs special consideration.

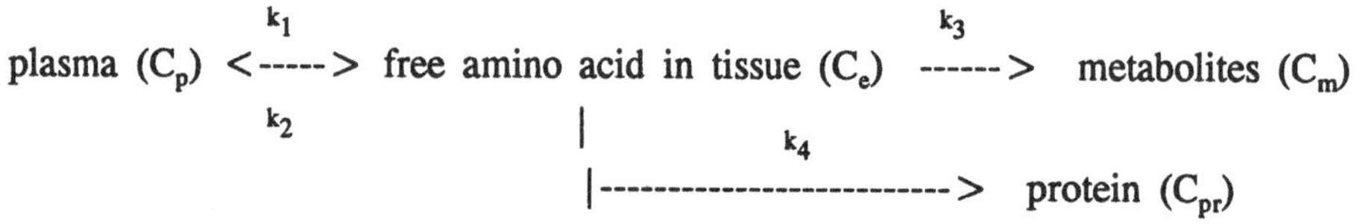

plasma (C_p) $\overset{k_1}{\underset{k_2}{\longleftrightarrow}}$ free amino acid in tissue (C_e) $\overset{k_3}{\longrightarrow}$ metabolites (C_m)

free amino acid in tissue (C_e) $\overset{k_4}{\longrightarrow}$ protein (C_{pr})

Fig. 1 Compartment model for measuring the PSR, using L-[1-^{11}C]tyrosine

The transport processes of the labeled tracer are assumed to be first order. The differential equations describing this system are:

$$dC_e(t)/dt = k_1.C_p(t) - (k_2+k_3+k_4).C_e(t) \qquad (1)$$

$$dC_m(t)/dt = k_3.C_e(t) \qquad (2)$$

$$dC_{pr}(t)/dt = k_4.C_e(t) \qquad (3)$$

The solutions of the above set of equations are, assuming the initial concentrations of radioactivity are zero:

$$C_e(t) = k_1\ e^{-(k_2+k_3+k_4)t} \int_0^t e^{(k_2+k_3+k_4)\tau}.C_p(\tau).d\tau \qquad (4)$$

$$C_m(t) = k_3 \int_0^t C_e(\tau).d\tau \qquad (5)$$

$$C_{pr}(t) = k_4 \int_0^t C_e(\tau).d\tau \qquad (6)$$

The estimation of the concentration of radioactivity in the different

compartments has to be based on measurable quantities. The measurable quantities in the case of in vivo PET are the tissue concentration, as measured by the PET scanner, and the free plasma concentration of the non-protein bound fraction as measured in the well counter after the proper chemical separation.
Following the same procedure as was done for the derival of the operational equation for the glucose consumption in the case of ^{14}C-deoxyglucose [19] an operational equation for the PSR can be derived:

$$PSR = \frac{C_i - k_1 . e^{-(k_2+k_3+k_4)t} \int_0^t e^{(k_2+k_3+k_4)\tau} . C_p(\tau) . d\tau}{\int_0^t C_p(\tau)/C_{pc} . d\tau - e^{-(k_2+k_3+k_4)t} \int_0^t e^{(k_2+k_3+k_4)T} . C_p(\tau)/C_{pc} . d\tau} \qquad (7)$$

In this operational equation only measurable quatities are referred and the protein synthesis rate in terms of nmol/(ml.min) can be calculated if also the cold tyrosine level (C_{pc}) is known.

Description of plasma concentration

Since it was not possible to establish exactly the input profile in the period just after the injection, a standardized procedure was used. By this a standardized input profile was assumed. All rats were injected via a catheter and it was established that the blood activity reached its maximum between 8 and 12 s after injection as was established by separate experiments. The non-protein bound radioactivity in the plasma was fitted with a multi-exponential function. For the adjustment of the parameters in the multi-exponential function the program MINUIT, obtained from the CERN-library, was used [20]. For the calculation of the integrated non-protein bound radioactivity the fitted multi-exponential function was extrapolated downward to the 10 s point, where the maximum in activity was defined based on the separate experiments mentioned above. Between 0 and 10 s a linear increase from zero to the maximum value was assumed for the calculation of the of the integrated plasma activity, see also the results and discussion section.

Description of tissue concentration

The tissue concentration, measured as described in the data processing part, was fitted as a combination of activity of the precursor pool, the protein bound pool, the metabolite pool and of course some activity from the plasma pool. The amount of activity due to the plasma pool was limited to a maximum of 5% of the actual plasma level at each moment. For the minimalization of the difference between measured and calculated tissue concentration again the MINUIT program was used [20]. The program was run on a VAX 8300 computer system. In order to localize probable local minima in χ^2-space the different data sets were object of extensive grid searches in which all possible starting parameters were varied over a wide range of values.

Results and discussion

Using a multi-exponential function for the tyrosine concentration in the plasma results in a good fit of the data. Due to the fast sampling in the beginning, 3 exponentials describe the data points. One for the early phase, one for the late phase and one exponential which describes the intermediate part. To obtain a correct integrated plasma activity the fitted acivity was extrapolated downward to the 10 s point and between 0 s and 10 s a linear increase in activty was assumed. From separate experiments it was found that the maximum in plasma acitivity, using our injection technique, was reached between 8 and 12 s. Extending an artery of the rat with a polyethelene tubing, and using this external tubing for a continuous monitoring of the blood activity was not successful, the amount of blood was to low for a reliable measurement. Extrapolation of the fitted blood curved to $t=0$ s induces a large overestimation of integrated plasma activity. Assuming a zero level of activity between 0 and 10 s leads to a underestimation. For this reason we set the maximum at 10 s with a linear increase between 0 and 10 s. With this procedure we have a standardized procedure leading to the same small error in all integrated plasma activities.

The clearance of tyrosine from the blood is rapid and consistent with ^{14}C- as well as ^{11}C-carboxylic labeled tyrosine experiments, as described by Ishiwata et al. [14] and Wiesel et al. [21].

Using the mathematical model in combination with the plasma data resulted in a unique fit of each set of tissue data. Independent of the starting values of the different parameters the same protein synthesis rate was found after the fit. The effect of compensating k-values, as has been observed in the FDG model, was also evident in these fits. This will occur because in the model the precursor pool in the tissue (C_e) is described by a combination of k-values as can be seen in the exponential in eq. 4. Since the variations in the individual k-values, $k_2+k_3+k_4$, are correlated, the effect on the calculated protein synthesis rate is minimal.

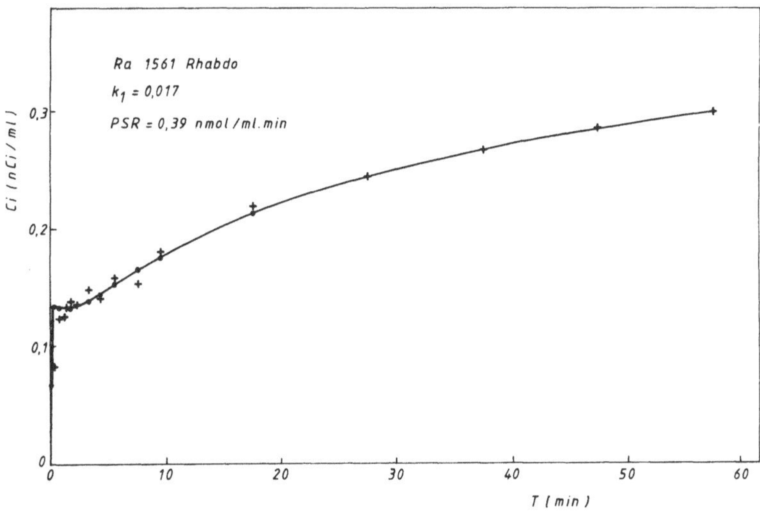

Fig. 2 Experimental tissue data (+) and fitted curve for the Rhabdomyosarcoma

In fig. 2 and 3 fits to the tumor data are shown. The crosses represent the tissue data points. The error in the rhabdmyosarcoma data is 15-20% while the error in the Walker 256 carcinosarcoma data amounts to 25-30%. The larger error in the latter tumor type is due to the fact that necrotic areas can be present in this rather fast growing tumor type. When $C_i(t)/C_p(t)$ is plotted versus $\int C_p(t)dt/C_p(t)$, the so called Patlak plot [22], a linear behaviour is observed after an initial rise, indicating an irreversable metabolic trapping. The small hump at very early times in fig. 3 is due to the plasma activity. The PSR found for the rhabdomyosarcoma tissue amounted to 0.4 ± 0.1 nmol/(g.min) while Walker 256 carcinosarcoma, the faster growing tumor, had a PSR of 0.8 ± 0.2 nmol/(g.min). These values for the PSR are of the same magnitude as found by Hawkins et al. [23] using L-[1-^{11}C]leucine for the measurement of the local cerebral protein synthesis rate in monkeys and humans, 0.5 nmol/(g.min).

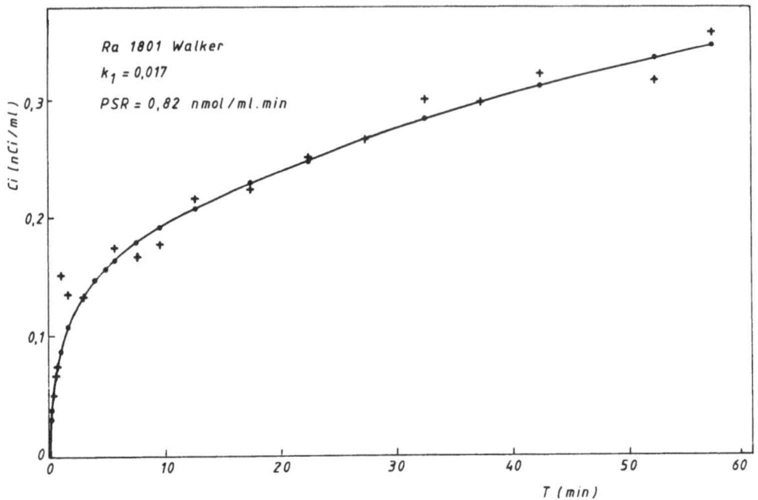

Fig. 3 Experimental tissue data (+) and fitted curve for the Walker 256 Carcinosarcoma

For amino acids labeled in the carboxylic position the main metabolic route yielding a non-protein metabolite, is decarboxylation. The $^{11}CO_2$ generated is cleared from the tissue by diffusion and blood flow. The idea of a subsequent removal of the carbon dioxide from the plasma by ventilation in the lungs has been questioned recently by Shields et al. [24]. Based on their findings the amount of carbon dioxide remaining in the body is rather high and an adjustment of the model may be necessary. In the validation of the model for human data this compartment should have proper attention. Of the described caboxylic labeled amino acids L-[1-^{11}C]tyrosine, L-[1-^{11}C]methionine and L-[1-^{11}C]leucine seem to have a comparable potential for measuring the protein synthesis rate in vivo [1]. For measuring the PSR these amino acids certainly have to be preferred above the frequently used methyl-labeled methionine because of their much lower amount of metabolites [25]. Also several ^{18}F-labeled amino acids have been prepared [1]. Of these fluoro-amino acids L-[2-^{18}F]fluorotyrosine is a promising tracer for quantitation of the protein synthesis rate by PET [25] although the yet unknown necessary lumped constant, accounting for differences between natural tyrosine and fluorotyrosine, may inhibit an absolute quantitation [26].

To establish an absolute calibration of the protein synthesis rate as measured by the method described in this contribution it may be necessary to assess other parameters in the protein synthesis process by autoradiographic methods in order to study e.g. the influence of the specific amino acid used and the recycling of amino acids derived from protein degradation [28]. According to experiments performed with L-[1-^{14}C]leucine the values obtained are probably an underestimation of the true rates of the incorporation of L-leucine into protein but the results of these experiments show changes in specific regions of the brain in response to altered function [28, 29].

Conclusions

We could measure the protein synthesis rate in rats using with L-[1-^{11}C]tyrosine. The faster growing Walker 256 carcinosarcoma has a protein synthesis rate twice the PSR of the rhabdomyosarcoma. The value of the PSR is of the same magnitude as the PSR values reported in literature measured by using different carboxylic labeled amino acids. To extend the model to human data it has to be tested and evaluated separately. Especially the metabolite profile in humans has to be measured to see wether the amounts of metabolites is as low as in the animal model. Also the investigation of the amount of $^{11}CO_2$ exhaled in comparison to remnant carbon dioxide in the human tissue has to have special attention.

References

1. Vaalburg W, Coenen HH, Crouzel C, PhH Elsinga, Langstrom B, Lemaire C and Meyer GJ, 1992, Amino acids for the measurement of protein synthesis in vivo by PET, Nucl Med Biol 19, 227-237.
2. Lilja A, Bergstrom K, Hartvig P, Spannare B, Halldin C, Lundqvist H and Langstrom B, 1985, Dynamic study of supratentorial gliomas with L-methyl-^{11}C-methionine and positron emission tomography, Am. J. Neurol. Rad. 6, 505-514.
3. Dunzendorfer U, Schmall B, Bigler RE, Zanzonico, Conti PS, Dahl JR, Kleinert E and Whitmore WF, 1981, Synthesis and body distribution of alpha-aminoisobutyric acid-L-^{11}C in normal and prostate cancer bearing rats after chemotherapy, Eur J Nucl Med 6, 535-538.
4. Daemen BJG, Paans AMJ, Elsinga PhH, Wieringa RA, Konings AWT and

Vaalburg W, 1989, Hyperthermia induced suppression of protein synthesis in tumors measured by PET, in: Nuclear Medicine, Trends and possibilities in Nuclear Medicine, eds. Schmidt HAE and Buraggi GL, Schattauer Verlag, Stuttgart-New York, 77-80.

5. Daemen BJG, Elsinga PhH, Paans AMJ, Wieringa RA, Konings AWT and Vaalburg W, 1989, The effect of radiotherapy on L-[1-^{11}C]tyrosine and ^{18}FDG metabolism of tumors as measured by PET, J Nucl Med 30, 789.
6. Schober O, Meyer G-J, Gaab MR, Muller JA and Hundeshagen H, 1986, Grading of brain tumors by C-11-methionine PET, J Nucl Med 27, P890.
7. Phelps ME, Barrio JR, Huang S-C, Keen R, Chugani H and Mazziotta JC, 1984, Criteria for the tracer kinetic measurements of cerebral protein synthesis in humans with positron emission tomography, Ann Neur 15 (suppl), S192-S202.
8. Kubota K, Matsuzawa T, Ito M, Ito K, Fujiwara T, Abe Y, Yoshioka S, Fukuda H, Hatazawa J, Iwata R, Watanuki S and Ido T, 1985, Lung tumor imaging by positron emisson tomography using C-11 L-methionine, J Nucl Med 26, 37-42.
9. Bergstrom M, Lundqvist H, Ericson K, Lilja A, Johnstrom P, Langstrom B, von Holst H, Eriksson L and Blomqvist G, 1987, Comparison of the accumulation kinetics of L-(methyl-^{11}C)-methionine and D-(methyl-^{11}C)-methionine in brain tumors studied with positron emission tomography, Acta Radiol, 28, 389-393.
10. Schober O, Duden C, Meyer G-J, Muller JA and Hundeshagen H, 1987, Non selective transport of [^{11}C-methyl]-L- and D-methionine into malignant glioma, Eur J Nucl Med 13, 103-105.
11. Jones RM, Cramer S, Sargent T and Budinger TF, 1985, Brain protein synthesis rates measured in vivo using methionine and leucine, J Nucl Med 26, P168.
12. Ishiwata K, Vaalburg W, Elsinga PhH, Paans AMJ and Woldring MG, 1988, Comparison of L-[1-^{11}C] Methionine and L-Methyl-[^{11}C]Methionine for measuring in vivo protein synthesis rates with PET, J Nucl Med 29, 1419-1427.
13. Banker G and Cotman CW, 1971, Characteristics of different amino acids as protein precursors in mouse brain: advantages of certain carboxyl-

labeled amino acids, Arch Biochem Biophys 142, 505-573.

14. Ishiwata K, Vaalburg W, Elsinga PhH, Paans AMJ and Woldring MG, 1988, Metabolic studies with L-[1-^{14}C]Tyrosine for the investigation of a kinetic model to measure protein synthesis rate with PET, J Nucl Med 29, 524-529.
15. Keen RE, Barrio JR, Huang S-C, Hawkins RA, Phelps ME, 1989, In vivo cerebral protein synthesis rates with leucyl-tranfer RNA used as precursor pool: determination of biochemical parameters to structure tracer kinetic models for positron emission tomogarphy, J Cereb Blood Flow Metab 9, 432-448.
16. Bolster JM, Vaalburg W, Paans AMJ, van Dijk ThH, Elsinga PhH, Zijlstra JB, Piers DA, Mulder NH, Woldring MG and Wynberg H, 1986b, Carbon-11 labelled tyrosine to study tumor metabolism by positron emission tomography, Eur J Nucl Med 12, 321-324.
17. Paans AMJ, de Graaf EJ, Welleweerd J, Vaalburg W and Woldring MG, 1982, Performance parameters of a longitudinal tomographic imaging system, Nucl Instr Meth 192, 491-500.
18. Smith CB, Crane AM, Kadekaro M, Agranoff BW and Sokoloff L, 1984, Stimulation of protein synthesis and glucose utilization in hypoglossal nucleus induced by axotomy, J Neuroscience 4, 2489-2496.
19. Sokoloff L, Reivich M, Kennedy C, Des Rosiers MH, Patlak CS, Pettigrew KD, Sakureda O and Shinohara M, 1977, The [^{14}C]deoxyglucose method for the measurement of local cerebral glucose utilization: theory, procedure, and normal values in the conscious and anaesthetized albino rat, J Neurochem 28, 897-916.
20. James F, Roos M, 1975, MINUIT-a system for function minimization and analysis of the parameter error and correlations, Comp Phys Comm 10, 343-367.
21. Wiesel F-A, Blomqvist G, Halldin C, Sjogren I, Bjerkenstedt L, Venizelos N and Hagenfeldt L, 1991, The transport of tyrosine into the human brain as determined with L-[1-^{11}C]tyrosine and PET, J Nucl Med 32, 2043-2049.
22. Patlak CS, Blasberg RG, Fenstermacher JD, 1983, Graphical evaluation of blood-to-brain transfer constants from multiple-time uptake data, J

Cereb Blood Flow Metab, 3, 1-7.

23. Hawkins RA, Huang S-C, Barrio JR, Keen RE, Feng D, Maziotta JC and Phelps ME, 1989, Estimation of local cerebral protein synthesis rate with L-[1-^{11}C]leucine and PET: Method, model and results in animal and human, J Cereb Blood Flow Metab 9, 446-460.
24. Shields AF, Graham MM, Kozawa SM, Kozell LB, Link JM, Swenson ER, Spence AM, Bassingthwaigthe JB and Krohn KA, 1992, Contribution of labeled carbon dioxide to PET imaging of carbon-11-labeled compounds, J Nucl Med 33, 581-584.
25. Ishiwata K, Vaalburg W, Elsinga PhH, Paans AMJ and Woldring MG, 1988, Comparison of L-[1-^{11}C]methionine and L-methyl-[^{11}C]methionine for measuring in vivo protein synthesis rates with PET, J Nucl Med 29, 1419-1427.
26. Coenen HH, Kling P, Stocklin, 1989, Cerebral metabolism of L-[2-^{18}F]fluorotyrosine, a new PET tracer of protein synthesis, J Nucl Med 30, 1367-1372.
27. Wienhard K, Herholz K, Coenen HH, Rudolf P, Kling P, Stocklin G and Heis W-D, 1991, Increased amino acid transport into brain tumors measured by PET of L-(2-^{18}F)fluorotyrosine, J Nucl Med 32, 1338-1346.
28. Smith CB, Deibler GE, Eng N, Schmidt K and Sokoloff L, 1988, Measurement of local cerebral protein in vivo: Influence of recycling of amino acids derived from protein degradation, Proc Natl Acad Sci USA, vol 85, 9341-9345, Neurobiology.
29. Sokoloff L, Cerebral circulation, energy metabolism, and protein synthesis: general characteristics and principles of measurement, In: Phelps ME, Mazziotta JC and Schelbert HR edittos. Positron emission tomography and autoradiography, New York, Raven Press, 1986: 2-71.

KINETIC MODELLING OF CARBON-11 LABELLED METHIONINE

H. LUNDQVIST

ABSTRACT. A three parameter compartment model is usually accepted to be adequate for describing the transport of ^{11}C-L-methionine across biological membranes. This is found to be true also in the placenta transport of the rhesus monkey. The placenta is an excellent experimental model since one can inject and sample on both sides of the membrane. Assumptions in the kinetic model like unidirectional transfer of radioactivity can thus be tested and verified.

1. INTRODUCTION

1.1. Kinetic model of ^{11}C-methionine in general

Although the biochemical pathway of methionine [1] should ask for a kinetic model with four rate constants or more there seems to be a general acceptance of the three parameter model first introduced by Bustany [2] and recently studied and applied in brain tumours by Hatazawa et al. [3].

There are many reasons why it is difficult to obtain detailed quantitative information on methionine from PET-data. Methionine is rapidly metabolized [4] why one has to correct for labelled metabolites 10-15 minutes after injections. Kinetic modelling has so far mainly been applied in the brain with its special way to handle amino acids by facilitated diffusion across the blood brain barrier (BBB) and facilitated and active transport across the cell membranes. Methionine is special here since it is the only physiological amino acid which has a high affinity to both systems which makes the interpretation more difficult. Other amino acids tends to compete

B. M. Mazoyer et al. (eds.), PET Studies on Amino Acid Metabolism and Protein Synthesis, 175–182.

and block the transport routes why changes in concentrations of other amino acids changes the transport kinetic of methionine.

2. TISSUE HETEROGENITY

One special problem is the tissue heterogenity i.e. the tissue is composed by different sub-tissues which can not be resolved by the resolution of PET and which have quite different kinetics. This problem is accentuated in the brain due to the large differencies in kinetics of white and gray matter. We studied the problem in a rhesus monkey. ^{14}C-*L*-methionine was given to the monkey. One hour after injection the monkey was killed, the brain sliced and the autoradiografic technique was used to obtain the radioactivity distribution in different parts of the brain as shown in fig. 1 [5]. The radioactivity concentration in the grey matter was found to be a factor of four higher than in the white matter. The difficulty to get "pure" regions of interest in a PET investigation is also clearly demonstrated in fig. 1.

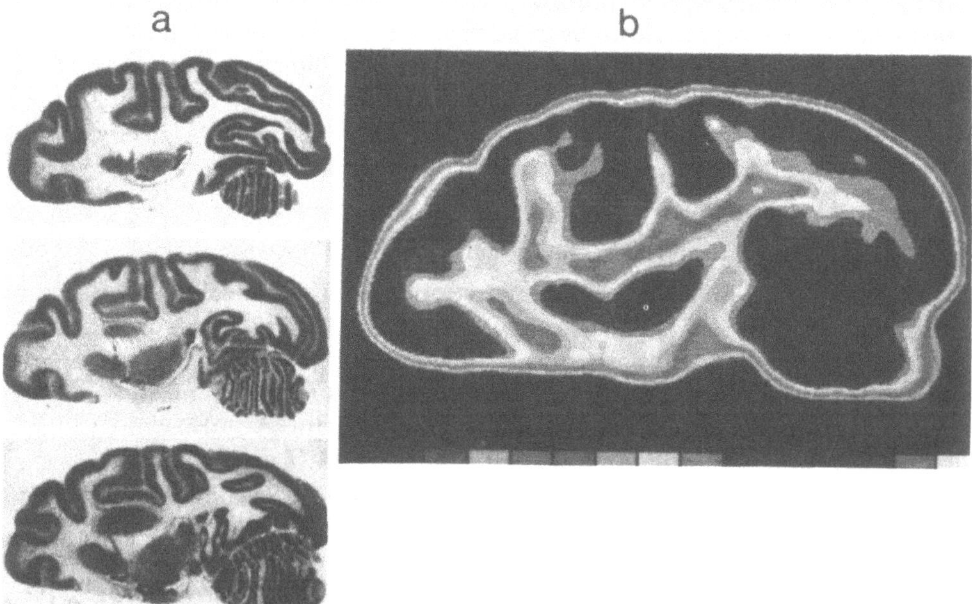

Fig. 1. (a) Three images of ^{14}C-*L*-methionine distribution in the rhesus monkey brain obtenaid by autoradiography is shown. The distance between the slices are about 1 cm.
(b) The three slices in (a) is filtered and added in order to simulate a PET-image with spatial resolution of 8 mm and a slice thickness of 14 mm.

What is then the information you are getting when applying your three parameter model to a data set composed of components with different kinetics? To get a rough idea one can simplify the problem to only two tissues which each obeys the three parameter model but with different rate constants. By simulations one can then obtain the results shown in figs 2.

To try to model pathological conditions like tumours where cell differentiation, necroses, BBB-rupture etc adds to the heterogenity is of course even more difficult. Further difficulties which adds to the problems is input function disturbances like smearing and metabolites. In our clinical work with methionine we have due to these reasons avoided to use kinetic modeling to analyze results but have used more simple methods like uptake

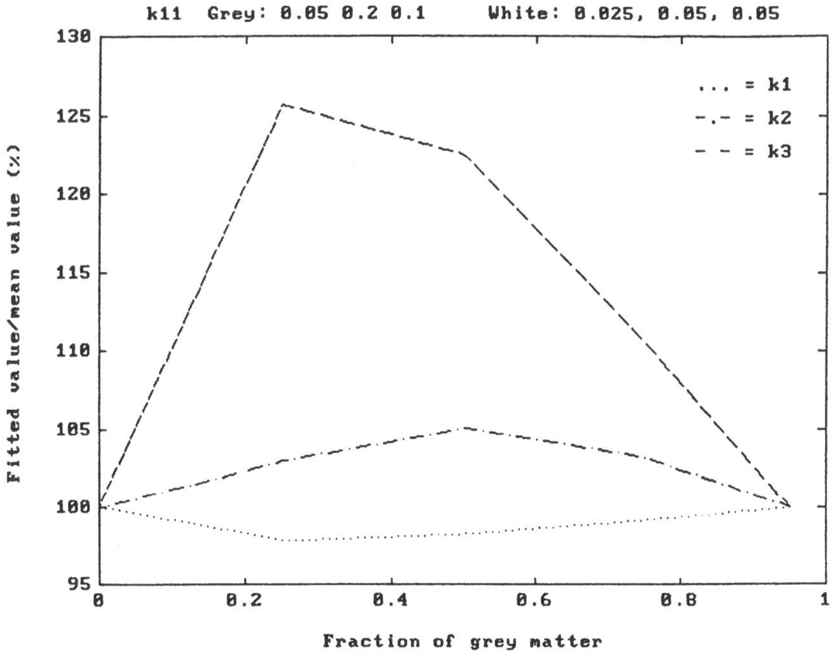

Fig. 2. The three parameter model for methionine is used to fit simulated data. The data is composed by two tissues, grey (k1=0.05 1/min, k2=0.2 1/min and k3=0.1 1/min) and white (k1=0.025 1/min, k2=0.05 1/min and k3=0.05 1/min) matter, which both are assumed to take up methionine according to the three parameter model. The fitted parameters are divided by a wheighted mean of the true parameters and plotted as a function of fraction of grey matter.

estimates and comparison with reference areas. We have restricted our use of kinetic modeling to transport only and in a time scale of 10-15 minutes where metabolization is a minor problem [4]. We will here present the work we have done in the placenta of the rhesus monkey, an excellent experimental model to study transport of various ligands since one can inject and sample on both sides of a membrane.

3. METHIONINE TRANSPORT ACROSS THE PLACENTA

3.1. Anatomy of the placenta

The placenta is a special organ with a large blood volume (40-60 %) and a structure which is species dependent. We have chosen to study the rhesus monkey placenta which has about the same structure and function as the human placenta. The purpose to model methionine transport across the placenta was to create a model which could inform about the nutrition condition of the fetus during different situations like starving of the mother or at drug influence.

The blood supply of the placenta is rather special. There is no capillaries but the arterial blood is pumped into small sacks called the cotyledons. The volumes of these cotyledons are probably varying with the blood pressure and the blood flow. Blood volume can thus vary rapidly and is in the range of 30-60 %. Placenta it self is nutrified from the cotedylon blood. Villous walls of the cotyledones contains the fetal blood system and will be well exposed to the maternal blood. Amino acids are then transported across a number of membranes to reach the fetal blood circulation. The fetus liver is supplied by the blood in the cord and the amino acid, peptides and proteins is further transported from the fetal liver to the rest of the fetus body.

In most studies with PET one can only measure the input function (arterial blood or plasma) and tissue. In the placenta one have three measurable compartments, arterial blood, placenta and the fetus which adds to the stiffness in the model. One can also inject on both sides, both in the blood system of the mother and of the fetus which is a way to verify if back transport of radioactivity takes place. We did that in one experiment. ^{11}C-*L*-Methionine was injected into the fetal heart and the radioactivity distribution in the mother was studied. For 20 minutes no radioactivity distribution was seen in the mother. After about 30 minutes radioactivity concentrations were seen in the kidneys of the mother. This was concluded that no back transport of ^{11}C-*L*-methionine took place but after 30 minutes labelled metabolic products were transported from the fetus across the placenta end then excreted by the mother.

3.2 Kinetic model of ^{11}C-methionine in the placenta

The placenta-fetus system is described by the compartment model presented in fig 3. The mathematical description of the model is described by Berglund et al. [6]. An example of fit to measured data is seen in fig 4. The

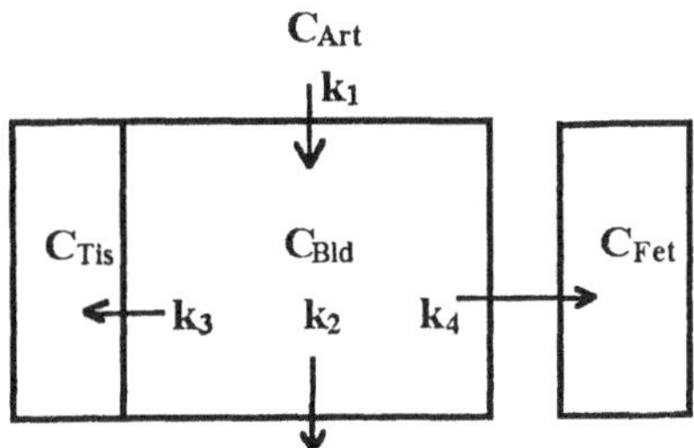

Fig 3. A compartment model for kinetic analysis of ^{11}C-*L*-methionine across the placenta. C_{Art} is the arterial blood concentration comming into the cotylodons of the placenta. C_{Bld} is the arterial blood concentration in the placenta. C_{Tis} is the trapped radioactivity concentration in placenta tissue. The radioactivity to the fetus is transported by the cord and the fetal radioactivity concentration, C_{Fet}, can thus be measured in a separate region of interest. k_1, k_2, k_3 and k_4 are rate constants.

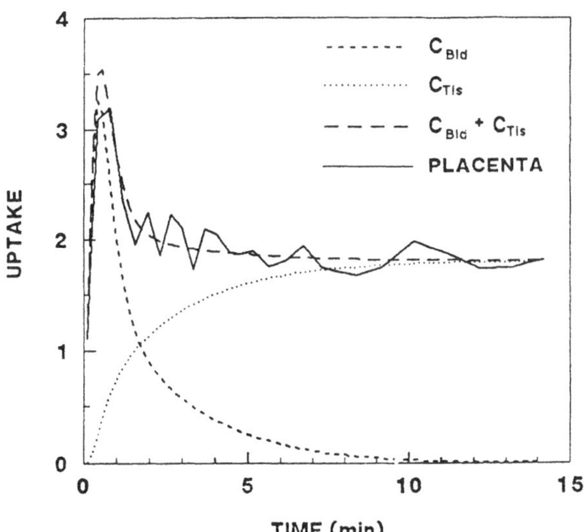

Fig. 4. Uptake in a placental region of interest of ^{11}C-*L*-methionine. In the figure is shown measured data and the mathematical fit to the model as a function of time.

TABLE 1. Kinetic parameters obtained in ten experiments with placenta transport of ^{11}C-*L*-methionine. k_1, k_3 and k_4 are explained in fig. 2. V_{Plac} is the volume om total placenta and V_{Fet} is the total volume of the fetus. V_{Plac} and V_{Fet} is introduced in order to couple the radioactivity concentrations measured in the placenta and the fetus. $k_{Com} = k_1*\varepsilon + k_3 + k_4$ and where ε is the blood volume of placenta. From this relation ε can be calculated and is also given in the table.

k_1 [min^{-1}]	k_3 [min^{-1}]	k_4*V_{Plac}/V_{Fet} [min^{-1}]	$k1/k_{Com}$	ε [%]
0.51	0.24	0.102	0.29	42
0.41	0.14	0.068	0.43	64
0.84	0.31	0.098	0.26	31
1.20	0.35	0.092	0.38	46
1.14	0.36	0.111	0.42	55
1.07	0.15	0.071	0.51	61
1.28	0.21	0.114	0.35	40
0.73	0.16	0.108	0.40	53
0.51	0.21	0.068	0.41	61
0.34	0.17	0.083	0.34	55

derived parameters from eleven experiments are given in table 1. The blood flow in placenta, k_1, varies with a factor of three while the transport of methionine to the placenta, k_3, and to the fetus, k_4*V_{Plac}/V_{Fet}, varies less. The placenta blood volume, V_{Bld}/V_{Plac}, can be calculated from the results and is also given if table 1. The calculated values agrees very well with measured values [7] and support that the model is consistent and well behavioured .

The model has so far been applied in two clinical experimental approaches

Placental transfer of methionine during protein starvation [6, 8]

Placental transfer of methionine during influence of nicotine [9]

The placenta system and PET has also been used to study placenta transport of opioids [10].

4. CONCLUSIONS

The placenta system in rhesus monkey is an excellent experimental system to study amino acid transport across membranes with PET since it is possible to

inject ligand on both sides of the membrane

sample and measure blood radioactivity on both sides of the membrane [11]

two independent region of interest are available, the placenta and the fetus

The dynamics of the system allows the experiments to be finished within 10-15 minutes during which time the metabolism of methionine and most other amino acids is negligible in plasma and tissue [4,12]

The usual three parameter kinetic model for ^{11}C-methionine do apply well in the placenta.

5. REFERENCES

1. Meyer G-J, Schober O, Hundeshagen H. Uptake of 11C-L- and D-methionine in brain tumors. Eur J Nucl Med 1985; 10:373-376.

2. Bustany P, Sargent T, Saudubray JM, Henry JF, Comar D. Regional human brain uptake and protein incorporation of [11C]L-methionine studied in vivo with PET.
J Cereb Blood Flow Metab (suppl) 1981; 1:I9-I10.

3. Hatazawa J, Ishiwata K, Itoh M et al. Quantitative evaluation of L-[methyl-C-11]methionine uptake in tumor using positron emission tomography. J Nucl Med 1989; 30:1809-1813.

4. Lundqvist H, Stålnacke C-G, Långström B, Jones B. Labelled metabolites in plasma after i.v. administration of 11-C-methyl-L-methionine.
In: Greitz T, Ingvar DH, Widén L, editors. The metabolism of the human brain studied with positron emission tomography.
New York: Raven Press, 1985: 233-239 .

5. d'Argy R. Short-lived radionuclides and image analysis in whole-body autoradiography. Thesis. Uppsala University (ISBN 91-554-2219-5) 1988.

6. Berglund L, Andersson J, Lilja A, Lindberg B S, Gebre-Medhin M, Antoni G, Bjurling P, L†ångström B, Lundqvist H. Amino acid transport across the placenta measured by positron emission tomography and analyzed by compartment modelling.
J Perinatal Medicine 1990; 18: 89-100.

7. Berglund L, Lilja A, Andersson J, Lindberg B, Ulin J, Långström B, Lundqvist H. Maternal Blood Volume in Placenta of the Rhesus Monkey Measured Invivo by Positron Emission Tomography.
Gynecol Obstet Invest 1991; 31:1-7.

8. Berglund L, Gebre-Medhin M, Lindberg B, Lilja A, Lå†ngström B, Lundqvist H. Placental transfer of methionine during protein restriction in the rhesus monkey studied by positron emission tomography.
Placenta 1989; 10: 387-397.

9. Berglund L, Lundqvist H, Nordberg A.,Lindberg B, Gebre-Medhin M. Nicotine and transplacental transfer of amino acids in the rhesus monkey studied by positron emission tomography (PET). Proceedings from "Sixth Annual Bristol-Myers Symposium on Nutr Res", 1987.

10. Lindberg B S, Berglund L, Gerdin E, Hartvig P, Lilja A, Lundqvist H, Långström B, Rane A. Feto-maternal kinetics of opioids in rhesus monkeys studied with positron emission tomography (PET).
Proceedings from "XI European Congress of Perinatal Medicine".
CIC Edizioni Intern (Rome) 1988; 713-719.

11. Lindberg B, Berglund L, Hartvig P, Lindgren P G, Lilja A, Lundqvist H, Långström B, Rane A. Positron emission tomography in experimental perinatology. J Perinat Med 1985; 13:277-286.

12. Berglund L, Halldin K, Lilja A, Lindberg B, Lundqvist H, Långström B, Malmborg P, Någren K, Stålnacke C-G. 11C-Methionine kinetics in pregnant rhesus monkeys studied by positron emission tomography:
A new approach to feto-maternal metabolism.
Acta Obstet Gynecol Scand 1984; 63:641-645.

Approaches to Quantitative Analysis of Amino Acid Transport and Metabolism

G.-J. MEYER, J. VAN DEN HOFF, W. BURCHERT, H. HUNDESHAGEN

Introduction

Unlike in the case of glucose utilization, there is an ongoing debate on which kinetic model might be best suitable for the description of amino acid metabolism, in order to facilitate a quantitative interpretation of the PET data obtained from the applications of amino acids. Although these PET data have been obtained mostly from investigations of tumors, basic questions on the uptake processes in normal brain can be addressed, using these data sets, by analyzing reference regions and non-tumorous brain areas. Some protocols have been especially designed to establish a quantitative model for protein synthesis in normal brain and to increase the knowledge about the transport mechanisms.

Despite a broad knowledge on the basic mechanisms of metabolic pathways of amino acids and protein synthesis (conf. [1, 2]), specific questions arising from the PET data under pathological conditions can only be answered by further basic investigations in animal experiments.

There is still a discussion about which amino acid would be most suitable for a given problem, as e.g. protein synthesis measurements. Since any essential amino acid should lead to the same protein synthesis rate, several amino acids are under investigation, especially in animal experiments. The most suitable amino acid for protein synthesis rate determination should be the one with the least metabolic side pathways, since such metabolic side chains complicate the interpretation of PET measurements tremendously. The degree of interference is dependent of the fate of the label, which may change with its position in the molecule. However, since all amino acids turn out to exhibit a series of interfering metabolic pathways, practical reasons, like ease of synthesis and purification of the amino acid begin to play an important role in the use of these compounds for clinical applications, especially in tumor diagnosis and follow-up of treatment. For a recent review of this problem cf. [3]

Furthermore it is not clear yet, whether a single parameter like protein synthesis or a less exactly defined mixture of metabolic processes are best suited to describe the metabolic

B. M. Mazoyer et al. (eds.), PET Studies on Amino Acid Metabolism and Protein Synthesis, 183–196.

activity of a tumor. Anyhow, transport phenomena have to be considered of predominant importance and especially their distortion under pathophysiological conditions [4].

Basic studies

Based on first pass extraction measurements of many amino acids, Oldendorf et al. [5,6] concluded that the transfer of amino acids from blood into brain tissue is an enzyme catalyzed transport, following Michaelis-Menten kinetics. He found that there are at least three groups of amino acids. Within the groups the amino acids compete with each other for the same carrier [7, 8]. Further experiments showed an individual stereospecificity for various amino acids, which ranged from high values for DOPA, Valine, Histidine, Lysine, and Tryptophane over medium values for Isoleucine and Tyrosine to amino acids with low stereospecificity like Phenylalanine, Leucine and Methionine[9, 10].

With respect to PET applications of amino acids as tools for the investigation of brain tumors, the mechanisms of uptake and the metabolic interpretation of the accumulation has been studied by many groups now. Although basic studies and modelling approaches can not be separated, since the former are necessary for the validation of the latter, the animal experiments and basic studies can be grouped by the different amino acids which have been used. ^{14}C- and/or ^{11}C -Leucine, which can be regarded as the most classic model amino acid with relatively simple metabolic pathways, has been used by the groups from NIH [11-16] and UCLA[17-20]. Most of their effort has been in finding quantitative models for a description of protein synthesis, with little attention to application in brain tumors, however.

On the other hand, ^{11}C-Methionine has been used most widely with a direct aim of tumor analysis, despite obvious difficulties in establishing quantitative models for the measurement of protein synthesis. The problems with methionine stem from the considerable fraction of metabolites which are formed in trans-methylation reactions. Although these difficulties have been underestimated in first reports by Bustany et al. [21, 22], and Lestage et al. [23], there is now a broad agreement on the usefulness of methionine for an analysis of the metabolic activity of brain tumors in general, regardless of the limitations that are associated with the interpretation of methionine data in terms of protein synthesis [24 - 31].

Some of these problems have been addressed by comparison of carboxylic- and methyl-labeled methionine, although both can not be considered especially suitable for protein synthesis measurements [32].

With respect to its lower abundance in brain protein and the decreased possibility of cross-contamination of the precursor pool from protein degradation, ^{11}C-labeled Valine has been suggested and used in animal experiments [33, 34]. ^{14}C-/^{11}C-labeled Phenylalanin has been used by several investigators [35-39], because it can be regarded as a model compound for aromatic amino acids, and has been well studied in its ^{14}C-labelled form, in numerous investigations [40]. Furthermore its own metabolic spectrum is simple, as long as the spectrum of its main metabolite Tyrosine can be neglected.

Finally the ^{11}C-label can be put in different positions in phenylalanine more easily than in other amino acids, thereby changing and simplifying the metabolite spectrum further .
The protein incorporation of ^{11}C-Tyrosine has been analyzed mainly in view of animal data which suggest a high protein incorporation rate for this amino acid [41].
Furthermore ^{11}C-glycine [42] and the common amino acid metabolite ^{11}C-pyruvate [43] have been used in preliminary studies for brain tumor analysis.
Several groups have analyzed the uptake patterns of D-amino acids [44-51]. Whereas the results of early investigators who reported a positive contrast of D-amino acid uptake in some non-brain tumors (10, 11) remained ambiguous, it is quite surprising that in brain tumors the uptake of the unphysiological enantiomer was in all investigations nearly as high as for the L-enantiomer or even better in some cases. Animal experiments as well as patient investigations have shown, that the similar uptake occurs despite strongly decreased protein incorporation [46-49]. This phenomenon has been used to attribute the uptake to diffusion processes following BBB damage. It was shown, however, that low grade tumors with intact BBB exhibit similar uptake characteristics for D - amino acids, thereby relating the accumulation of the D - amino acids to active transport phenomena also.
Few ^{18}F-labelled amino acids have been investigated, among them ^{18}F-p- and ^{18}F-o-fluoro-phenylalanine [52-55] and ^{18}F-m-/^{18}F-o- fluoro-tyrosine[54, 56]. Only the latter one has shown to be incorporated into brain protein to the same degree as natural amino acids. For this amino acid analog only few and negligible amounts of metabolites have been identified up to now. Therefore it seems to be suitable for the measurement of protein synthesis rates. Preliminary studies have been performed in dynamic mode, which allowed a kinetic analysis according to a 3-compartment model. The k-map analyses showed that most of the PET signal information can be attributed to the transport of the amino acid analogue into tumor tissue, whereas the metabolic rate map added little more information only [57].
It can be concluded from the various animal experiments and basic patient investigations that the uptake is surprisingly similar for various amino acids, including some D-enantiomeres, and especially D-methionine. This holds true for normal brain tissue as well as for brain tumors. In contrast to the uptake, the protein incorporation is a slow process, which is markedly different for each of the investigated amino acids. The plasma clearance curves for all investigated amino acids are relatively similar, but show differences with respect to the fractions of protein bound activity and low molecular weight components. Several of these results have been corroborated in dynamic PET investigations in patients already, as discussed in a different chapter of this book.
The conclusion, that the uptake of amino acids is a fast process, which in tumors seems to be triggered by demand, is further supported by the observation, that iodinated amino acids like p-Iodophenylalanine [58, 59] and Iodo-alpha-methyl-tyrosine [60-63] which have been developed for SPECT investigations of brain tumors according to the promising PET results discussed above, show a very similar accumulation in brain tumors, despite a total lack of incorporation into brain proteins.
Finally this interpretation is further corroborated by investigations which indicate that

the uptake can be blocked by saturation of the carrier in the BBB with other amino acids. [29, 64]

Modelling Approaches

Early modelling approaches in order to calculate brain amino acid uptake data in terms of protein synthesis rate (PSR) have been suggested for methionine by Bustany et al . [65-67] and Lestage et al. [23] and for leucine by Dienel et al. [68], Dweyer et al. [69], Mies et al. [70], Smith et al. [11], and Phelps et al. [17]. While it has become clear that a 3 compartment model for methionine can not be validated by experimental results [71], efforts have been intensified to validate a 4 compartment model for Leucine [18-20]. The only major degradation pathway for leucine labelled in carbon-1 position leads to a decarboxylation with loss of the label as CO_2, which then rapidly clears via the lungs. Therefore fast dynamic measurements in the early phase after uptake allow a calculation of the CO_2 loss in regions of interest.
A kinetic model for tyrosine has been discussed briefly only [72]. Despite high uptake values in tumors, the relative complicated metabolic pathways, which this amino acid enters and which potentially interfere with a conclusive model, have distracted many PET researchers up to now.
In principle this holds true for phenylalanine as well. However, a smaller fraction of the amino acid undergoes metabolic pathways other than protein incorporation or degradation, initiated by decarboxylation. This small fraction then is almost totally converted to tyrosine and may enter the various possible pathways for this amino acid [73,74].
An approach for the measurement of brain protein synthesis by a simplified 3 compartment model with ^{14}C-Valine has been reported by Kirikae et al. [33,34]. It has, however, not yet been applied to PET investigations.
In a study by Anders [71] 3- and 4-compartment systems have been compared on two sets of experimental data from leucine and methionine [48]. The deterministic analysis was generalized by describing the concentration changes in the compartments by matrices, which simplified the calculations upon variation of the compartment models. The results of this approach have shown, that a 3-compartment model is definitely insufficient to fit the data, while the generally accepted 4-compartment model can fit the data sufficiently for Leucine as well as for Methionine despite their different metabolic pathways. Since in the case of methionine the metabolic compartment can not be analyzed individually by PET, however, the analytical approach by non-linear fitting of a set of four differential equations remains unsolvable for PSR by PET with this amino acid.
For the calculation of PSR with ^{11}C-carboxyl labelled amino acids the loss of labelled $^{11}CO_2$ is the major complicating factor, which according to recent results can possibly be corrected sufficiently in actual PET experiments, provided this is the only major competing metabolic pathway. However, the contamination of the plasma curve by a large fraction of protein bound activity must be corrected for also.

For methyl-labelled ^{11}C-methionine the loss of activity by is nearly negligible, but the trans-methylated products, which seem to exhibit a small elimination and which can not be differentiated from protein bound material in the PET signal, make it impossible to calculate a protein synthesis rate. The plasma curve represents the input function better than in the case of many carboxylic labeled amino acids, however, it was shown that small amounts of metabolites contribute to the signal in the case of methionine also, requiring a chemical analysis of plasma samples for good fitting [75,76]. Despite of these difficulties and limitations, kinetic analytical approaches on methionine data by the simplified 3 compartment model have revealed some insight into the accumulation processes. As has been shown more recently for fluoro-tyrosine by Wienhard et al. [57], Ericson et al. [77] have shown that for methionine most of the information which can be obtained from a kinetic analytical approach is residing in the transport parameters. Although this analysis admittedly suffers from some oversimplification in the applied model, the authors conclude that the results are indicative of a specific physiological process in the tumors without a disruption of the BBB, where demand and transport are closely related.
The aspect that an increase of the transport into tumor tissue is more relevant than increased protein synthesis rates is corroborated by a finding of Meyer et al. [78]. In this study measurements of protein incorporation of methionine in human brain tumor samples suggest that the protein synthesis rate is similar in brain tumors of different malignancy, but that the uptake into tumor tissue increases with the tumor grade.

Implementation of the Gjedde Patlak Analysis

Under physiological conditions which include some metabolic sink of a labeled tracer a Gjedde-Patlak analytical approach [79-81] seems to be the method of choice. This modeling tool is able to describe the unidirectional transport into the irreversible compartment, disregarding whether this compartment is splitted or not, as in the case of methionine, where the transmethylated components can not be distinguished from the protein bound fraction. This approach has been evaluated for methionine by several investigators already [71, 27, 51].
Our recent approach to analyze dynamic PET data obtained from amino acid investigations makes use of the programming tool "Matlab" including parametric image processing. After reading calibrated image files and corrected blood data files into the "Matlab" processing window, the calculated image matrix is loaded back into the image analysis tool. Numerical values for Kmr, and the virtual distribution volume can be retrieved from either regions of interest or pixel by pixel in the image window. For judgement of the quality of the data, parametric standard deviation images for Kmr and Vd as well as the chi-square matrix for the fit of the regression line are displayed. A graphical display window in the "Matlab" tool allows to judge the fit on data from regions of interest. Examples for a tumor region and a reference tissue region from one patient with a high grade glioma are plotted in fig. 1 and 2 for data obtained with Methionine and FDG.

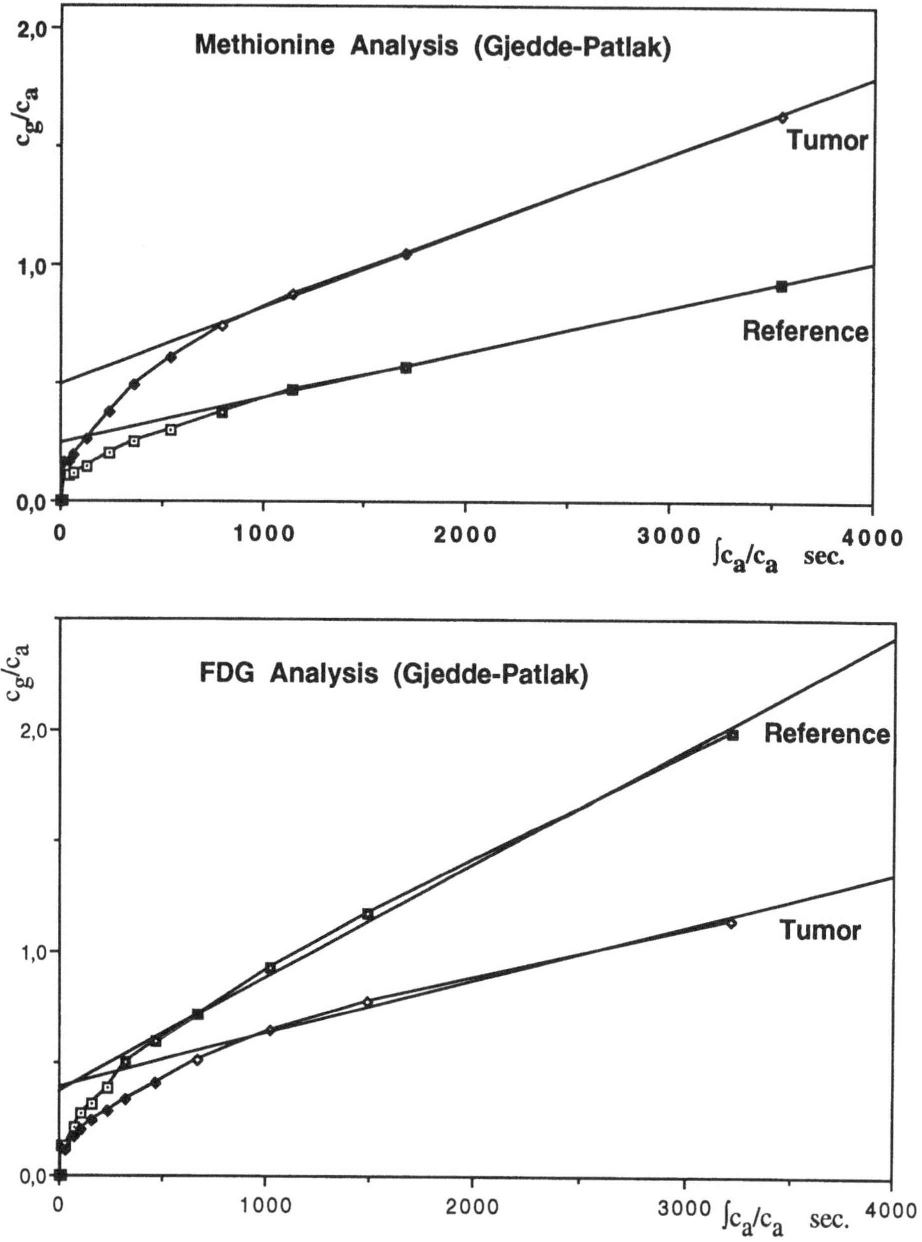

Fig. 1 a, b. Gjedde - Patlak analysis of an a) FDG- and b) Methionine- Study

The values on Kmr and Vd for normal tissue as obtained from non-tumor regions of patients with low grade tumors are shown in table 1. The reference values given by Bergström et al. [83] are somewhat higher than our values. On the other hand recent data presented by a Japanese group [84] on low grade tumors indicate rate constants in the order of 0.012 - 0.017 which is somewhat smaller than our data.

Table 1: Metabolic rate constants (Kmr) and distribution volume (Vd) for L-Methionine and Fluorodeoxyglucose in grey- and white-matter brain tissue as obtained by Gjedde - Patlak analysis.

		grey matter		white matter	
		This Study	Ref [83]	This study	Ref [83]
meth	Kmr	0.019 ±0.011	0.031 ±0.002	0.012 ±0.008	0.018 ±0.003
	Vd	0.37 ±0.08	0.44 ±0.17	0.21 ±0.07	0.34 ±0.12
		This Study	Ref [84]	This study	Ref [84]
f d g	Kmr	0.028 ±0.005	0.028	0.015 ±0.003	0.014

Since the data obtained by our procedure with FDG compare very well with the literature data, this may be taken as an indication for the validity of the approach. In summary, however, there are too few data available for methionine at the moment, in order to compare the data sets adequately.
A further evaluation of the data in terms of an individual fitting for K1, K2, and K3 in a linearized solution of the three compartment model, according to the general approach of Blomqvist [82] has been successful for FDG data. As can bee seen from the difference of the uptake characteristics of FDG and Methionine (as shown in fig 2.) it is more critical, however, to fit individual K-values from Methionine data, because of the rapid increase of the tissue signal.

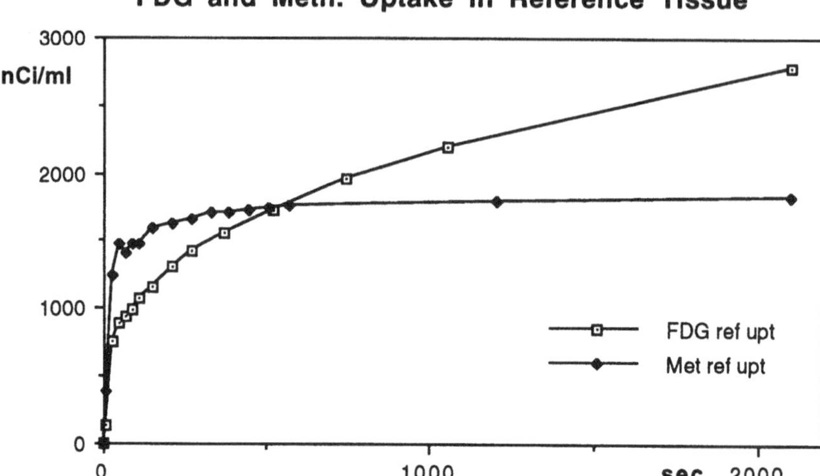

Fig. 2 Tissue response curves for FDG and Methionine

Since the additional information after about 300 sec is limited, the fitting procedure for the K values requires avery good sampling in the early phase. It is hoped that this problem can be overcome soon, by choosing a higher frame rate during the first minutes and by implementing an automatic and continuous blood sampling system. It must be considered, however, that the unstable fitting procedure may, as well or additionally, stem from insufficient assumptions on the 3 compartment model. Although this can be overcome by introducing one more rate-constant, the solution of the system then becomes arbitrary unless firm restrictions on the allowable values of the K-parameters are introduced.
Nevertheless, it can be seen from the uptake characteristics directly, that transport phenomena must play a dominant rule for the overall interpretation of the tissue signal.

References

1. Dunlop DS. Measuring protein synthesis and degradation rates in CNS tissue. In: Marks N, Rodnight R, Eds., Research Methods in Neurochemistry, Vol. 4, Plenum Publ. New York 1978, pp 91-141
2. Pardridge WM, OLdendorf WH. Kinetic analysis of blood brain barrier transport of amino acids. Biochim. Biophys. Acta 401, 128-136, 1975
3. Willem Vaalburg, Heinz H. Coenen, Christian Crouzel, Philip H. Elsinga, Bengt Langström, C. Lemaire, Geerd-J. Meyer. Amino Acids for in-vivo measurement of protein synthesis by PET. Nuclear Medicine and Biology 19, 227-237 (1992)

4. Steinwall O. Transport inhibition phenomena in unilateral chemical injury of blood brain barrier. In : Lajtha A., Ford D. (eds.) : Brain Barrier Systems. Elsevier Publ. Co., Amsterdam 1968, pp 357-366

5. Oldendorf WH. Brain uptake of radiolabeled amino acids, amines and hexoses after arterial injetion. Am. J. Physiol. 221, 1629-1639, 1971

6. Oldendorf WH, Szabo J. Amino acid assignment to one of three blood brain barrier amino acid carriers. Am. J. Physiol. 230, 94-98, 1976

7. Vahvelainen M.-L., Oja S.S. Kinetic analysis of phenylalanine-induced inhibition in the saturable influx of tyrosine, tryptophan, leucine and histidine into brain cortex slices from adult and 7-day-old rats. J. Neurochem. 24 , 885-892, 1975

8. Pardridge W.M. Kinetics of competetive inhibition of neutral amino acid transport across the blood-brain barrier. J. Neurochem. 28, 103-108, 1977

9. Lajtha, A., Toth J. The brain barrier system-V: Stereospecificity of amino acid uptake, exchane and efflux. J. Neurochem. 10, 909-920, 1963

10. Oldendorf WH. Stereo-specificity of blood-brain barrier permeability to amino acids. Am J. Physiol. 224, 967-969, 1972

11. Smith CB, Davidsen L, Deibler G, Patlak C, Pettigrew K, Sokoloff L. A Method for the determination of local protein synthesis in the brain. Trans. Am. Soc. Neurochem. 11-94, 1980

12. Smith CB, Deibler G, Eng N, Schmidt K, Sokoloff L. Measurement of local cerebral protein synthesis: Influence of recycling of amino acids in the precursor pool derived from protein degradation. Proc. Natl. Acad. Sci. USA 85, 9341-9345, 1988 (conf also: J. Cerebr. Blood Flow Metabol. 9, S201, 1989)

13. Smith Q.R., Takasato Y., Sweeney D.J., Rapoport S.I. Regional cerebrovascular transport of leucine as measured by the in situ brain perfusion technique. J. Cereb. Blood Flow Metab. 5, 300-311 (1985)

14. Smith Q.R., Takasato Y. Kinetics of amino acid transport at the blood-brain barrier studied using an in situ brain perfusion technique. Ann. NY Acad. Sci. 481, 186-201, (1986)

15. Smith Q.R., Momma S., Aoyagi M., Rapoport S.I. Kinetics of neutral amino acid transport across the blood-brain barrier. J. Neurochem. 49, 1651-1658, 1987

16. Smith QR, Takasato Y, Rapoport SI. Kinetic analysis of L-leucine transport across the blood brain barrier. Brain Res. 311, 167-170, 1984

17. Phelps ME, Barrio JR, Huang S.-C., Keen RE, Chugani H, Mazziotta JC. Criteria for the tracer kinetic measurement of cerebral protein synthesis in humans with positron emission tomography. Ann. Neurol. 15 (Suppl), S192-S202, 1984

18. Keen RE, Barrio, JR, Huang SC, Hawkins RA, Phelps ME. In vivo cerebral protein synthesis rates with Leucyl-transfer RNA used as precursor pool: Determination of biochemical parameters to structure tracer kinetic models for positron emission tomography. J. Cerebr. Blood Flow Metabol. 9, 429-445, 1989

19. Hawkins, RA, Huang S.-C., Barrio JR, Keen RE, Feng D, Mazziotta JC, Phelps ME. Estimation of local cerebral protein synthesis rates with L-[1-11C] Leucine and PET: Methods, model, and results in animals and humans. J. Cerebr. Blood Flow Metabol. 9, 446-460, 1989

20. Phelps M.E., Barrio J.R., Huang S.C., Keen R.E., Chugani H., Maziotta J.C. Measurement of cerebral protein synthesis in man with positron computerized tomography:: Model, assumption, and preliminary results. In : Greitz T. (ed) The Metabolism of the Human Brain Studied with Positron Emission Tomography. Raven Press, New York : 215-232 (1985)

21. Bustany P, Sargent T, Saudubray JH, Henry JF, Comar D. Regional human brain uptake and protein incorporation of 11C-L-Methionine studied in vivo with PET. J. Cerebr. Blood Flow Metabol. 1, 17-18, 1981

22. Bustany P, Chatel M, Derlon, JM, Darcel F, Sgouropoulos P, Soussaline F, Syrota A Brain tumor protein synthesis and histological grades: A study by positron emission tomography (PET) with 11C-L-Methionine. J. Neurooncol. 3, 397-404, 1986

23. Lestage P, Gonon, M, Lepetit P, Vitte PA, Debilly G, Rosatto C, Lecestre D, Bobillier P. An in vivo kinetic model with L-[35S]-Methionine for the determination of local cerebral rates for methioninbe incorporation into protein in the rat. J. Neurochem. 48, 352-363, 1987

24. Mosskin M. Diagnostic imaging in glioma,(Thesis), Karolinska Hospital Stockholm, Caslon Press, Stockholm 1987

25. Mosskin M, Ericson K, Hindmarsh T, von Holst H, Collins VP, Bergström M, Eriksson L, Johnström P. Positron emission tomography compared with MRI and CT in supratentorial gliomas using multiple stereotactic biopsies as reference. Acta Radiologica 30, 225-323, 1989

26. Ishiwata K, Ido T, Abe Y, Matsuzawa T, Iwata R. Tumor uptake studies of S-adenosyl-L-[methyl-11C]-methionine and L-[methyl-11C]-methionine. Nucl. Med. Biol. 15, 123, 1988

27. Hatazawa J, Ishiwata K, Itoh M, Kameyama M, Kubota K, Ido T, Matsuzawa, T, Yoshimoto T, Watanuki S, Seo S. Quantitative evaluation of [methyl-11C]-methionine uptake in tumor using positron emission tomography. J. Nucl. Med. 30, 1809-1813, 1989

28. Kameyama M, Shirane R, Itoh J, Sato K, Katakura R, Yoshimoto T, Hatazawa J, Itoh M, Seo S, Ido T. The accumulation of 11C-Methionine and histological grade in cerebral glioma studied with PET. CYRIC Annnual Report 1988, Cyclotron and Radioisotope Center Tohoku University, Sendai 1989, pp 228-237

29. O'Tuama LA., Guilarte TR, Douglass KH, Wagner Jr, HN, Wong, DF, Dannals R.F., Ravert, HT, Wilson AA, La France N.D., Bice AN, Links JM. Assesment of [11C]-L-Methionine transport into the human brain. J. Cerebr. Blood Flow Metabol. 9, 341-345 1988

30. Schober O., Meyer G.-J., Duden C., Lauenstein L., Niggemann J., Müller J.-A., Gaab M.R., Becker H., Dietz H., Hundeshagen H. Die Aufnahme von Aminosäuren in Hirntumoren mit der Positronen-Emissionstomographie als Indikator für die Beurteilung von Stoffwechselaktivität und Malignität. Fortschr. Röntgenstr. 147, 503-509, 1987

31. Derlon J-M, Bourdet C, Bustany P, Chatel M, Theron J, Darcel F, Syrota A. ^{11}C-L-Methionine uptake in gliomas. Neurosurgery 25, 720-728, 1989

32. Ishiwata K, Vaalburg W, Elsinga PH, Paans AMJ, Woldring MG. Comparison of L-[1-11C]-Methionine and L-Methyl-[11C]-Methionine for measuring in vivo protein synthesis rates with PET. J. Nucl. Med. 29, 1419-1427, 1988

33. Kirikae M, Diksic M, Yamamoto YL. Transfer coefficients for L-Valine and the rate of incorporation of L-[1-14C]-Valine into proteins in normal adult rat brain. J. Cerebr. Blood

Flow Metabol. 8, 598-605, 1988

34. Kirikae M, Diksik M, Yamamoto YL. Quantitative measurements of regional glucose Utilization and rate of valin incorporation into proteins by double tracer autoradiography in the rat brain tumor model. J. Cerebr. Blood Flow Metabol. 9, 87-95, 1989

35. Vahvelainen M.-L., Oja S.S. Kinetic analysis of phenylalanine-induced inhibition in the saturable influx of tyrosine, tryptophan, leucine and histidine into brain cortex slices from adult and 7-day-old rats. J. Neurochem. 24 , 885-892, 1975

36. Vahvelainen M.-L., Oja S.S. Kinetics of influx of phenylalanine, tyrosine, tryptophan, histidine and leucine into slices of brain cortex from addult and 7-day-old rats. Brain Research 40, 477-488, 1972

37. Pollay M. Regional transport of phenylalanine across the blood-brain barrier. J. Neurosci. Res. 2, 11-19, 1976

38. Casey D.L., Digenis G.A., Wesner D.A., Washburn L.C., Chaney J.E., Hayes R.L., Callahan A.P. Preparation and preliminary tissue studies of optically active D- and L-[11C]-phenylalanine. Int. J. Appl. Rad. Isotop. 32 , 325-330 ,1981

39. Momma S., Aoyagi M., Rapoport S.I., Smith Q.R. Phenylalanine transport across the blood-brain barrier as studied with the in situ brain perfusion technique. J. Neurochem. 48, 1291-1300 (1987)

40. Choi B., Pardridge W.M. Phenylalanine transport at the human blood-brain barrier. J. Biol. Chem. 261, 6536-6541, 1986

41. Bolster JM, Vaalburg W, Paans AMJ, vanDijk TH, Elsinga PH, Zijlstra JB, Piers DA, Mulder NH, Woldring MG, Wynberg H. Carbon -11 labelled Tyrosin to study tumor metabolism by positron emission tomography. Eur. J. Nucl. Med. 12, 321-324, 1986

42. Johnström P, Stone-Elander S, Ericson K, Mosskin M, Bergström M. 11C-labelled glycine Synthesis and preliminary report on its use in the investigation of intracranial tumors using positron emission tomography. Appl. Radiat Isot. 38, 729-734, 1987

43. Tsukiyama T, Hara T, Iio M, Kido G, Tsubokawa T. Preferential accumulation of 11C in-human brain tumors after intravenous injection of 11C-1-pyruvate. Eur. J. Nucl. Med. 12, 244-248, 1986

44. Takeda A., Goto R., Tamemasa O., Chaney J., Digenis G. Biological evaluation of radio-labelled D-methionine as a parent compound in potential nuclear imaging. Radioisotopes 33, 213-217, 1984

45. Tamemasa O., Goto R., Suzuki T. Preferential incorporation of some 14C-labelled D-amino acids into tumor-bearing animals. Gann. 69, 517-523 (1978)

46. Tamemasa O., Goto R., Takeda A., Maruo U. High uptake of 14C-labelled D-amino acids by various brain tumors. Gann. 73, 147-152, 1982

47. Lauenstein L., Meyer G.-J., Sewing K.-F., Schober O., Hundeshagen H. Uptake kinetics of 14C L-leucine and 14C L- and 14C D-methionine in rat brain and incorporation into protein. Neurosurg. Rev. 10, 147-150 (1987)

48. Lauenstein L. Aufnahme und Proteininkorporation von L-Leucin und L- und D-Methionin ins Gehirn der Ratte und in Hirntumoren. Thesis, Medizinische Hochschule Hannover (1988)

49. Schober O, Duden C, Meyer G.-J., Müller JA, Hundeshagen H. Non selective transport of [11C-methyl]-L-, and -D-methionine into a malignant glioma. Eur. J. Nucl. Med. 13, 103-

105, 1987

50. Meyer G.-J., Schober O., Hundeshagen H. Uptake of 11C-D- and L-methionine in brain tumors. Eur J. Nucl. Med. 10, 373-376, 1985

51. Bergström M, Lundqvist A, Ericson K, Lilja A, Johnström P, Langström B, von Holst H, Eriksson L, Blomqvist G. Comparison of the accumulation kinetics of L-[methyl-11C]-methionine and D-[methyl-11C]-methionine in brain tumors studied with positron emission tomography. Acta Radiol. 28, 225-229, 1987

52. Bodsch W., Coenen H.H., Stöcklin G., Takahashi K., Hossmann K.-A. Biochemical and autoradiographic study of cerebral protein synthesis with [18F]- and [14C]fluorophenylalanine. J. Neurochem. 50, 979-983 (1988)

53. Mineura K, Kowada M, Shishido F. Brain tumor imaging with synthesized 18F-fluorophenylalanine and positron emission tomography. Surg. Neurol. 31, 468-9, 1989

54. Murakami M, Takahashi K, Kondo Y, et al. 2-18F-phenylalanine and 3-18F-tyrosine: Synthesis and preliminary data of tracer kinetics. J. Labelled Comp. Radiopharm. 25, 773-782, 1988

55. Murakami M, Takahashi K, Kondo Y, Mizusawa S, Nakamichi H, Sasaki H, Hagami E, Iida H, Kanno I, Miura S, Itoh I, Uemura K. The slow metabolism of L-[2-18F]-Fluorophenylalanine. J. Labelled Comp. Radiopharm. 27, 245-255, 1989

56. Coenen H.H., Kling P., Stöcklin G. Cerebral metabolism of L-[2-18F]fluorotyrosine, a new tracer of protein synthesis. J. Nucl. Med. 30, 1367-1372 (1989)

57. Wienhard K, Herholz K, Coenen HH, Kling RP, Stöcklin G, Heiss WD. Increased Amino Acid Transport into Brain Tumors Measured by PET of L-[2-^{18}F]Fluorotyrosine. J. Nucl. Med. 32, 1338-1346, 1991

58. Orth F. Untersuchungen zur Aufnahme und zum Protein-Einbau von Phenylalanin und p-Jod-phenylalanin in das Gehirn der Ratte und in Hirntumoren der Ratte. Dissertation. Medizinische Hochschule Hannover 1990.

59. Meyer G.-J., Orth F, Coenen HH, Stöcklin, G, Hundeshagen H. Uptake and protein incorporation of some iodinated amino acids in brain tumors of rats. Eur. J. Nucl. Med. 8, 427, 1989

60. Biersack H.J., Coenen H.H., Stöcklin G., Kashab M., Reichmann K., Bockisch A.SPECT of brain tumors with L-3-[123I]iodo-alpha-methyl-tyrosine (IMT). J. Nucl. Med. 29, 911, 1988

61. Biersack H.J., Coenen H.H., Stöcklin G., Reichmann K., Bockisch A., Oehr P., Kashab M., Rollmann O. Imaging of brain tumors with L-3-[123I]iodo-alpha-methyl-tyrosine and SPECT. J. Nucl. Med. 30, 110-112, 1989

62. Kawai K., Fujibayashi Y., Saji H., Konishi J., Yokoyama A. New radioiodinated radiopharmaceutical for cerebral amino acid transport studies: 3-iodo-alpha methyl-L-tyrosine. J. Nucl. Med. 29, 778 , 1988

63. Langen KJ, Roosen N, Coenen HH, Kuikka JT, Herzog H, Stöcklin G, Feinendegen LE. Brain and Brain Tumor Uptake of -3-[^{123}I]Iodo-α-Methyl Tyrosine: Competition with natural L-Amino Acids. J. Nucl. Med. 32, 1225-1228, 1991

64. Bergström M, Ericson K, Hagenfeldt L. Mosskin M, von Holst H, Noren G, Eriksson L, Ehrin E, Johnström P. PET study of methionine accumulation in glioma and normal brain tissue: Competition with branched amino acids. J. Computer Assist. Tomogr. 11, 208-213,

1987

65. Bustany P., Henry J.F., Sargent T., Zarifian E., Cabanis E., Collard P., Comar D. Local brain protein metabolism in dementia and schizophrenia : in vivo studies with 11C-L-methionine and positron emission tomography. In : Heiss W.D., Phelps M.E. (eds) : Positron Emission Tomography of the Brain. Springer, Berlin, Heidelberg, New York, Tokyo 208-211 (1983)

66. Bustany P., Henry J.F., deRotrou J. Local cerebral metabolic rate of 11C-L-methionine in early stages of dementia, schizophrenia and Parkinson's disease. J. Cereb. Blood Flow Metab. 3, 492-493 (1983)

67. Bustany P., Comar D. Protein synthesis evaluation in brain and other organs in humans by PET. In : Reivich M., Alavi A. (eds) : Positron Emission Tomography. Liss., New Y-ork 183-201 (1985)

68. Dienel GA, Pulsinelli WA, Duffy TE. Regional protein synthesis in rat brain following acute hemispheric ischemia. J Neurochem 35, 1216-1226, 1980

69. Dweyer, BE, Donatoni P, Wasterlain CG. A quantitative autoradiographic method for the measurement of local rates of brain protein synthesis. Neurochem. Res. 7, 563-576, 1982

70. Mies G, Bodsch W, Paschen W, Hossmann KA. Experimental application of triple labeled quantitative autoradiography for measurement of cerebral blood flow, glucose metabolism and protein biosynthesis. In: Heiss WD, Phelps ME, Eds., Positron Emission Tomography of the Brain. Springer Berlin 1983, pp 19-28

71. Anders B. Anwendung von kinetischen Modellen zur quantitativen Berechnung der Proteinsynthese im Gehirn und Optimierung einer Messanordnung zur Aufnahme von Plasma-Aktivitätskurven. Diplomarbeit, MedizinischeHochschule Hannover andTechnische Hochschule Hannover, 1988

72. Ishiwata K, Vaalburg W, Elsinga PH, Paans AMJ, Woldring MG. Metabolic studies with L-[1-14C]-Tyrosine for the investigation of a kinetic model to measure protein synthesis rates wit PET. J. Nucl. Med. 29, 524-529, 1988

73. Diamondstone T.I. Amino acid metabolism II. Metabolism of the individual amino acids. In : Derlin T.M. (Ed.) Textbook of Biochemistry with Clinical Correlations New York 563-626 (1982)

74. Young S.N. The significance of tryptophan, phenylalanine tyrosine, and their metabolites in the nervous system. In : Lajtha A. (ed) : Handbook of Neurochemistry. Plenum Press, New York (1982) Vol. 3, 559-581

75. Ishiwata K, Hatazawa J, Kubota K, Kameyama M, Itoh M, Matsuzawa T, Takahashi T, Iwata R, Ido T. Metabolic fate of L-[methyl-11C]methionine in human plasma. Eur. J. Nucl. Med. 15, 665-669, 1989

76. Hatazawa J., Ishiwata K., Itoh M., Kameyama M., Kubota K. Ido T, Matsuzawa T, Yoshimoto T, Watanuki S, Seo S. Quantitative Evaluation of L-[Methyl-C-11]Methionine Uptake in Tumor Using Positron Emission Tomography. J. Nucl. Med. 30, 1809-1813 (1989)

77. Ericson, K, Blomqvist G, Bergström M, Eriksson L, Stone -Elander S. Application of a kinetic model on the methionine accumulation in intracranial tumors studied with positron emissionm tomography. Acta Radiol. 28, 505-509, 1987

78. Meyer G.-J., Harre R., Orth F., Gaab M.R., Dietz H., Hundeshagen H. In vivo protein

synthesis in human brain tumors measured with 11C-L-methionine. Eur. J. Nucl. Med. 15, 506, 1989

79. Patlak CS, Blasberg RG, Fenstermacher JD. Graphical evaluation of blood to brain transfer constants from multiple time uptake data. J. Cerebr. Blood Flow Metabol. 3, 1-7, 1983

80. Gjedde A. A high and low affinity transport of D-glucose from blood to brain. J. Neurochem. 36, 1463 , 1981

81. Lassen NA, Gjedde A. Kinetic analysis of the uptake of glucose and some of its analogs using the single capillary model. Comments on some points of controversy. In: Lambrecht RM, Rescigno A. Eds., Tracer Kinetics and Physiologic Modelling. Springer, Berlin 1983, pp348-407

82. Blomqvist G. On the construction of functional maps in positron emission tomography. J. Cereb. Blood Flow Metab. 4, 629-632, 1984

83. Bergström M, Muhr C, Ericson K, Lundqvist H, Lilja A, Erikcson L, Blomquivst G, Langström B, Johnström P. The normal pipituary examined with positron emission tomography and (methyl-11C)-L-methionine and (methyl-11C)-D-methionine. Neuroradiol. 29, 221-225, 1987

84. Wienhard K, Wagner R, Heiss WD. PET: Grundlagen und Anwendungen der Positronen-Emissions-Tomographie. Springer Verlag, Heidelberg 1989 p. 38

85. Sato K, Kameyama M, Ishiwata K, Hatazawa J, Katakura R, Yoshimoto T. Dynamic study of methionine uptake in glioma using positron emission tomography. Eur. J. Nucl. Med. 19, 426-430, 1992

^{11}C-METHIONINE AND ^{82}RUBIDIUM UPTAKE IN HUMAN BRAIN TUMORS: COMPARISON OF CARRIER DEPENDENT BLOOD-BRAIN BARRIER TRANSPORT

U. Roelcke, EW. Radü, KL. Leenders

ABSTRACT. In 14 patients with brain tumors uptake of ^{11}C methionine and ^{82}Rb into the tumor tissue showed a significant correlation. This suggests that methionine uptake mainly reflects transport alteration from blood into tumor.

1. Introduction

Using ^{11}C and ^{18}F labeled amino acids and PET, differentiation of amino acid transport from specific metabolism is required to quantitate brain protein synthesis. Several authors suggest that a measure of tracer transport and metabolism can be derived by means of Multiple Time Graphical Plotting [1,2]. On the other hand, it has been demonstrated that in normal brain and brain tumor tissue, uptake of labeled amino acids can be competitively inhibited by unlabeled amino acids. From this it was concluded that tracer uptake mainly reflects transport across the blood-brain barrier [3,4]. Studies combining ^{68}Ga-EDTA and amino acids failed to show a correlation of both tracers [1,5]. However, ^{68}Ga-EDTA enters the brain by passive influx when the barrier is disrupted.

In the present study we compared ^{11}C-methionine and ^{82}Rubidium uptake into brain tumors. ^{82}Rubidium, like potassium, is actively transported across the blood-brain barrier (BBB) only and not further metabolized.

2. Subjects and Methods

We have studied 14 patients (9 gliomas, 1 brain secondary, 1 lymphoma, 3 meningioma) using PET, ^{11}C-methionine (MET) and ^{82}Rubidium (RUB). MET was administered intraveneously over 1 minute (3-10 mCi). Scan duration was 32 mins. RUB was administered over 30 secs (30-40 mCi) and its subsequent brain uptake measured over 6 minutes. Normalized uptake (NU) values were calculated from brain tumor regions as: ROI activity over applied radioactivity per body weight [Bq/ml per Bq/g]. The values were obtained after

B. M. Mazoyer et al. (eds.), PET Studies on Amino Acid Metabolism and Protein Synthesis, 197–199.

tracer uptake had reached a steady state level (MET: from 5 to 32 mins; RUB: from 1 to 6 mins).

3. Results

Tracer uptake into tumor tissue took place shortly after administration (MET: within a few mins; RUB: within 1 min) and remained mostly at a stable level thereafter (MET global ROI: 1.20 +/- 0.11 NU; MET tumor ROI: 1.21 to 7.17 NU. RUB global ROI: 0.21 +/- 0.08 NU; RUB tumor ROI: 0.19 to 6.58 NU). Both tracers showed an uptake in accordance with glioma grades. As expected, highest values were obtained in the meningioma group. A dot diagram (Figure) of NU_{MET} and NU_{RUB} showed a highly significant correlation (NU_{MET} = 0.73 NU_{RUB} + 1.06, n = 14, r = .937, p < .001). When excluding the meningioma patients, the regression was similar (NU_{MET} = 0.71 NU_{RUB} + 1.01) and also significant (n = 11, r = .81, p < .01).

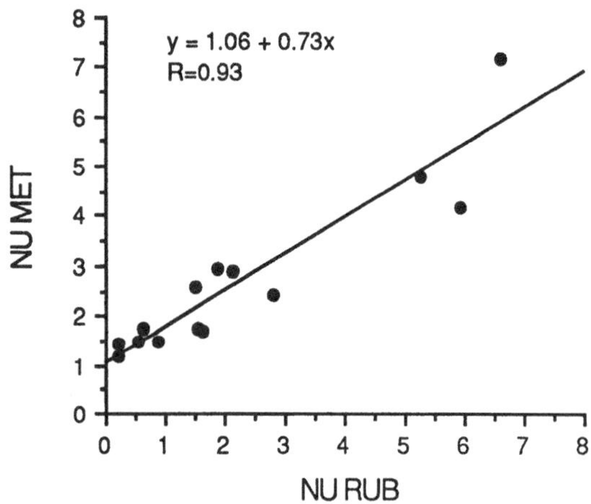

Figure. RUB versus MET uptake in 14 human brain tumors. NU values are expressed in Bq/ml per Bq/g.

4. Conclusions

Our results show, that, irrespective of the presence of cellular BBB disruption, RUB and MET influx from plasma into tumor tissue increase in parallel. Uptake values for both tracers are highest in meningiomas in which no blood-brain barrier is present. Since RUB undergoes BBB transport only and is not further metabolized within the brain, the similar behaviour of MET suggests that RUB and MET uptake into brain tumor tissue is dominated by alterations of the BBB or changes of transport systems.

On the basis of our preliminary findings it is questionable whether protein synthesis in brain tumors can be assessed applying techniques like the Multiple Time Graphical Plotting method, when using the tracer ^{11}C methionine.

5. References

1. Wienhard K, Herholz K, Coenen HH et al. Increased amino acid transport into brain tumors measured by PET of L-(2-^{18}F) fluorotyrosine. J Nucl Med 1991; 32:1338-1346.
2. Hatazawa J, Ishiwata K, Itoh M et al. Quantitative evaluation of L-[methyl-C-11] methionine uptake in tumor using positron emission tomography. J Nucl Med 1989; 30: 1809-1813.
3. Bergström M, Ericson K, Hagenfeldt L et al. PET study of methionine accumulation in glioma and normal brain tissue: Competition with branched chain amino acids. J Comput Assist Tomogr 1987; 11: 208-213.
4. O'Tuama LA, Phillips PC, Strauss LC et al. Two-phase [^{11}C]L-methionine PET in childhood brain tumors. Pediatr Neurol 1990; 6: 163-170.
5. Ericson K, Liljia A, Bergstroem M et al. Positron emission tomography with ([^{11}C]methyl)-L-methionine, [^{11}C]D-glucose, and [^{68}Ga]EDTA in supratentorial tumors. J Comput Assist Tomogr 1985; 9: 683-689.

DISCUSSION

D. Heiss With Rubidium you only show the breakdown of the blood brain barrier. If you get the same results with methionine, you only have investigated tumor with damaged blood brain barrier. I do not believe that leakage of contrast media is a slower process than transport of aminoacids. The leakage you see very fast with Ga-EDTA, as fast as you see the accumulation of aminoacids. If the blood brain barrier is damaged you also have leakage of contrast medium with the first pass during angiography.

A. Planas In the case of a tumor with no brain barrier disruption, can an increase in number of capillaries be sufficient to explain an increase in aminoacid uptake?
D. Heiss There is a dependency to the histological type of the tumors. With a high portion of fibrillary components, the aminoacid accumulation might be lower than in protoplasmatic astrocytomas with high cellular ratio. In low grade astrocytomas the capillary density is not increased compared to normal grey brain tissue.

B. M. Mazoyer et al. (eds.), PET Studies on Amino Acid Metabolism and Protein Synthesis, 201.

TRACERS FOR CLINICAL EVALUATION OF GLIOMAS: A NEUROLOGIST´S VIEW

K.Herholz

Introduction

Positron-labeled amino acids, e.g. ^{11}C-methionine or ^{18}F-fluorotyrosine, may be used for clinical evaluation of gliomas. In that context, their potentials and limitations, as far as they are currently known must been seen in comparison with other PET tracers and other tomographic imaging techniques. With regard to spatial resolution - which is always the most important issue when surgical or stereotactic therapy is considered - PET is clearly superior to other isotope imaging techniques, but cannot compete with computed tomography (CT) and magnetic resonance imaging (MRI). Thus, the clinician needs to know, which of the local functional alterations that can be demonstrated with PET add clinically relevant information to what can already be seen with CT and MRI.

The clinically most important issues are diagnostic sensitivity and specificity, tumor grading, therapy planning and monitoring. Progress with PET may be expected mostly for the last three points. The glucose analogue ^{18}F-2-fluoro-2-deoxy-D-glucose (FDG) has been studied most extensively and seems to be able to improve tumor grading even beyond the limitations of histological grading, but does often not allow to delineate tumors. Amino acids may contribute significantly to all issues, but appropriate clinical studies are still rare, and basic mechanisms of tracer uptake and accumulation are poorly understood. Other tracers, e.g. positron-labeled nucleosides, are also of high interest but in a very early stage of clinical evaluation. In the following chapters the current state of knowledge is reviewed and possibilities for further progress are indicated.

Basic problems of data interpretation

A problem common to most metabolically active tracers, in particular to FDG and ^{11}C-methionine, is the choice of appropriate reference tissue, since they are also taken up by normal brain tissue, more by gray than by white matter, and the uptake may depend on age [1,2]. Thus, because of the variable background, the contrast between tumor and

B. M. Mazoyer et al. (eds.), PET Studies on Amino Acid Metabolism and Protein Synthesis, 203–214.

normal tissue depends on its location and is often less than with tracers (or contrast agents) that simply reflect blood-brain barrier (BBB) damage, because for them background is uniformly low. The literature on FDG data in brain tumors is confounded by the variability that is introduced by using different reference tissues. DiChiro and Brooks [3] advocated the use of white matter, because gliomas arise from white matter and increased metabolic activity due to malignant degeneration is best seen in contrast to the low metabolic rates of normal white matter. However, the delineation of normal white matter on FDG PET images is not trivial, therefore other investigators have used contralateral mirror regions, gray matter regions, or whole brain metabolism. Obviously, the definition of "hypermetabolism" and the value of tumor-to-brain ratios depend on that choice, and comparisons of data from different laboratories must take that into account.

Visual image interpretation of a "hot spot" suggesting highly specific uptake, and mere calculation of tumor-to-brain ratios may also be misleading in quantitative terms. E.g., in malignant gliomas, tumor-to-brain ratios for the metabolically inert tracer ^{68}Ga-EDTA, that merely reflects the BBB leakage, are much higher than for ^{18}F-fluorotyrosine and FDG, although absolute transport rates for the latter are higher [4].

For these reasons, absolute quantification is clearly desirable whenever possible. It requires adequate blood sampling (arterial, or arterialized venous if peripheral extraction rates are low enough to be neglected), scanner calibration, corrections for randoms, scatter and attenuation. Clinical interpretation and interindividual comparisons are facilitated if the physiological variables that determine uptake can be estimated by quantitative models. Use of appropriate models is therefore preferable to reporting mere countrates, even if normalized on body weight and injected dose. However, validation of such models is required and issues of tissue heterogeneity must be solved (see ref. 5 for review). Unfortunately, an ideal tracer, whose uptake is determined by one single physiological process, has not yet been found. We therefore have to deal with complicated situations which make compromises for practicability of clinical application inevitable.

Sensitivity: Is there a lesion ?

For detection of brain lesions, in particular of those mainly located in white matter, is MRI currently the most sensitive method. T_2-weighted images show tumors, focal edema and most other types of focal brain damage as signal-intense areas with high spatial resolution. Tracer methods, in particular blood flow and FDG studies, can detect focal functional impairment that is not seen with MRI. However, functional changes depend on many complex influences, and tumor-to-brain contrast is often low in blood flow and FDG studies. Increased uptake of amino acids has been seen in most, but not all gliomas (Ericson, this volume), and therefore seems to be very sensitive marker of gliomas, although probably not as sensitive as MRI. Amino acids are clearly superior to tracers like ^{68}Ga-EDTA, which is taken up only by tumors with BBB damage, but not in most grade 2 gliomas [4].

From a clinical point of view, MRI fulfils virtually all currently perceived needs in terms of sensitivity to detect a potentially tumorous lesion. Its spatial resolution, which also affects the sensitivity to detect small lesions is clearly superior to all tracer methods. Development of radiotracers for this purpose is therefore of little interest.

Diagnosis: Is it a tumor ?

In quite a number of cases, it is not possible with current imaging techniques to be sure before surgery that a brain lesion seen on CT or MRI is a tumor. For instance, it may be difficult to distinguish between a glioblastoma (or metastasis) and a brain abscess. Both may present as a rim-shaped contrast enhancing lesion with central necrosis on CT and MRI (each with application of contrast material). With PET, I am aware only of FDG studies which demonstrated that both processes may show variable, often enhanced metabolic activity, usually in the contrast enhancing rim [6], which is obviously not helpful in the differential diagnosis. An experimental study [7] indicated, that amino acids may also not be able to differentiate, whereas thymidine was taken up by tumor but no by a brain abscess. In some cases, it may even be difficult to differentiate such a tumor from a recent brain infarct, which may also show central necrosis surrounded by a contrast-enhancing rim. However, that problem is usually solved pragmatically by a control study after a few weeks, since contrast enhancement disappears in infarcts.

Another typical problem is the distinction between a low-grade glioma without BBB damage and a chronic ischemic lesion. Both may present as low density lesions on CT and as signal-intense lesions on T2-weighted MRI and contrast enhancement may be absent. The mass effect, which is the usual clue to diagnose a tumor, may not be seen clearly enough in small tumors. Blood flow and FDG uptake are usually low in both types of lesions. Amino acids are potential candidates to approach that diagnostic problem. As yet, I am not aware of appropriate clinical studies. Thus, the issue of tumor specificity is unsolved, and most previous experience indicates that it will be difficult to achieve significant progress because reactive glia proliferation may also occur after ischemia. It also would require to clarify the exact mechanism of increased uptake of amino acids in most gliomas.

After radiation therapy it is sometimes impossible to distinguish between necrosis and recurrent tumor with CT and MRI, because both types of lesions may show contrast enhancement due to BBB damage and a mass effect. FDG has been favoured for this distinction [8-11] because its uptake is consistently low in necrosis, whereas increased uptake of ^{11}C-methionine has occasionally been seen also in necrotic areas [12].

Diagnosis: What kind of tumor ?

This even more difficult question has repeatedly been attacked with CT and MRI, but has not been completely solved. Current criteria for differentiation are largely derived from location, delineation, and contrast enhancement characteristics. E.g., meningiomas are usually located in continuity with the meninges, have a relatively sharp demarcation, and show intense and rather homogeneous contrast enhancement, whereas gliomas are located within brain tissue, are poorly delineated and show variable and inhomogeneous contrast enhancement. There is abundant literature on atypical appearances that may cause diagnostic problems. So far, attempts to improve diagnostics by measuring relaxation times with MRI [13,14] have not been clearly successful and there are no tumor type-specific radiotracers available. It is unlikely that amino acids will improve this situation substantially.

Grading: How malignant ?

Histological tumor grading relies mainly on cell density and anaplasia, mitotic activity, and other morphological characteristics. Several grading systems have been proposed and the WHO classification [15] is widely used as a standard. Although histology proper obviously cannot be replaced by tomographic imaging, functional correlates of structural changes might be accessible.

For gliomas, the most frequently used criterion of malignancy is BBB damage as visualized by application of contrast agents with CT and MRI. Apparently, this is a phenomenon secondary to malignant dedifferentiation, but it also occurs in non - malignant lesions, such as pilocytic astrocytomas (grade 1), in tumors not of neuroepithelial origin, e.g. meningiomas, which are devoid of a BBB, and in recent inflammatory and vascular lesions. BBB damage also influences cerebral uptake of many radiotracers including ^{99m}Tc-pertechnetate, ^{68}Ga-EDTA, ^{82}Rb, and most probably also of amino acids and nucleic acids. E.g. α-aminoisobutyric acid [16] is widely used as a sensitive indicator of BBB damage in experimental studies. Of course, BBB damage does not influence uptake of freely diffusible tracers, and its influence on glucose transport is probably small compared to the high activity of the glucose transporter enzyme GLUT-1 resulting in K_1 values for FDG in the order of 0.1 ml/100g/min. We have shown that there is little difference in FDG transport rates between gliomas and normal brain [17]. In an evaluation of the use of a radiotracer for grading of gliomas it is important to determine to what extent its uptake depends on BBB damage, which may cause a trivial correlation with glioma grade.

Amino acid transport across the BBB is facilitated by several carriers (see contribution by Lajtha in this volume). The L-system for large neutral amino acids has the highest capacity and is therefore most interesting for clinical applications. L-[1-^{11}C]leucine, L-[1-^{11}C]tyrosine, L-[methyl-^{11}C]- and L-[1-^{11}C]-methionine, and L-2-[^{18}F]fluorotyrosine are transported by this system and may subsequently be incorporated into proteins [see ref. 18 for review]. In most brain tumors, increased transport of amino acids can be inhibited competitively [19]. There is no close correlation of amino acid transport and ^{68}Ga-EDTA diffusion, and distribution volumes of amino acids may well exceed unity [4]. Thus, there is ample evidence that carrier-mediated transport mechanisms are prevailing over passive diffusion in brain tumors, but their relation to tumor grading is not entirely clear. Experimental studies indicate, that amino acid transport is not energy-dependent but may be driven by intra-extracellular amino acid or sodium gradients [20].

Quantitation of protein metabolism is still hampered by the complexity of amino acid metabolism and difficulties in estimating intracellular amino acid concentrations. Thus, although tumor proliferation is most probably associated with increased protein synthesis, that has not yet been demonstrated in human brain tumors in vivo. By measuring uptake of ^{18}F-fluorotyrosine dynamically, we differentiated between transport and metabolism (presumably protein incorporation, as indicated by experimental studies [21]) and found preliminary evidence that the latter may correlate with glioma grade [4]. Studies with ^{11}C-methionine [12,22,23] had also suggested that amino acid uptake is related to tumor grade.

In principle, nucleoside incorporation as a measure of DNA synthesis rate is probably the most interesting parameter for grading. [2-^{11}C]-thymidine [24] appears to be a promising tracer for that purpose. Other nucleosides, like thymidine labeled at the methyl group [25] and ^{18}F-5-fluoro-2´-deoxyuridine [26,27] have also been introduced. DNA incorporation has been demonstrated in-vitro, but significant systemic and local metabolism occurs in-vivo, currently preventing quantitation of DNA synthesis rates [28]. The normal BBB has only a low transport capacity for nucleoside compounds [29] and the influence of BBB alterations on the differential uptake of nucleosides in tumors and brain has not been fully elucidated. It should also been kept in mind that even glioblastomas usually have a relatively low growth fraction compared with metastases from other malignant tumors [30], and signal to noise ratios may therefore not be favourable for measuring DNA synthesis rates.

The most extensively studied tracer for tumor grading is FDG. A correlation between tumor grade and FDG uptake was reported by DiChiro and coworkers [31,32], but another group found no correlation [33]. We saw a significant but not very close correlation between FDG phosphorylation, which mainly determines uptake, and glioma grade (Kendall´s τ_B=0.40, p=0.01 in 26 patients [17]). Since increased FDG uptake in gliomas correlates with lactate production [34], it may be regarded as an indicator of non-oxidative glycolysis, which is increased in malignant tissue [35]. Histological tumor grading has its own limitations, due to some subjectivity in the evaluation of anaplastic features and the examination of only a small subset of tissue, that may not be representative for the whole, often inhomogeneous tumor. Its clinical relevance is based on the correlation with prognosis, since the term "malignant" basically indicates rapid fatal outcome if therapy is not available. Therefore, the correlation between tracer uptake and prognosis is clinically most relevant. Several studies indicated that the relation between FDG uptake in tumor and normal brain is of high prognostic significance [36-38]. In our own experience [39] the ratio of highest tumor metabolism to contralateral brain metabolism is significantly related to prognosis within homogeneous groups of low grade and high grade gliomas. Thus, FDG PET apparently adds prognostic information to histopathological examination. The prognostic value of tumor-to-brain metabolic ratios is not only related to tumor metabolism, but is also due to the decline of brain metabolism that occurs when the tumor infiltration or edema impairs brain function [40,41].

FDG PET has some important limitations: first differentiation between tumor and normal brain tissue is often very diffucult because metabolic rates may be very similar; thus CT or MRI scans are needed for comparison. Second, the exact biochemical links between increased glycolysis and malignant tissue dedifferentiation are not known, although the association has already been described several decades ago by Warburg (reviewed in ref. 35). Third, the accuracy of measuring glucose metabolic rates with FDG in tumors is uncertain because the lumped constant needed for that calculation may vary in tumors (see ref. 5 for review). In spite of these limitations is FDG currently the only tracer of broadly documented clinical prognostic significance while for all other tracers, including amino acids, this challenge is still to be met.

Therapy planning: How large? Where are the borders ?

This is an extremely difficult question because most gliomas infiltrate normal brain diffusely and do not exhibit sharp borders. Cellular conglomerates suspect of incipient tumor growth can even be found beyond apparent tumor borders in histopathological specimens (for details see ref. 42). For that reason, probably, gliomas of grade 2 or higher can generally not be cured even with radical surgical procedures. Consequently, defining tumor borders with neuroimaging techniques is a task impossible to be solved perfectly. The issue really is, to delineate a tumor so that a surgical or local radiotherapeutic therapy following this outline will maximize the relapse-free interval without destroying intact or functionally indispensable brain tissue. This is a complex issue that has not yet been adressed with any positron tracer. The current state of knowledge is, that CT and MRI are not very reliable in terms of tumor extent. Contrast enhancing masses (if no previous therapy was applied) generally reliably indicate compact tumor tissue, but non-enhancing hypodense or signal-intense areas may represent edema or tumor infiltration, and the latter may also extend beyond all abnormalities seen on CT and MRI [43]. Experience at the Karolinska Hospital in Stockholm [12,44] indicates that tumor delineation with ^{11}C-methionine is better than with CT and Mineura et al. [45] report correct autopsy-confirmed tumor delineation in a case of gliomatosis cerebri. Good correspondence was also seen between the extent of malignant tumors, Ki-67 labeling, and fluorodeoxyuridine uptake in stereotactic biopsies [26].

The spatial relation of tumors to functionally important areas, such as primary motor cortex and language related areas, may sometimes be difficult to determine with CT or MRI because of anatomical variability and its distortion by tumor and edema or unclear hemispheric dominance. Focal functional activiation by appropriate tasks and PET examinations of blood flow or glucose metabolism may answer such questions. This potential has, however, not yet been demonstrated in systematic clinical studies of gliomas. Most activation studies (see ref 46 for review) show effects by averaging over several subjects, and the reliability and accuracy of localization of specific brain functions with these methods in individual patients remains to be demonstrated.

In summary, within the inherent limitations of spatial resolution with PET amino acids appear promising with respect to tumor delineation, although the issue has not yet been studied in all relevant aspects. Similarly, more clinical functional activiation studies are needed to explore their potential in glioma patients.

Therapy monitoring: Does it respond ?

Radiotherapy and chemotherapy with certain alkylating agents, e.g. ACNU, BCNU and CCNU, is of proven, but limited effectivity in malignant gliomas. Median survival times are at best only few years, and only about 10% of patients survive up to 5 years in most studies (see refs. 47 and 48 for review). By current means, success or failure of treatment is often evident only after completion of a full course of radiotherapy (with typically 50 to 60 Gy tumor dose in fractions of 2 to 3 Gy) and several cycles of chemotherapy. A technique to predict tumor response before or early during treatment would permit to tailor therapy to individual responsiveness.

Work by Bergström et al. [19,49] in pituitary adenomas indicates that reduced uptake of ^{11}C-methionine occurs early during effective drug treatment. Comparable data in gliomas are not available. FDG has been used several times to monitor therapeutic effects in gliomas. Rozental et al. [50,51] report an increase of glucose metabolism on the first day after interstitial irradiation or intensive polychemotherapy, followed later by a decline to or below baseline. Langen et al. [52] found highly variable responses after intra-arterial therapy with ACNU. We saw a distinct decrease of glucose metabolism (by 29 to 64%) few weeks after radio- and chemotherapy in some clinically responding, metabolically highly active medulloblastomas in children, whereas poorly responding tumors showed much less decrease (Holthoff et al., submitted). Ogawa et al. [53] described a decline of glucose metabolism by 14% on average in gliomas within one month after radiochemotherapy. In patients with non-progressing glioblastomas, we observed generally little changes in tumor glucose metabolism in FDG PET follow-up studies at intervals of approx. 6 months, whereas progressing tumors were characterized by increasing tumor and declining brain metabolism (unpublished data).

In summary, the issue of how to monitor therapy in gliomas is not yet settled. Available data indicate that amino acids may be well suited for that purpose, whereas FDG is probably only useful in tumors with initially high metabolism. An experimental study [54] indicates that methionine and thymidine demonstrate response to radiotherapy more clearly than FDG.

Perspectives

Amino acids are promising agents with regard to tumor delineation and for monitoring of therapeutic response. They may also be useful for glioma grading, although current evidence favours FDG for evaluation of prognosis. More data are needed to elucidate the significance of changes at the blood-brain barrier for tracer uptake and its relation to tumor metabolism. For FDG, current data indicate that BBB changes are of little relevance and increased uptake is mainly due to increased non-oxidative glycolysis, whereas amino acid uptake appears to be governed by changes at the BBB. Very little is known about the factors determining the uptake of nucleosides in gliomas, which could possibly be used to determine DNA synthesis rate as a measure of tumor proliferation. In addition to experimental studies, more detailed clinical studies with histopathological confirmation of tumor type, standardized therapy protocols and clinical follow-up are urgently needed to evaluate the role of PET in neuro-oncology.

References

1. Chugani HT, Phelps ME, Mazziotta JC. Positron emission tomography study of human brain functional development. Ann Neurol 1987; 22: 487-497.

2. O'Tuama LA, Phillips PC, Smith QR et al. L-methionine uptake by human cerebral cortex: Maturation from infancy to old age. J Nucl Med 1991; 32:16-22.

3. Di Chiro G, Brooks RA. PET - FDG of untreated and treated cerebral gliomas. J Nucl Med 1988; 29:421-422.

4. Wienhard K, Herholz K, Coenen HH et al. Increased amino acid transport into brain tumors measured by PET of L-(2-^{18}F)fluorotyrosine. J Nucl Med 1991; 32: 1338-1346.

5. Herholz K, Wienhard K, Heiss WD. Validity of PET studies in brain tumors. Cereb Brain Metab Rev 1990; 2:240-265.

6. Sasaki M, Ichiya Y, Kuwabara Y et al. Ringlike uptake of [F-18] fdg in brain abscess - a PET study. J Comp Assist Tomogr 1990; 14:486-487.

7. Grossman SA, Eller S, Dick J, Burch PA, Yang K. Thymidine (T), Leucine (L) and 2Deoxyglucose (2DG) incorporation in brain-tumors and abscesses - a quantitative autoradiographic (QAR) study with implications for PET scans. Proc Am Assoc Cancer Res 1988; 29:516-516.

8. Patronas NJ, Di Chiro G, Brooks RA, Delapaz RL, Kornblith PL, Smith BH, Rizolli HV, Kesseler R, Manning RG, Channing M, Wolf AP, O'Conner, CM. Work-in-progress: FDG and PET in the evaluation of radiation necrosis of the brain. Radiology 1982; 144:885-889.

9. Doyle WK, Budinger TF, Valk PE, Levin VA, Gutin PH. Differentiation of cerebral radiation necrosis from tumor recurrence by 18-FDG and Rb-82 positron emission tomography. J Comput Assist Tomogr 1987; 11:563-570.

10. DiChiro G, Oldfield E, Wrigth DC et al. Cerebral necrosis after radiotherapy and/or intraarterial chemotherapy for brain tumors - PET and neuropathologic studies. American J Roentgenology 1988; 150:189-197.

11. Kowada M. Clinical-value of PET with F-18-fluorodeoxyglucose and L-methyl-C-11-methionine for diagnosis of recurrent brain-tumor and radiation-injury. Acta Radiologica 1991; 32:197-202

12. Mosskin M, Vonholst H, Bergstrom M et al. Positron emission tomography with C-11-Methionine and computed tomography of intracranial tumors compared with histopathologic examination of multiple biopsies. Acta Radiologica 1987; 28:673-681.

13. Chatel M, Darcel F, Certaines De J, Benoist L, Bernard AM. T1 and T2 proton-NMR relaxation-times in-vitro and human intracranial-tumors. J Neuro-Oncology 1986; 3:315-321.

14. Komiyama M, Yagura H, Baba M et al. MR imaging: possibility of tissue characterization of brain tumors using T1 and T2 values. AJNR 1987; 8:65-70.

15. Zülch KJ. Histology typing of tumours of the central-nervous-system. World Health Organization, Geneva, 1979.

16. Blasberg RG, Fenstermacher JD, Patlak CS. Transport of α-aminoisobutyric acid across brain capillary and cellular membranes. J Cereb Blood Flow Metab 1983; 3:8-32.

17. Herholz K, Rudolf J, Heiss WD. FDG transport and phosphorylation in human gliomas measured with dynamic PET. J Neuro-Oncology 1992; 12:159-165.

18. Vaalburg W, Coenen HH, Crouzel C et al. Amino-acids for the measurement of protein-synthesis invivo by PET. Nuclear Medicine and Biology-International Journal of Radiation Applications and Instrumentation Part B 1992; 19:227-237.

19. Bergström M, Muhr C, Lundberg PO et al. Rapid decrease in amino acid metabolism in prolactin-secreting pituitary adenomas after bromocriptine treatment: a PET study. J Computer Assisted Tomography 1987; 11(5):815-819.

20. Greenwood J, Hazell AS, Pratt OE. The transport of leucine and aminocyclopentane carboxylate across the intact, energy-depleted rat blood-brain-barrier. J Cereb Blood Flow Metab 1989; 9:226-233.

21. Coenen HH, Kling P, Stöcklin G. Cerebral metabolism of L-[2-F-18]fluorotyrosine, a new PET tracer of protein-synthesis. J Nucl Med 1989; 30:1367-1372.

22. Bustany P, Chatel M, Delon JM, Darcel F, Sgouropoulos F, Syrota A. Brain tumor protein synthesis and histological grades: a study by positron emission tomography with C-11-L-Methionine. J Neuro-Oncology 1986; 3:397-404.

23. Derlon JM, Bourdet C, Bustany P et al. [C-11]-l-methionine uptake in gliomas. Neurosurgery 1989; 25:720-728.

24. Vanderborght T, Labar D, Pauwels S, Lambotte L. Production of [2-C-11]thymidine for quantification of cellular proliferation with PET. Applied Radiation and Isotopes-International Journal of Radiation Applications and Instrumentation Part A 1991; 42:103-104.

25. Poupeye E, Counsell RE, Deleenheer A, Slegers G, Gocthals P. Synthesis of C-11-labelled thymidine for tumor visualization using positron-emission-tomography. Applied Radiation and Isotopes-International Journal or Radiation Applicationa and Instrumentation Part A 1989; 40:57-61.

26. Thomas DGT, Gill SS, Wilson CB, Darling JL, Parkins CS. Use of relocatable stereotaxic-frame to integrate positron-emission-tomography and computed-tomography images - application in human-malignant brain-tumors. Stereotactic and Functional Neurosurgery 1990; 54-5:388-392.

27. Tsurumi Y, Kameyama M, Ishiwata K et al. F-18-fluoro-2'-deoxyuridine as a tracer of nucleic acid metabolism in brain tumors. J Neurosurgery 1990; 72:110-113.

28. Shields AF, Lim K, Grierson J, Link J, Krohn KA. Utilization of labeled thymidine in DNA synthesis - studies for PET. J Nucl Med 1990; 31:337-342.

29. Pardridge WM. Recent advances in blood-brain-barrier transport. Annual Review of Pharmacology and Toxicology 1988; 28:25-39.

30. Burger PC, Shibata T, Kleihues P. The use of the monoclonal antibody Ki-67 in the identification of proliferating cells: application to surgical neuropathology. Am J Surg Pathology 1986; 10:611-617.

31. DiChiro G, DeLaPaz RL, Brooks RA et al. Glucose utilization of cerebral gliomas measured by 18-F-fluorodeoxyglucose and positron emission tomography. Neurology 1982; 32:1323-1329.

32. DiChiro G. Positron-emission-tomography using [F-18]-fluorodeoxyglucose in brain tumors - a powerful diagnostic and prognostic tool. Investigative Radiology 1987; 22:360-371.

33. Tyler JL, Diksic M, Villemure JG et al. Metabolic and hemodynamic evaluation of gliomas using positron emission tomography. J Nucl Med 1987; 28:1123-1133.

34. Herholz K, Heindel W, Luyten PR et al. In-vivo imaging of glucose consumption and lactate concentration in human gliomas. Ann Neurol 1992; 31:319-327.

35. Warburg O. On the origin of cancer cells. Science 1956; 123:309-314.

36. Patronas NJ, DiChiro G, Kufta C et al. Prediction of survival in glioma patients by means of PET. J Neurosurg 1985; 62:816-822.

37. Alavi JB, Alavi A, Chawluk J et al. Positron emission tomography in patients with glioma - a predictor of prognosis. Cancer 1988; 62: 1074-1078.

38. Kim CK, Alavi JB, Alavi A, Reivich M. New grading system of cerebral gliomas using positron emission tomography with F-18-fluorodeoxyglucose. J Neuro-Oncology 1991; 10:85-91.

39. Herholz K, Friedrichs B, Jeske J, Heiss WD. Prognostic significance of positron emission tomography with F-18-fluorodeoxyglucose in gliomas. J Cancer Res Clin Oncol 1992; 118 (Suppl.): R119.

40. Delapaz RL, Patronas NJ, Brooks RA et al. PET study of suppression of gray-matter glucose-utilization by brain tumors. AJNR 1983; 4:826-829.

41. Jeske J, Herholz, K, Heindel W, Heiss, WD. Stoffwechseluntersuchungen an Gliomen mit der Positronen-Emissions-Tomographie und der Phosphor-31-MR-Spektroskopie in Diagnostik und Therapieplanung. Onkologie 1989; 12 (Suppl.1):42-45.

42. Russell DS, Rubinstein LJ. Pathology of tumours of the nervous-system. 5th Edition, Edward Arnold, London, 1989:421-448.

43. Greene GM, Hitchon PW, Schelper RL, Yuh W, Dyste GN. Diagnostic yield in ct-guided stereotactic biopsy of gliomas. J Neurosurgery 1989; 71:494-497

44. Ericson K, Lilja A, Bergström M et al. Positron emission tomography with 11-C-methyl-L-methionine, 11-C-D-Glucose and 68-Ga-EDTA in supratentorial tumors. J Comput Assist Tomogr 1985; 9:683-689.

45. Mineura K, Sasajima T, Kowada M, Uesaka Y, Shishido F. Innovative approach in the diagnosis of gliomatosis cerebri using carbon-11-l-methionine positron emission tomography. J Nucl Med 1991; 32:726-728.

46. Fox PT. Functional brain mapping with positron emission tomography. Seminars in Neurology 1989; 9:323-329.

47. Stewart DJ. The role of chemotherapy in the treatment of gliomas in adults. Cancer Treatment Reviews 1989; 16: 129-160

48. Mahaley MS.
Neuro-oncology index and review (adult primary brain tumors): Radiotherapy, chemotherapy, immunotherapy, photodynamic therapy. J Neuro-Onc 1991; 11:85-147

49. Bergström M, Muhr C, Lundberg PO, Langstrom B. PET as a tool in the clinical-evaluation of pituitary adenomas. J Nucl Med 1991; 32:610-615.

50. Rozental JM, Levine RL, Nickles RJ, Dobkin JA. Glucose uptake by gliomas after treatment. A positron emission tomographic study. Arch Neurol 1989; 46:1302-1307.

51. Rozental JM, Levine RL, Mehta MP et al. Early changes in tumor metabolism after treatment: the effects of stereotactic radiotherapy. Int J Rad Oncol Biol Phys 1991; 20:1053-1060.

52. Langen KJ, Roosen N, Kuwert T et al. Early effects of intra-arterial chemotherapy in patients with brain-tumors studied with PET - preliminary-results. Nuclear Medicine Communications 1989; 10:779-790.

53. Ogawa T, Uemura K, Shishido F et al. Changes of cerebral blood flow and oxygen and glucose metabolism following radiochemotherapy of glioma - a PET study. J Comput Assist Tomography 1988; 12:290-297.

54. Kubota K, Ishiwata K, Kubota R et al. Tracer feasibility for monitoring tumor-radiotherapy - a quadruple tracer study with fluorine-18-fluorodeoxyglucose or fluorine-18-fluorodeoxyuridine, l-[methyl-C-14]methionine, [6-H-3]thymidine, and Ga-67. J Nucl Med 1991; 32:2118-2123.

PET STUDIES OF AMINO ACID METABOLISM: INTEGRATION IN CLINICAL ROUTINE AND CURRENT RESEARCH ON INTRACRANIAL TUMOURS

Kaj Ericson

It is at present not possible to measure the cerebral rate of protein synthesis. This is dependent on the fact that amino acids not only are involved in the protein synthesis but also in other metabolic processes. These processes are not always entirely known, and many of the pathways and metabolic products cannot be quantified. This makes the kinetic models unnecessarily complex and inexact for many amino acids. However, models have been developed for methionine, leucine and phenylalanine (1-5). At Karolinska Hospital, we have been working with methionine since 1983 and have no experience with the other two tracers.

Methionine is quite simple to label, at least in the methyl group. It was therefore natural to test this substance for brain tumour studies with positron emission tomography (PET). Already from the very start 11-C-methionine proved to be a very useful tracer for studies of intracranial tumours (6). It was hoped that this tracer would make it possible to measure the rate of protein synthesis in normal and pathologic brain tissue. Most investigators used methionine labelled in the methyl group. However, methionine is a powerful methyl donor, and the label may therefore be lost to other metabolic processes than protein synthesis. Methionine may also be labelled in the carboxyl group, but this is a more complicated process. Further studies have shown that methionine is transported from the blood to the brain cells by stereospecific facilitated diffusion (7-9). It has been claimed that a proper determination of the protein synthesis cannot be performed with methionine labelled in the methyl group (4, 10). Regardless of this fact methionine remains the most effective tracer for brain tumours, so far. About 80-90 % of brain tumours show an increased accumulation of methionine as compared with normal brain tissue (11). There is a rough correlation with the degree of malignancy. Tumours without an increased accumulation are always astrocytomas grade II. It has been claimed that 18-F-fluorodeoxyglucose (FDG) is a better tracer for tumours and particularly for malignancy grading (12-16). This is not entirely true. It is, however, probably true that FDG is superior to methionine in the differentiation between radiation necrosis and malignant tumour. The beauty of methionine as a tracer (like

B. M. Mazoyer et al. (eds.), PET Studies on Amino Acid Metabolism and Protein Synthesis, 215–221.

phenylalanine and leucine) is that the accumulation is not hindered by the blood brain barrier.

We can measure the metabolic rate of methionine and other kinetic parameters. This is a time-consuming procedure which requires adequate data of the plasma concentration of the cold tracer and of the radioactivity in plasma. Blood samples, preferably arterial, must be obtained for a good input function. It is doubted if these calculations really are necessary in the clinical routine use of PET for tumour studies.

The ratio between activity in tumour tissue and white matter has been measured in static images accumulated during the whole measuring period and in metabolic maps of the metabolic rate (Fig.1). For clinical purposes the metabolic maps add little information. Even when comparing PET studies of the same patient the calculation of ratios seems better than using metabolic maps due to the rather large variation of the metabolic values from one time to the other.

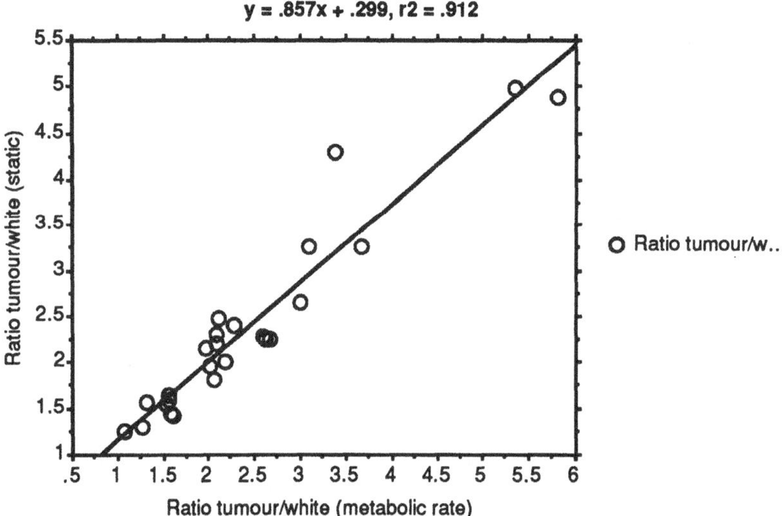

Fig, 1: Correlation between uptake ratios measured in static images and in metabolic maps.

A "static" PET image with methionine as a tracer typically conists of a series of measurements to cover the whole examination time of 20-50 minutes. This series of measurements may then be manipulated to obtain some dynamic information. A high blood volume will, for example, be visualized in the first minutes of the study while the slower incorporation will show in the later parts.

The accumulation of methionine in the cells is dependent on the concentration of the tracer in the extracellular fluid. Thus, if there is a disrupture of the blood-brain barrier large amounts of the tracer leaks out to the extracellular space and large amounts may also be transported into the intracellular space. This is one explanation to the increased accumulation in the more malignant tumours. However, there must also be a difference in the real metabolism of methionine. The presence of a blood-brain barrier disrupture makes the kinetic models even more complicated. It is very unlikely that a kinetic model for normal brain tissue will be valid for tumour tissue with a pathological barrier.

We have followed a large number of tumours before and after treatment. The natural course of astrocytomas grade II seems to be a change over to an astrocytoma grade III or in some cases maybe a glioblastoma. This change may take long time but may also occur rapidly. This change in malignancy grade is reflected in the methionine accumulation. Thus, PET is an excellent tool to diagnose malignification. FDG can also be used for this purpose. A change from low uptake in the grade II astrocytoma to a high uptake would suggest malignification. When examining irradiated tumours there is another factor to consider. The irradiation may cause a disrupture of the blood-brain barrier. This results in leakage of methionine to the extracellular space and enables an increased methionine accumulation in the cells. It is difficult or impossible to differentiate this uptake from that occuring in a malignant tumour. This is true even when a radiation necrosis has developed. The necrotic area does not accumulate methionine to any great extent, but increased accumulation will be seen in the tissue surrounding the necrosis mimicking a malignant tumour with central necrosis. FDG is probably a better tracer for this differentiation showing decreased accumulation in radiation necrosis and increased in malignant tumour (17).

Generally, the metabolic rate of methionine is reduced towards more normal values 1-2 years after irradiation. This seems to reflect the radiation effect on the metabolism. This effect may last for many years. We have patients with astrocytomas grade II that have been followed since 1983 and still are without signs of recurrency or malignant transformation. In most cases, however, the metabolic rate is eventually again increased, often signifying malignification. The effect of stereotactic irradiation with the multi-cobalt unit (the Gamma knife) on the methionine accumulation has also been studied in a few cases. A reduction of the metabolic rate of 30-50 % was seen at reexamination 2-3 years after irradiation.

PET is of value also in the follow-up of operated tumour patients. Provided the tumour has an increased methionine uptake at the preoperative PET scan, residual tumour may be detected with a high degree of accuracy in the early post-operative period when findings at CT and MR are equivocal.

For reasons unknown oligodendroglioms show a very high uptake of methionine.

Biopsy is often performed of brain tumours to ascertain the histologic character before determining the treatment strategy. Stereotactic biopsies guided by CT or MR are not rarely inconclusive. This is particularly true for low-grade astrocytomas and is partly due to difficulties in delineating the tumours. With PET it is often much easier to determine the metabolically most active tumour part and direct the biopsy needle to this area (18).

A few words should be mentioned on the histological examination of brain tumours. It seems that the grading of tumours is to some extent subjective. The distribution of different tumour types vary from one country to another in a way that cannot be explained by demographic differences. This is a serious problem since treatment results cannot readily be compared. PET may be of value also in this context. A diagnosis based on the metabolic characteristics on PET might be a better correlate to prognosis.

Little is known of the methionine uptake in non-tumourous lesions. Cavernous hemangiomas, for example, show a normal or slightly reduced methionine accumulation (19).

Another amino acid has been tested. Glycine (20) is a small amino acid which lacks a transport system over the blood-brain barrier. Well over the barrier, however, the tracer is most eagerly accumulated intracellularly . The accumulation reflects both the barrier disrupture and the intracellular uptake, and glycin is therefore a much more effective tracer for barrier disruptures than, for example, Ga-EDTA. However, for delieation of tumours, methionine is far superior to glycine.

There are, no doubt, many tracers behaving in very much the same way as methionine. They can be used for tumour detection and delineation regardless of minor differences in the distribution. The use of PET and amino acids represents a break-through in the evaluation of brain tumours. Tmours may now be evaluated with regard to their metabolic characteristics rather than to morphology. This will influence also the treatment since response to treatment can be monitored in a new way. Although improved diagnosis in itself is an advantage, it seems even more important to find new treatment strategies for brain tumours. PET may help us in this endeavour.

References

1. Barrio J, Keen R, Chugani H, Ackerman R, Chugani D, Phelps M (1983) L-(1-C-11)phenylalanine for the determination of cerebral protein synthesis rates in man with positron emission tomography. J Nucl Med 24: P70

2. Bustany P, Henry J, Sargent T, Zarifian E, Cabanis E, Collard P, Comar D (1983) Local brain protein metabolism in dementia and schizophrenia: In vivo studies with 11-C-L-methionine and positron emission tomography. *In:* Heiss WD Phelps ME. (eds) Positron Emission Tomography of the Brain. Springer-Verlag, New York, 208-11

3. Bustany P, Henry J, de Rotrou J, Signoret P, Cabanis E, Zarifian E, Ziegler M, Derlon J, Crouzel C, Soussaline F, Comar D (1985) Correlations between clinical state and positron emission tomography measurement of local brain protein synthesis in Alzheimer´s dementia, Parkinson´s disease, schizophrenia, and gliomas. *In:* Greitz T. (eds) The Metabolism of the Human Brain Studied with Positron Emission Tomography. Raven Press, New York, 249-51

4. Phelps ME, Barrio JR, Huang SC, Keen RE, Chugani H, Mazziotta JC (1984) Criteria for the tracer kinetic measurement of cerebral protein synthesis in humans with positron emission tomography. Ann Neurol

5. Smith C, Davidsen L, Deibler G et al (1980) A method for the determination of local rates of protein synthesis in brain. Trans Am Soc Neurochem 11: 94

6. Bergström M, Collins P, Ehrin E, Ericson K, Eriksson L, Greitz T, Halldin C, von Holst H, Långström B, Lilja A, Lundqvist H, Någren K (1983) Discrepancies in brain tumor extent as shown by computed tomography and positron emission tomography using 68-Ga-EDTA, 11-C-glucose, and 11-C-methionine. J Comput Assist Tomogr 7: 1062-66

7. Bergström M, Lundqvist H, Ericson K, Lilja A, Johnström P, Langström B, von Holst H, Eriksson L, Blomqvist G (1987) Comparison of the accumulation kinetics of L-(methyl-11-C)methionine and D-(methyl-11-C)methionine in brain tumors studied with positron emission tomography. Acta Radiol 28: 225-29

8. Bergström M, Ericson K, Hagenfeldt L, Mosskin M, von Holst H, Norén G, Eriksson L, Ehrin E, Johnström P (1987) Methionine accumulation in glioma and normal brain tissue studied with positron emission tomography - competition with branched amino acids. J Comput Assist Tomogr 11: 208-213

9. Ericson K, Blomqvist G, Bergström M, Eriksson L, Stone-Elander S (1987) Application of a kinetic model on the methionine accumulation in intracranial tumours studied with positron emission tomography. Acta Radiol 28: 505-509

10. Phelps M, Barrio J, Huang S, Keen R, Chugani H, Mazziotta J (1985) Measurement of cerebral protein synthesis in man with positron computerized tomography: Model, assumptions, and preliminary results. *In:* Greitz T. (eds) The Metabolism of the Human Brain Studied with Positron Emission Tomography. Raven Press, New York, 215-32

11. Ericson K, Bergström M, Collins V, Ehrin E, Eriksson L, von Holst H, Mosskin M, Norén G (1986) Positron emission tomography in neuroradiology . Integration of a research tool in clinical routine. *In:* Valk J (ed) XIII Congress of the European Society of Neuroradiology. Elsevier Science Publishers, Amsterdam, 245-251

12. Patronas N, Di Chiro G, Brooks R, De Laz R, Kornblith P, Smith B, Rizzoli H, Kessler R, Manning R, Channing M, Wolf A, O'Connor C (1982) Work in progress: 18-F-fluorodeoxyglucose and positron emission tomography in the evaluation of radiation necrosis of the brain. Radiology 144: 885

13. Di Chiro G, De La Paz R, Brooks R (1982) Glucose utilization of cerebral gliomas measured by ^{18}F-fluorodeoxyglucose and PET. Neurology 32: 1323-1329

14. Di Chiro G, Brooks R, Sokoloff L (1983) Glycolytic rate and histologic grade of human cerebral gliomas: a study with ^{18}F-fluorodeoxyglucose and positron emission tomography. *In:* Heiss W-D Phelps M (eds) Positron emission tomography of the brain. Springer-Verlag, Berlin-Heidelberg-New York, 181-191

15. Di Chiro G (1987) Positron emission tomography using [18-F]fluorodeoxyglucose in brain tumors - a powerful diagnostic and prognostic tool. Invest Radiol 22: 360-371

16. Di Chiro G (1991) Which PET radiopharmaceutical for brain tumors? J Nucl Med 32: 1346-8

17. Di Chiro G, Oldfield E, Wright D (1988) Cerebral necrosis after radiotherapy and/or intraarterial chemotherapy for brain tumors: PET and neuropathologic studies. AJR 150: 189-197

18. Mosskin M, von Holst H, Bergström M, Collins V, Eriksson L, Johnström P, Noren G (1987) Positron emission tomography with 11-C-methionine and computed tomography of intracranial tumours compared with histopathologic examinations of multiple biopsies. Acta Radiol 28: 673-81

19. Ericson K, von Holst H, Mosskin M, Bergström M, Lindqvist M, Norén G, Eriksson L (1986) Positron emission tomography (PET) in the examination of cavernous angiomas of the brain. Case reports. Acta Radiol (Diagn) 27: 379-383

20. Johnström P, Stone-Elander S, Ericson K, Mosskin M, Bergström M (1987) 11-C-labelled glycine: Synthesis and preliminary report on its use in the delineation of intracranial tumors using positron emission tomography. In J Appl Instrum 38(9): 729-734

UPTAKE OF [^{11}C]METHIONINE IN NON-BRAIN TUMORS

S. LESKINEN-KALLIO

ABSTRACT. L-[methyl-^{11}C]Methionine ([^{11}C]methionine) has been successfully used to image brain and lung tumors. In this study, the uptake of [^{11}C]methionine in lymphoma, breast cancer and head and neck cancer was studied by positron emission tomography (PET). [^{11}C]Methionine accumulated avidly in 57/61 (93%) tumors. The accumulation of [^{11}C]methionine appears to associate with the proliferative rate of breast cancer. Four different analysis methods of the uptake of [^{11}C]methionine in cancer were studied. The kinetic analysis of the uptake of [^{11}C]methionine in cancer can be substituted with 5 minutes emission scanning 25-40 minutes after the injection, when the radioactivity concentration in the tumor is adjusted for the injected dose and body surface area.

1. Introduction

L-[methyl-^{11}C]Methionine ([^{11}C]methionine) has been reported to image human brain tumors and lung cancer [1,2]. Furthermore, there is some evidence that its uptake may be associated with the histological grade of cancer [1,2]. Lymphoma, breast cancer, and head and neck cancer are often potentially curable malignant diseases. If the treatment of cancer is to be successful, the extent of the primary tumor and metastases and the degree of malignancy needs to be known with accuracy for planning of appropriate treatment strategies.

Assessment of the uptake rate of [^{11}C]methionine from the blood to the tumor may be the most precise way of analysis. If the assumptions of the Patlak model [3] are met, an influx constant (K_i) can be determined for tracer transfer from blood to tumor. This kinetic analysis requires the determination of the concentration of [^{11}C]methionine in plasma for the input function and dynamic emission scanning for 40 minutes. If feasible, simpler studies would increase the capacity of the PET-camera, reduce the radiation exposure to the personnel, and reduce the inconvenience of long scanning to the patient and the personnel.

The aim of this study was to evaluate the potential of [^{11}C]methionine-PET for the imaging of non-brain tumors and to compare the uptake of [^{11}C]methionine in malignant tissue with the grade of malignancy and proliferative rate of cancer. Four different analysis methods of [^{11}C]methionine accumulation were compared.

B. M. Mazoyer et al. (eds.), PET Studies on Amino Acid Metabolism and Protein Synthesis, 223–225.

2. Material and Methods

The investigations involved 14 patients with lymphoma, 14 with breast tumor, and 23 with head and neck cancer. All patients were off cancer therapy at the time of the investigation. Furthermore, 14 patients with breast cancer metastases were included in the study, where the different quantitation formulas were compared.

An ECAT PET scanner type 931/08-12 was used for imaging. The device acquires 15 contiguous slices simultaneously with a slice thickness of 6.7 mm; the full width of half maximum is 6.1 mm transaxially in the center of the field of view. Transmission scanning was performed with a removable ring source containing ^{68}Ge immediately before each emission scan. Following the injection, dynamic scanning was carried out for 40-60 minutes.

Immediately after injection of [^{11}C]methionine, repeated venous blood sampling was started from an antecubital vein contralateral to the injection site. The hand and arm were warmed up with a pad. The radioactivity concentration of the plasma was determined. The low molecular weight fraction of the plasma taken at 20, 40 and 60 minutes after the injection was separated by fast gel filtration using Sephadex PD-10 columns (Pharmacia Fine Chemicals, Uppsala, Sweden) for radioactivity measurements. The radioactivity concentration of the low molecular weight fraction of the plasma was used as the input function in the graphical analysis according to Patlak.

The regions of interest (ROI) were drawn on the last frame image on the hot spots in the tumor so that the standard deviation of the average counts in the ROI was less than 15% in the last frames. The ROI was copied to all frames. Several ROIs with high accumulation were selected from the total tumor tissue, because [^{11}C]methionine accumulation was clearly heterogenous in several cases. Of all the tumor time activity curves in each study, the time activity curve with the maximum average counts in the last frame from 35 to 40 minutes after injection was chosen to represent the [^{11}C]methionine uptake of the tumor for further analysis.

Four different formulas were used for the calculations of the uptake of [^{11}C]methionine: graphical analysis according to Patlak et al. [3], standardized uptake value (SUV), radioactivity concentration in tumor per injected dose (RD), and radioactivity concentration in tumor per injected dose and body surface area (RDA).

3. Results

[^{11}C]Methionine accumulated avidly in 57/61 (93%) malignant non-brain tumors (lymphoma, breast cancer, and head and neck cancer). The reason why 4 tumors were not visualized is their small size (diameter$<$20 mm) and small proliferation rate. There was some accumulation of [^{11}C]methionine in an abscess, but no accumulation in an inflammatory lesion or a benign breast tumor. A high accumulation of [^{11}C]methionine in breast cancer correlated well with a large S-phase fraction ($r=0.77$ and $p=0.01$, $N=9$).

There was no correlation between the accumulation of [^{11}C]methionine and the histopathological grade of head and neck cancer. The uptake of [^{11}C]methionine was generally higher in the head and neck cancer (mean K_i 0.147 ± 0.070 min^{-1}, mean SUV 8.5 ± 3.5) than in the bone marrow (K_i 0.091 ± 0.029 min^{-1}, $N=17$; SUV 5.5 ± 1.4; $N=17$), parotid gland (K_i 0.072 ± 0.022 min^{-1}, $N=21$; SUV 4.4 ± 1.0, $N=26$), submandibular gland (K_i 0.078 ± 0.022 min^{-1}, $N=28$; SUV 4.5 ± 1.0, $N=30$), lacrimal gland (K_i 0.044 ± 0.006 min^{-1}, $N=4$; SUV 2.7 ± 0.3; $N=4$), and cerebellum (K_i 0.027 ± 0.010, $N=3$; SUV 2.2 ± 0.7, $N=5$). There was a strong correlation between the K_i and SUV values of [^{11}C]methionine accumulation in normal tissues ($r=0.79$, $p<0.0001$, $N=73$).

The uptake rate (K_i) of [^{11}C]methionine from the plasma to the tumor can be measured according to a graphical method of Patlak et al. However, the radioactivity accumulation in tumor adjusted for the injected dose and body surface area (RDA) correlates strongly with the uptake rate K_i ($r=0.92$, $p<0.0001$). The correlation between K_i and RD was 0.84 ($p<0.0001$), and between K_i and SUV 0.90 ($p<0.0001$).

4. Discussion

[^{11}C]Methionine-PET can be used to image lymphoma, breast cancer and head and neck cancer. The accumulation of [^{11}C]methionine appears to associate with the proliferative rate of breast cancer, but further studies are needed to confirm this finding.

For clinical purposes, RDA can substitute the kinetic uptake measurements of [^{11}C]methionine from the plasma to the tumor. RDA is simpler to derive, since blood samples are not needed and the dynamic scanning time is reduced from 40 minutes to 5 minutes. The radiation exposure to the personnel is reduced when frequent blood samples need not to be taken and the capacity of the PET unit is improved. The correlation between the K_i values and the SUVs was also highly significant. If the patient is of normal weight, the SUV can well substitute the K_i, but especially in patients who have a body mass index that differs greatly from the average, the SUV is unreliable.

5. Acknowledgements

I thank Dr Heikki Minn for presenting this paper in Lyon, Feb 1992. The radiopharmaceuticals were produced at the Turku Medical Cyclotron Laboratory. I thank the personnel of the Nuclear Medicine Department for pleasant cooperation. The study was financially supported by grants from the Finnish Cancer Society, and the Arvo and Inkeri Suominen Foundation.

6. References

1. Derlon J-M, Bourdet C, Bustany P et al. [^{11}C]L-Methionine uptake in gliomas. Neurosurgery 1989; 25:720-728.

2. Fujiwara T, Matsuzawa T, Kubota K et al. Relationship between histologic type of primary lung cancer and carbon-11-L-methionine uptake with positron emission tomography. J Nucl Med 1989; 30:33-37.

3. Patlak CS, Blasberg RG, Fenstermacher JD. Graphical evaluation of blood-to-brain transfer constants from multiple-time uptake data. J Cerebr Blood Flow Metabol 1983; 3:1-7.

Utilization of Amino Acid Transport Rates for the Differential Diagnosis of Brain Tumors

G.-J. MEYER, W.BURCHERT, K.-F. GRATZ, H. HUNDESHAGEN

Introduction

Tumor growth is associated with the increase of tissue mass, a large fraction of which are protein containing structures. Since these are built up from amino acids it can be assumed that these building blocks accumulate in tumorous tissue to a larger extent, than in the surrounding tissue.
This approach has proved to be extremely successful especially in the case of brain, where the surrounding tissue has a relatively low amino acid turn over rate. By now amino acids seem to be the most promising agents to study the metabolic activity of brain tumors. As shown in a study of the Karolinska group, the amino acid image delineates the metabolically active tumor tissue over a quite larger area than would have been expected from all other imaging modalities, and more exactly when compared with stereotactic biopsies [1].
Clinical results on the usefulness of amino acid uptake in brain tumors for analysis of metabolic activity are as numerous as those reported on glucose metabolism. The main focus of Swedish researchers was first to establish a comparison of tumor extension by morphologically oriented methods like CT and MRT with functional PET imaging [2-9]. Most of their results have been corroborated by investigators from France [10, 11], Germany [12-19], USA [20, 21]] and Japan [22-24].
In general amino acids have been used successfully in three areas of clinical applications.

Identification of brain tumors

Low grade gliomas are difficult to delineate by conventional radiological methods. Although significant improvements have been demonstrated by MRT, which is likely to detect small lesions in T_2 weighted images, differential diagnosis of these lesions in terms of tumor, scar, and oedema is often difficult. Reports, in which amino acid uptake measurement could differentiate tumors from other malformations, or could identify areas of increased metabolic activity despite the absence of delineated morphological signs have been reported by several groups. [cf 22, 25].

B. M. Mazoyer et al. (eds.), PET Studies on Amino Acid Metabolism and Protein Synthesis, 227–236.

Monitoring of therapy response

Furthermore amino acid uptake in tumors can be used as a tool for therapy control [7]. Functional changes of the target tissue, which are the goal of any tumor therapy can not be visualized by CT and MRT directly. All therapeutic interventions, - surgery, radiation, and chemotherapy -, induce necrosis and morphological changes at the tumor location which complicate the interpretation of CT and MRT Images. ^{11}C-methionine PET has shown to be suitable to differentiate necrotic changes and scar from residual tumor tissue and recurrencies [26]. The possibility to evaluate metabolic activity on a quantitative basis allows furthermore to measure the direct response of the tumor tissue to radiation and chemotherapy. Whereas the morphologically oriented methods allow a judgement of the tissue response after several weeks or months only, PET measurements provide quantitative information on the metabolic response directly after the treatment. A special example has been demonstrated by Bergström and Muhr [27], who reported on the control of bromocryptine therapy of a prolactinoma with ^{11}C-L-methionine. In their study the therapeutic response was detectable 3 h after treatment already, whereas tumor regression was not detectable with CT and MRT before several weeks.

Metabolic Activity and Grading of Brain Tumors

The comparison of CT and MRT with PET has revealed that the active tumor tissue is most accurately delineated by the ^{11}C-methionine PET. This has been verified by controlling the PET data with multiple stereotactic biopsie examinations [5, 8, 9]. Although the spacial resolution of MRT is not reached by PET, PET images allow a clear differentiation of perifocal oedema from active tumor tissue, which is difficult by MRT. Besides these visual aids for therapy planning, the quantitative or even semi-quantitative analysis bears more advantages for the PET method. In high grade tumors (grade IV) which are usually well characterized in CT and MRT also, PET offers the advantage of demonstrating the most metabolically active areas better than the former methods. The growth direction can be determined by one ^{11}C-methionine PET investigation, whereas this may require sequential images over a longer time span with other methods. Despite the extremely bad prognosis of high grade brain tumors this information can influence therapeutic approaches significantly.

While most high grade tumors are easy to identify by all radiological methods, forecasts on their metabolic activity can usually be extracted only from the dynamics of clinical parameters. Furthermore grading of moderate malignant gliomas is often difficult to specify even from biopsy, because of the limited sampling possibilities. Former investigations which tried to grade the tumors by quantifying the uptake of amino acids in tumors by calculating tumor over non-tumor accumulation ratios have shown that despite a good correlation of the uptake with the histological grade, there is some overlap between the histological grades II and III as well as between grades III and IV [15-19, 24]. New approaches which analyze the uptake data in a more quantitative way, provide a somewhat better discrimination of grades, but still leave some overlap in grade II and grade III glioma. The relative large variability especially in grade III tumors suggests, that histological grading may not be the ultimate answer on the metabolic state of these tumors. The amino acid PET data suggest that especially in the histological class III, at least two subgroups can be defined according to their metabolic activity. Until now, however, too few PET measurements have been performed in follow-up studies, in order to correlate these findings with clinical data like survival times, treatment response, and prognosis. It should be mentioned here, that recent findings in lung cancer investigations with ^{11}C-L-methionine indicate a good correlation of differential uptake ratios with the histological type of tumor [28].

Multi-Parameter Analyses

The physiological status of an organ can not be described by any single parameter. Usually a set of parameters including perfusion, energy consumption, and some organ specific metabolic functions need to be analyzed. The same holds true for brain tumors. Because of the large inter-patient variability of tumors, even within one histological class, multi-parameter studies on one individual tumor are of special interest for investigations of a correlation of these parameters. Several studies for the analysis of such possible correlations have been undertaken.

The results from these can be summarized by the statements that the generally decreased oxygen consumption of brain tumors is accompanied by a variable flow pattern, and that glucose utilization in brain tumors is often related to the perfusion characteristics. The correlation coefficient of the latter pair is weak, however, because of some exceptions, stemming especially from the class of meningiomas, which sometimes exhibit low glucose utilization despite high flow rates.

Multi parameter studies which compare amino acid uptake and metabolism with the other parameters have been few up to now. Experimental three parameter studies on brain tumors which compared blood flow, glucose utilization and amino acid uptake have been reported by Mies et al. [29]; Kirikae et al. [30] compared glucose utilization and valine uptake in a rat brain tumor model, and Abe et al. [31] measured blood flow and methionine uptake in a non brain rat tumor model. All studies agree that amino acid uptake is a sensitive and reliable parameter for increased metabolic activity, as analyzed by histological examination of tumor samples. Kirikae et al. identified valine uptake as being seven fold more sensitive than glucose utilization.

Several multi parameter investigations in patients [2, 4, 32, 33] have shown that glucose utilization may be inferior to amino acid uptake in identifying the extent and the metabolic activity of brain tumors. This was corroborated in a study from Montreal [34] on a grade IV glioma, in which glucose metabolism, flow, oxygen extraction, and oxygen utilization, were below normal values, whereas blood volume, as well as tissue pH were both increased.

In our studies, the physiological function of individual brain tumors has been analyzed by the three parameters: blood flow [15O-Water], glucose utilization [18F- FDG] and amino acid uptake [^{11}C-Methionine].

In previous investigations, which were performed with a two head positron camera [Cyclotron Corp. Mod. 4200] all three parameters were recorded by steady state methods. The data were analyzed in four relative ways using a) tumor region over contralateral region ratios, b) tumor region over average slice minus tumor region, c) tumor region over average white matter and, d) tumor region over average grey matter ratios.

Using the steady state protocol 112 patients were analyzed with ^{11}C-methionine. All measurements were carried out prior to surgery, which was carried out in 82 patients, thus leading to the same number of histologically proven samples. 31 patients were stud-

ied with two parameters, 15 were studied with all three modalities.
The results can be summarized as follows: All tumors showed ratios >> 1 in terms of amino acid uptake by all analytic methods. The majority of tumors of grade III showed a ratio < 1 for glucose utilization by method a) b) and d). This was matched in all cases by a deceased blood flow as well. Using method c) about half of these tumors showed a ratio < 1 for glucose utilization. This again was matched individually by the blood flow ratios obtained with the corresponding method. Only some high grade gliomas of grade IV exhibited glucose utilization ratios > 1 by all analytical methods. These cases showed matched high blood flow ratios as well. Only meningioma exhibited a mismatch of flow and glucose utilization. In meningiomas, however, flow and amino acid uptake correlated significantly.

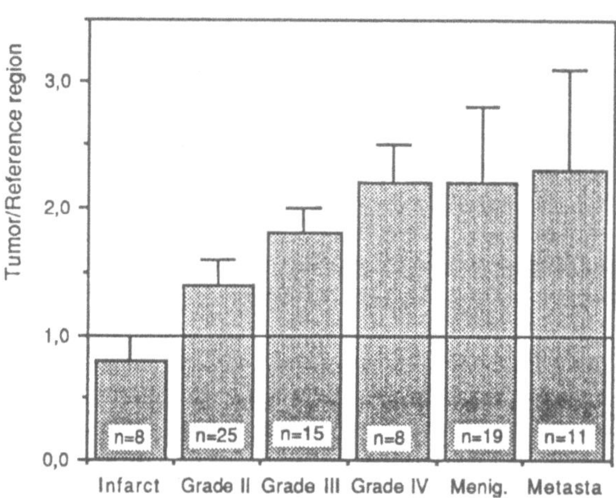

Fig. 1. Grading according to the ratio of tumor /non-tumor on the contralateral reference region by ^{11}C-methionine.

A new positron camera system (Siemens ECAT 951/31) is in use since two years. 16 patients with brain tumors or suspected lesions have been investigated with both, ^{11}C-methionine and ^{18}F-FDG. In 4 of these patients flow was additionally measured with ^{15}O-water. All data were analyzed quantitatively according to a dynamic aquisition protocol with blood sampling and metabolite analysis. Data analysis for ^{11}C-methionine and ^{18}F-FDG was carried out in terms of the Gjedde-Patlak approach as described in another contribution to this volume.
Two examples of dynamic investigations with ^{11}C-methionine and ^{18}F-FDG are shown in fig.2 (a-f) and 3 (a-d). In both cases CT (not shown) was normal, while MRT detected lesions in T_2 weighted images.

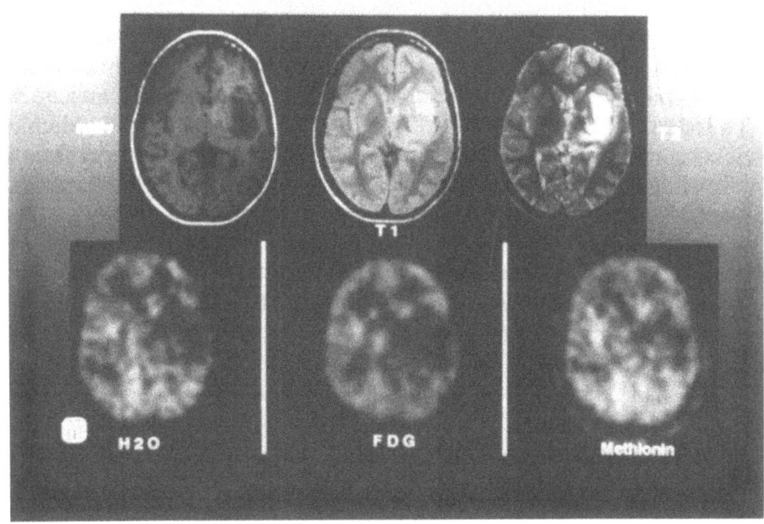

Fig. 2 MRT and PET images of a patient with a suspect lesion which turned out to result from an infarction (see text)

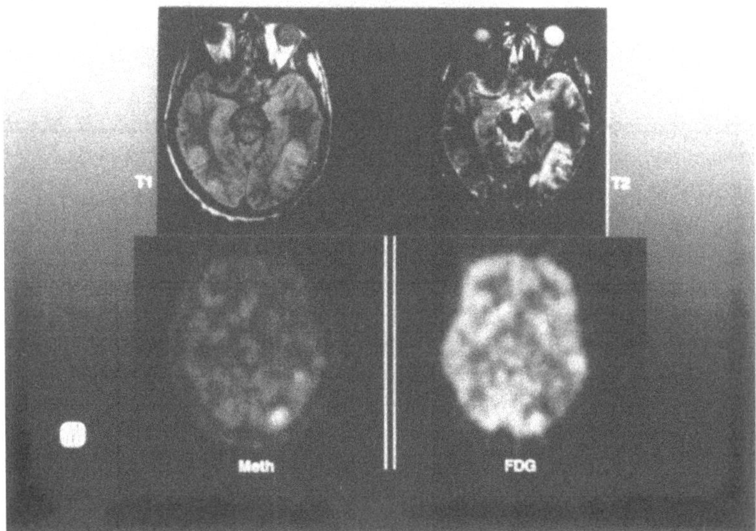

Fig 3. MRT and PET images of a patient with a suspect lesion which was proven to be a low grade glioma

In fig. 2 MRT and PET images of a patient are shown who's clinical symptoms were not clearly indicative for an infarction. Although MRT (proton density-, T_1- and T_2-images) revealed a lesion which probably results from an infarction, clinical history and symptoms remained unclear. In order to investigate the possibility of an underlying tumorous process, PET was performed with ^{15}O-water, ^{11}C-methionine and ^{18}F-FDG. This investigation showed that the lesion is unlikely to include a tumor.
In fig. 3 a lesion can be identified by MRT in . The PET scan with L-Methionine (fig. c) reveals an accumulation of the tracer in small areas which are suspect from the MRT. The unidirectional transport rate in these areas are consistent with low grade tumors. A summary of the quantitative results is shown in table 1.

Table 1: Unidirectional transport rates K_{mr} for ^{11}C- L-methionine and ^{18}F-FDG in brain and brain tumor

brain structure / tracer	grey matter	white matter	grade II	grade III	grade IV
methionine	0.019 ±0.011	0.012 ±0.008	0.022 ±0.012	0.033 ±0.011	0.050 0.016
f d g	0.028 ±0.006	0.015 ±0.005	<0.015 ±0.010	0.017 ±0.010	0.025 ±0.008

In all tumors methionine yields metabolic rate constants which are higher than in grey and white matter. In contrast the rate constant for FDG is much smaller than in grey matter for most tumors and is comparably high only for grade IV tumors. Only when compared with white matter, FDG metabolic rate constants can discriminate high grade tumor tissue from normal brain tissue.
In comparison with the results obtained with the steady state protocol and semi-quantitative data-analysis, a better discrimination of grades may be obtained with the quantitative analysis. However, the group of patients analyzed so far is still too small to allow distinct or definite conclusions.
The methodological limitation of any analytical approach is, of course, highly dependent on the accuracy of region of interest specification. Especially in the case of low contrast of the tumor region over adjacent brain structures, as in the case of those FDG images, where the tumor is embedded in gray matter structures or within the borders of grey and white matter, errors in the definition of tumor region influence the result. In low grade tumors orientation may become difficult if morphological information by MRT or CT is missing. In these cases it is therefore necessary to compare the PET images with MRT

and/or CT images. However, quite often the orientation of these images does not match the orientation of the PET images exactly. Although software for the reorientation of three dimensional image sets is under development, most of these programs have difficulties in either retaining quantitative pixel information under reorientation processing, or to reorient data sets under routine application conditions. Furthermore they fail generally if the data quality is poor. An interesting approach to overcome some of these problems has been reported recently [35].

In summary it has been coroborated that the amino acid uptake is the only parameter which gives a positive signal in nearly all types of brain tumors. The small fraction of low grade tumors with no uptake or even a a slightly decreased amino acid accumulation results from calcified astrocytomas II.

It is concluded from this study that the unidirectional transport rate of amino acid into brain tumors is a sensitive parameter for grading of these tumors.

The biochemical interpretation of the uptake data in terms of more precise quantitative functional parameters as e.g. protein synthesis rate remains difficult. Nevertheless amino acid uptake has been shown to be a reliable measure for metabolic activity.

Since broad multi parameter studies are far beyond the scope of a clinical routine application for positron emission tomography, amino acid uptake measurements alone have the potential to become a standard application for diagnostic brain tumor studies with PET.

References

1. Ericson K, Bergström M, Eriksson L, Hatam A, Greitz T, Soederström CE, Widen L. Positron emission tomography with 68Ga-EDTA compared with transmission computedtomography in the evaluation of brain infarcts. Acta Radiol. 22, 385-389, 1981

2. Bergström M, Collins VP, Ehrin E, Ericson K, Eriksson L, Greitz T, Halldin C, Holst H v, Langström B, Lilja A, Lundquist H, Nagren K. Discrepancies in brain tumor extent as shown by computed tomography and positron emission tomography using 68Ga-EDTA, ^{11}C-Glucose, and ^{11}C-Methionine. J. Comput. Assist. Tomogr. 7, 1062-1066, 1983

3. Mosskin M, von Holst H, Ericson K, Noren G. The blood tumor barrier in intracranial tumors studied with X-ray computed tomography and positron emission tomography using 68-Ga EDTA. Neuroradiol. 28, 259, 1986

4. Ericson, K., Lilja A, Bergström M., Collins VP., Eriksson, L., Ehrin E., v.Holst, H., Lundqvist H., Langström B., Mosskin, M. Positron emission tomography with ^{11}C-methyl-L-methionine, 11C-D-glucose, and 68-Ga-EDTA in supratentorial tumors. J. Comput. Assist. Tomography 9, 683-689, 1985

5. Mosskin M, Ericson K, Hindmarsh T, von Holst H, Collins VP, Bergström M, Eriksson L, Johnström P. Positron emission tomography compared with MRI and CT in supratentorial gliomas using multiple stereotactic biopsies as reference. Acta Radiologica 30, 225-323, 1989

6. Lilja A, Bergström K, Hartvig P, Spännare B, Halldin C, Lundquist H, Langstrom B Dynamic Study of Supratentorial Gliomas with L-methyl-11-C-Methionine and Positron Emission Tomography. Amer. J. Neuro. Radiol. 6, 505-514 [1985]

7. Lilja A, Lundqvist H, Olsson Y, Spännare B, Gullberg P, Langström B. Positron emission tomography and computed tomography in differential diagnosis between recurrent or residual glioma and treatment induced brain lesions. Acta Radiol. 30, 121-128, 1989

8. Mosskin M, von Holst H, Bergström M, Collins VP,Eriksson L, Johnström P, Noren G.. Positron emission tomography with ^{11}C-methionine and X-ray computed tomography of intracranial tumors compared with histopathologic examination of multiple biopsies. Acta Radiol. 28, 673-681, 1987

9. Mosskin M, Ericson K, Hindmarsh T, von Holst H, Collins VP, Bergström M, Eriksson L, Johnström P. Positron emission tomography compared with magnetic resonance imaging and computed tomography in supratentorial gliomas using multiple stereotactic biopsies as reference. Acta Radiol. 30, 225-232, 1989

10. Bustany P, Sargent T, Saudubray JH, Henry JF, Comar D. Regional human brain uptake and protein incorporation of ^{11}C-L-Methionine studied in vivo with PET. J. Cerebr. Blood Flow Metabol. 1, 17-18, 1981

11. Bustany P, Chatel M, Derlon, JM, Darcel F, Sgouropoulos P, Soussaline F, Syrota A. Brain tumor protein synthesis and histological grades: A study by positron emission tomography [PET] with ^{11}C-L-Methionine. J. Neurooncol. 3, 397-404, 1986

12. Meyer, G.J., Schober O, Gaab, MR, Dietz, H., Hundeshagen H. Multi parameter studies in brain tumors. In: Beckers C, Goffinet A, Bol A, Eds., Positron Emission Tomography in Clinical Research and Clinical Diagnosis. Kluwer Acad. Publ., Dordrecht,1989, pp 229-248

13. Schober O, Duden C, Meyer G.-J., Müller JA, Hundeshagen H. Non selective transport of [^{11}C-methyl]-L-, and -D-methionine into a malignant glioma. Eur. J. Nucl. Med. 13, 103-105, 1987

14. Meyer G.-J., Schober O., Hundeshagen H. Uptake of ^{11}C-D- and L-methionine in brain tumors. Eur J. Nucl. Med. 10, 373-376, 1985

15. Schober O., Creutzig H., Meyer G.-J., Becker H., Schwarzrock R., Dietz H., Hundeshagen H. ^{11}C-Methionin-PET, IMP-SPECT, CT und MRI bei Hirntumoren. Fortschr. Roentgenstr. 143, 133-136, 1985

16. Schober O., Meyer G.-J., Stolke D., Hundeshagen H. Brain tumor imaging using ^{11}C-labelled L-methionine and D-methionine. J. Nucl. Med. 26, 98-99, 1985

17. Schober O., Meyer G.-J., Gaab M.R., Müller J.A., Hundeshagen H. Grading of brain tumors by L-[^{11}C]methionine PET. J. Nucl. Med. 27 : 890-891, 1986

18. Schober O., Meyer G.-J., Duden C., Lauenstein L., Niggemann J., Müller J.-A., Gaab M.R., Becker H., Dietz H., Hundeshagen H. Die Aufnahme von Aminosäuren in Hirntumoren mit der Positronen-Emissionstomographie als Indikator für die Beurteilung von Stoffwechselaktivität und Malignität. Fortschr. Röntgenstr. 147, 503-509, 1987

19. Schober O., Meyer G.-J., Gaab M.R., Dietz H., Hundeshagen H. Multi-Parameter Studies in Brain Tumors by PET. J. Nucl. Med. 29 , 853 1988

20. O'Tuama LA., Guilarte TR, Douglass KH, Wagner Jr, HN, Wong, DF, Dannals R.F., Ravert, HT, Wilson AA, La France N.D., Bice AN, Links JM. Assesment of [^{11}C]-L-Methionine transport into the human brain. J. Cerebr. Blood Flow Metabol. 9, 341-345 1988

21 O'Tuama LA, La France ND, Dannals RF, Douglass KH, Links JM, Bice AN, Williams JA, Villemagne V, Wagner Jr. HN. Quantitative imaging of neutral amino acid transport by human brain tumors. J. Cerebr. Blood Flow Metabol. 7, S517, 1987

22. Mineura K, Sasajima T, Yoshitaka S, Kowada M, Shishido F, Uemura K. Early and accurate detection of primary cerebral glioma with interfibrillary growth using ^{11}C-L-methionine positron emission tomography. J. Med. Imaging 3, 192-193, 1989

23. Hatazawa J, Ishiwata K, Itoh M, Kameyama M, Kubota K, Ido T, Matsuzawa, T,Yoshimoto T, Watanuki S, Seo S. Quantitative evaluation of [methyl-^{11}C]-methionine uptake in tumor using positron emission tomography. J. Nucl. Med. 30, 1809-1813, 1989

24. Kameyama M, Shirane R, Itoh J, Sato K, Katakura R, Yoshimoto T, Hatazawa J, Itoh M, Seo S, Ido T. The accumulation of ^{11}C-Methionine and histological grade in cerebral glioma studied with PET. CYRIC Annnual Report 1988, Cyclotron and Radioisotope Center Tohoku University, Sendai 1989, pp 228-237

25. Schober O., Meyer, G.-J., Klinische Anwendung der Positronen Emissions Tomographie. In: Handbuch der medizinischen Radiologie Band XV/1B, Diethelm L, Heuck F, Olsson O, Strnad F., Vieten H., Zuppinger A., Eds. Springer Verlag Berlin 1988 pp315-469.

26. Eriksson L. personal communications, cf. also: Proc. Eur Symp. on Positron Emission Tomography in Cancer, Hammersmith Hospital, London 5.-6. July 1989

27. Bergström M, and Muhr C. personal communications. cf. also: Proc. Eur Symp. on Positron Emission Tomography in Cancer, Hammersmith Hospital, London 5.-6. July 1989

28. Fujiwara, T, Matsazuwa T, Kubota K, Abe Y, Itoh M, Fukuda H, Hatazawa J, Yoshioka S, Yamaguchi K, Ito K, Watanuki S, Takahashi T, Ishiwata K, Iwata R, Ido T. Relationship between histologic type of primary lung cancer and carbon -11-L-methionine uptake with positron emission tomography. J. Nucl. Med. 30, 33-37, 1989

29. Mies G, Bodsch W, Paschen W, Hossmann KA. Experimental application of triple labeled quantitative autoradiography for measurement of cerebral blood flow, glucose metabolism and protein biosynthesis. In: Heiss WD, Phelps ME, Eds., Positron Emission Tomography of the Brain. Springer Berlin 1983, pp 19-28

30. Kirikae M, Diksik M, Yamamoto YL. Quantitative measurements of regional glucose Utilization and rate of valin incorporation into proteins by double tracer autoradiography in the rat brain tumor model. J. Cerebr. Blood Flow Metabol. 9, 87-95, 1989

31. Abe Y, Matsuzawa T, Itoh M, Ishiwata K, Fujiwara T, Sato T, Yamaguchi K, Ido T. Regional coupling of blood flow and methionine uptake in an experimental tumor assessed with autoradiography. Eu. J. Nucl. Med. 14, 388-392, 1988

32. Meyer, G.J., Schober O, Gaab, MR, Dietz, H., Hundeshagen H. Multi parameter studies in brain tumors. In: Beckers C, Goffinet A, Bol A, Eds., Positron Emission Tomography in Clinical Research and Clinical Diagnosis. Kluwer Acad. Publ., Dordrecht,1989, pp 229-248

33. Kameyama M, Tsurumi Y, Shirane R, Katakura R, Suzuki J, Itoh M, Fukuda H, Matsuzawa T, Watanuki, S, Ido T. Multi parametric analysis of brain tumor with PET. J. Cereb. Blood Flow Metabol. 7, S466 [1987]

34. Yamamoto YL, Thompson CJ, Meyer E, Robertson SJ, Feindel W. Dynamic positron emission tomography for study of cerebral hemodynamics in a cross section of the head using positron-emitting ^{68}Ga-EDTA and ^{77}Kr. J. Comp. Assist. Tomogr. 1, 43-56, 1977

35. Kapoulesas I, Alavi A, Alves WM, Gur RE, Weiss DW. Registration of Three-dimensional MR and PET Images of the human brain without Markers. Radiol.181, 731-739, 1991

USE OF AMINO ACID UPTAKE AND PROTEIN SYNTHESIS RATES FOR TUMOUR DIAGNOSIS

H. LUNDQVIST

ABSTRACT. The Uppsala PET-group has used ^{11}C-L-methionine for tumour diagnosis during 10 years. Our experience is that this technique gives an excellent delineation of the tumour and in many cases a good differential diagnostic information. We prefer not to use modelling in brain tumours but a simple ratio between the tumour area and a symmetric normal area in the opposite hemisphere. Changes in this ratio is used to evaluate changes due to e.g. therapy. In tumours in the rest of the body the Patlac technique is used. We have recently started to use amino acids which do not enter the protein synthesis for special type of tumours. An example is ^{11}C-5-hydroxy-tryptophan to find serotonine producing carcinoid-metastasis in the liver.

1. INTRODUCTION

Ten years ago the first PET-scanner was installed in Uppsala. The very first investigation was made with ^{11}C-methionine [1]. The patient was suffering from a brain tumour, gliomas, and since then ^{11}C-methionine has been the main tracer in our clinical work especially for tumour diagnostics and for the evaluation of treatment effects [2, 3, 4, 5, 6, 7, 8, 9, 10, 11, 12, 13].

The new PET-centre in Uppsala has now been operating since sept-91 and it is interesting to see that about 50 % of the investigations are still made with ^{11}C-methionine and for tumour diagnosis.

B. M. Mazoyer et al. (eds.), PET Studies on Amino Acid Metabolism and Protein Synthesis, 237–241.

2. PROBLEMS TO MODEL ^{11}C-L-METHIONINE

We and others has tried to derive quantitative physiological information from PET and methionine, e.g. tried to model the tracer [14]. There are some problems to be considered:

L-methionine is transported to a high degree both by the L- and the A-transport system. To our knowledge it is the only physiological amino acid which has this ability.This is an advantage in brain tumours since L-methionine passes efficiently an intact BBB and is still a good marker of increased active transport of the tumour cells.But it also indicates that the transport is rather complicated and is dramatically changed especially in pathological conditions.

Some general problems for amino acids are that the input function is disturbed by a rather slow equilibration between plasma and red cells. In order to compensate for this correctly one has to make a fast separation to obtain radioactivity concentration in plasma. The membrane transport is disturbed by the plasma concentration of cold methionine but also the cold concentration of all other amino acids competing of the same transport system. When incorporated into the brain the heterogenity is large with a factor of four between white and grey matter uptake which makes quantitative modeling difficult [14]. In tumours the heterogenity will be even larger and hence the modeling more difficult.

Special problems for L-methionine is a complex metabolism which is well described in other articles of this volume. Beside the protein synthesis route there are routes giving yield to low molecular metabolites which will appear both in plasma and in tissue and which severely disturbs the quantitative evaluation of methionine kinetics.

3. THE CLINICAL USE OF ^{11}C-L-METHIONINE

With all these problems in mind we have been rather restrictive to use modeling when evaluating tumour uptake of ^{11}C-L-methionine. Our protocol in clinical practise in the brain and brain tumours is:

scan dynamically for 30-40 minutes
no blood samples
make a ratio between the tumour area and a symmetric normal area
in the opposite hemisphere and use this as an index.

We have found this to be much more stable and reliable measure than e.g. the Patlac-slope. By this measure we can see significant changes less than 10 %. Parameterization will give model parameters which may give more physiological relevant information but which also will be more noisy.

We are just now doing our first experiences in the whole body. For these studies we can not use this simple approach since there is no obvious reference tissue which can be used. For these studies we take blood, either arterial or arterialized venous blood and use the Patlac-technique i.e. having the plasma as reference.

To use ^{11}C-L-methionine to measure protein synthesis rates in normal tissue is difficult (see other contributors to this volume). It is even more difficult in a pathological situation like tumours. This is well illustrated in a in vitro system we have developed. Cultivated human tumour cells are grown as spheroids in order to simulate small tumour nodules with well and less well nutrified cells and where parts of the nodules are necrotic [15]. This system is fed by ^{11}C-L-methionine and the uptake of the spheroids is measured in a well counter. The system is used to study the ^{11}C-L-methionine uptake dependence of different drugs. Protein incorporation, free amino acid and metabolites are measured. The results from these studies is then a help to interpret the similar situation in vivo and PET. One can in this system clearly see that high utake values can be due to protein incorporation or due to transport and that it is difficult to separate inbetween these two.

4. THE USE OF OTHER AMINO ACIDS

We have recently started to use amino acids which do not enter the protein synthesis for special type of tumours. An example is the use of ^{11}C-5-hydroxy-tryptophan to find serotonine producing carcinoid-metastasis in the liver. The advantage is that there is no back-ground from protein synthesis and the uptakes correlated to serotonine production give a significant indication of the spread of metastasis. So far the evaluation is made by just measuring the radioactivity uptakes or by using a reference Patlac-technique with normal liver tissue as the reference.

5. REFERENCES

1. Bergström K, Lilja A, Spännare B, Aquilonius S-M, Jung B, Hartvig P, Halldin C, Långström B, Stålnacke C-G, Lundqvist H. Dynamic study of supratentorial gliomas with C-L-methionine and positron emission tomography (PET). Am J Neurorad 1983; 4:1139-1140.

2. Bergström M, Collins P, Ehrin E, Ericson K, Eriksson L, Greitz T, Halldin C, von Holst H, Långström B, Lilja A, Lundqvist H, Någren K. Discrepancies in brain tumor extent as shown by computer tomography and positron emission tomography using (68Ga)EDTA, (11C)Glucose, and (11C)Methionine. J Comp Ass Tomog 1983; 7:1062-1066.

3. Ericson K, Lilja A, Bergström M, Collins P, Eriksson L, Ehrin E, von Holst H, Lundqvist H, Långström B, Mosskin M. Positron emission tomography with (11C-methyl)-L-methionine, (11C)D-glucose, and (68Ga)EDTA in supratentorial tumors. J Comp Ass Tomog 1985; 9(4):683-689.

4. Lilja A, Lundqvist H, Olsson Y, Spännare B, Gullberg P, Långström B. Positron emission tomography and computed tomography in differential diagnosis between recurrent or residual glioma and treatment-induced brain lesions. Acta Radiol 1989; 30:121-128.

5. Lilja A, Bergström K, Hartvig P, Spännare B, Halldin C, Lundqvist H, Långström B. Dynamic study of supratentorial gliomas with L-methyl-11C-methionine and positron emission tomography. AJNR 1985; 6:505-514.

6. Muhr C, Bergström M, Lundberg P O, Bergström K, Hartvig, P Lundqvist H, Antoni G, Långström B. Dopamine receptors in pituitary adenomas: PET visualization with 11C-N-methylspiperone. J Comp Ass Tomography 1986; 10:175-180.

7. Bergström M, Muhr C, Lundberg P O, Bergström K, Lundqvist H, Antoni G, Fasth K-J, Långström B. In vivo study of amino acid distribution and metabolism in pituitary adenomas using positron emission tomography with 11C-D-methionine and 11C-L-methionine. J Comput Assist Tomogr 1987; 11:384-389.

8. Bergström M, Muhr C, Lundberg P O, Bergström K, Gee A D, Fasth K-J, Långström B. Rapid decrease in amino acid metabolism in prolactin-secreting adenomas after bromocriptine treatment.
J Comput Assist Tomogr 1987; 11:815-819.

9. Bergström M, Lundqvist H, Ericson K, Lilja A, Johnström P, Långström B, v Holst H, Eriksson L, Blomqvist G. Comparison of the accumulation kinetics of L-(methyl-11C)-methionine and D-(methyl-11C)-methionine in brain tumors studied with positron emission tomography.
Acta Radiol 1987; 28:225-229.

10. Bergström M, Muhr C, Ericson K, Lundqvist H, Lilja A, Eriksson L, Blomquist G, Långström B, Johnström P. The normal pituitary examined with positron emission tomography and (methyl-11C)-D-methionine.
Neuroradiology 1987; 29:221-225.

11. Lundberg P O, Muhr C, Bergström M, Bergström K, Lundqvist H, Långström B.The use of Positron Emission Tomography (PET) with 11C-labelled L- and D-methionine in diagnosis and follow-up of pituitary adenomas treated with bromocriptine.
Advances in the Biosciences 1988; 69:155-162.

12. Muhr C, Bergström M, Lundberg P O, Hartman M, Bergström K, Pellettieri L, Långström B. Malignant prolactinoma with multiple intracranial metastases studied with positron emission tomography.
Neurosurgery 1988; 22:374-379.

13. Bergström M, Muhr C, Lundberg P O, Långström B. PET as a Tool in the clinical evaluation of pituitary adenomas. J Nucl Med 1991; 32:610-615.

14. Lundqvist H. (1992) 'Kinetic modelling of carbon-11 labelled methionine' in Comar D, Heiss W-D and Mazoyer B (eds.), PET studies on aminoacid metabolism and protein synthesis, Kluwer Academic Publisher, Dortrecht.

15. Gáti I, Bergström M, Muhr C, Långström B, Carlsson J. Application of (methyl-^{11}C)-methionine in the multicellular spheroid system.
J Nucl Med 1991; 32:2258-2265.

^{11}C-METHIONINE UPTAKE IN BRAIN TUMORS MEASURED BY PET: EARLY CLINICAL RESULTS

B. SADZOT, B. KASCHTEN, G. DELFIORE, J.M. PETERS,
A. STEVENAERT, G. FRANCK and D. COMAR

Summary

CT scan and magnetic resonance imaging are excellent for detecting brain tumors but they suffer from a number of limitations. We studied with PET the uptake of ^{11}C-methionine in 6 patients with newly diagnosed brain tumors and in 2 patients who were suspected of tumor recurrence. They were also studied with CT and MRI. Four patients were also studied with ^{18}FDG. Six patients were later operated, one patient underwent a biopsy while one patient had been operated of a grade IV astrocytoma three months before. The uptake of ^{11}C-methionine was increased in each of these brain tumors, allowing a clear delineation of their extent. The tumor/controlateral cortex ratio ranged from 1.6 to 3.5 and seemed related to tumor grade. Glucose uptake was increased in 2 cases of grade IV astrocytoma and decreased in 2 grade I-II astrocytoma. ^{11}C-methionine might be a more interesting tracer for imaging brain tumors than ^{18}FDG and appears useful for the evaluation of tumor activity.

1. Introduction

CT scanner and magnetic resonance imaging (MRI) are excellent imaging modalities for detecting brain tumors. They provide crucial morphological informations. These techniques however suffer limitations, for example in their ability to delineate tumors surrounded by oedema or to differentiate persistant tumor from the effects of surgery in the early post-operative period. The differential diagnosis between tumor recurrence and radionecrosis is a common and difficult problem, as well as the evaluation of response to radiotherapy and/or chemotherapy. Besides some obvious cases, the degree of malignancy of the neoplasic lesion can be difficult to appreciate. For that particular purpose, it is always possible to carry out a biopsy but the value of this approach is limited by its small spatial sampling. It does not allow to fully appreciate the histological heterogeneity of the tumor.

Positron emission tomography has been proved to be a unique tool to measure specific biochemical parameters in the human brain. Depending on the radiotracer administered, it is possible to measure in vivo glucose metabolism,

B. M. Mazoyer et al. (eds.), PET Studies on Amino Acid Metabolism and Protein Synthesis, 243–254.

neuroreceptor density and affinity, amino acid uptake and to obtain parametric images.

The metabolic and biochemical characterization of brain tumor may help answer these questions and provide unique informations on prognosis. Glucose metabolism has been extensively studied in brain tumors. Di Chiro et al. [1] found on a large series of 100 patients a significant relationship between histological grade and metabolic activity. More variable results were obtained by Tyler et al. [2] on 16 untreated patients who showed variable and often low glucose metabolic rates that did not correlate with tumor grade. ^{18}FDG-PET has also been used to identify early brain tumor recurrence [3].

Protein synthesis is another important aspect of tissue activity. The rate of amino acid incorporation in tissue is believed to reflect the protein synthesis rate [4]. This idea gave impetus to the developpement of natural or analog labeled amino acids for use with PET.

Amongst those that have been synthetized, L-[methyl-^{11}C]methionine has been and remains the most widely used. Early [5] and more recent [6] reports have stressed the interest of this tracer for tumor imaging. A significant correlation between the intratumoral/normal tissue activity ratio and the histological grade was found by Derlon et al. [6]. The number of PET centers carrying out these studies and the number of published cases is still limited. We report our recent experience with this tracer and we present here the first cases we have studied in the last few months.

2. Methods

2.1. Radiochemistry

^{11}C-methionine was synthetized by a reaction between $^{11}CH_3$ I and L-homocysteine thiolactone according to Comar et al. [7] and subsequent HPLC separation. More than 100 mCi of ^{11}C-methionine is routinely obtained at the end of a 45 minutes synthesis time. Specific activity is usually higher than 500 mCi/mmole (E.O.S.). The radiosynthesis is fully automatized.

2.2. Data acquisition

The PET measurements were performed with a two ring NeuroEcat PET scanner (EG&G, Ortec) [8] operating in the high resolution mode, providing images with a spatial resolution of 8.6 mm in plane (FWHM) and a slice thickness of 12 mm.

Because only two direct planes can be simultaneously acquired with this tomographic system, it is necessary to select planes of interest prior to dynamic PET studies. For that purpose, a thermolabile plastic face mask (Runtech, Liège) was molted on the face of each subject. A limited X-ray computed tomography (CT) was used to identify a plane passing through the largest portion of the tumor. Once the plane of interest was identified, an alignment line was drawn on the face mask. In our experience, this system ensures an inter and intra individual, standardized and reproducible positioning.

A ^{68}germanium transmission scan was first obtained for photon attenuation correction. Twenty mCi of ^{11}C-methionine were then injected intravenously as a bolus. Dynamic PET data acquisition was started immediately for 75 minutes according to the following protocol: 3x1 min, 3x2 min, 3x5 min, 2x10 and 1x20 minutes. Scan duration was increased to account for the radioactive decay.

During the study, blood samples were withdrawn via a humeral artery catheter previously inserted under local anesthesia. To characterize the plasma time activity curves, a ≈ 2 ml arterial blood sample was obtained every 10 sec during the first 3 minutes, and then at 4, 5, 7.5, 10, 12.5, 15, 20, 30, 45, 60, 75 minutes.

Four subjects were studied on another day with the ^{18}FDG method, in order to study glucose consumption according to a standardized protocol developed in our laboratory [9]. Arterial blood samples (± 25 times 2 ml) were withdrawn at a decreasing rate during the study. Scan data collection was started 45 minutes after injection. Six to 12 planes (8 mm apart) were acquired to image most of the brain structures. Cerebral metabolic rates for glucose were calculated pixel by pixel, using the operational equation of Phelps et al. [10], as well as the rate and lumped constant proposed by these authors for the adult man, resulting in parametric images.

2.3. Data analysis

^{11}C-methionine studies: the attenuation corrected images collected from 15 to 55 minutes post-injection were summed to reduce statistical fluctuations and enhance the anatomical details. These summed images have higher counts and less noise resulting in better images of the tumors. Regions-of-interest (ROI; 2x2 pixels) were then placed on several parts of the tumor and in the normal cortex. These ROI were then automatically transfered to the images obtained at each time point. The counts/pixel were transformed into percent of injected dose per liter of brain (% ID/l). In each patient, the tumor region with the highest uptake was selected. The plasma time-activity curves were also expressed as % ID/l but were not corrected for labeled metabolites.

^{18}FDG studies: For the purpose of this study, parametric images of glucose consumption were visually evaluated. Glucose consumption in the area of the tumor was compared to glucose consumption in the normal grey and white matter.

3. Results

Eight patients with brain tumors were studied with CT scan, MRI, and ^{11}C-methionine-PET. In adition, 4 patients had also a ^{18}FDG-PET. Relevant clinical informations and some results are detailed in Table I. Six patients had newly diagnosed brain tumors and were never treated before the PET study, while two patients known to have a brain tumor had worsening of their clinical neurological deficit and were suspected of tumor recurrence. They had been operated of glioblastoma within the 3 previous months.

Table I. Clinical data and PET results T = Tumor; CTX = cortex; New Diagn.: newly diagnosed brain tumor. The ^{11}C-methionine ratios were calculated with data acquired from 35 to 55 minutes post-injection.

					^{11}C-Methionine		
Pt N°	Type	Treatment	Histologic grade	Contrast enhanc.	T/CTX Ratio	T/PLASMA Ratio	Glucose consumption
1	New Diagn	Surgery	IV	+	3.43	5.2	
2	New Diagn	Surgery	IV	+	2.74	3.52	Increased
3	New Diagn	Surgery	IV	+	2.65	3.91	Increased
4	Recurrence ?	No Surgery	IV	+	2.43	6.16	
5	New Diagn	Surgery	III	+	1.75	3.06	
6	Recurrence ?	Surgery	III	+	1.64	3.00	
7	New Diagn	Biopsy	II	-	1.66	4.43	Decreased
8	New Diagn	Surgery	I	-	1.66	4.44	Decreased

Six patients were operated and one patient had a stereotactic tumor biopsy. A recent histological diagnosis based on the Kernohan scale is therefore available for seven patients. One patient with a recurrence of a grade IV glioblastoma was not reoperated.

In each patient, except in patient 8, CT scan clearly showed the tumor lesion and MRI did not provide substantial information of clinical importance. Patient 8 had a small cortical tumor, measuring 20 by 20 mm, that was missed on CT and seen only on MRI. All the high grade tumors (III and IV) had irregular areas of contrast enhancement. By contrast, no uptake of contrast agent was observed in the low grade tumors.

A striking increase in ^{11}C-methionine uptake was observed in each tumor lesion relative to the surrounding background activity in the normal brain (Figures 1 and 2). Even the small cortical tumor of patient 8 appeared as a hot spot with ^{11}C-methionine (Figure 1). The tumor limits were well defined with ^{11}C-methionine as indicated by the comparison between CT and PET images. In 7 of the 8 cases, ^{11}C-methionine uptake in the tumor was not uniform but heterogeneous. Outside the tumor, ^{11}C-methionine uptake was uniformly low and no anatomical structure could be recognized.

Four patients had a ^{18}FDG-PET study in addition to the ^{11}C-mehionine scan. In the 2 patients with a high grade tumor, ^{18}FDG uptake was increased in some part of the tumor relative to the adjacent neocortex. By contrast, glucose consumption was decreased in the 2 lower grade tumors (pts 7 and 8). Tumor limits were poorly or not distinguishable on the ^{18}FDG scan. Examples of the combined ^{18}FDG and ^{11}C-methionine scans are shown in Figure 1 and 2.

Regions of interest were placed on the ^{11}C-methionine scans in various parts of the tumor and on the normal cortex. Representative time-activity curves from one patient are shown in Fig. 3 and 4. The kinetics of the activity, expressed as the % injected dose, is shown in Fig. 3.

The penetration of ^{11}C-methionine into the brain was rapid, the peak activity being reached before 15 minutes post-injection. After the peak, the uptake remained stable with little or no wash-out. The heterogeneity of the uptake in the tumor is visible at the inspection of the curves. Interestingly, the kinetics of the tumor/cortex ratio and the kinetics of tumor/plasma ratio were strikingly different (Figure 4).

The tumor/cortex ratio was retained as the best mean to quantify ^{11}C - methionine incorporation.

Figure 1. Clinical example 1. This 25 years old woman had complex partial seizures for 6 years. Surface EEG showed interictally some right temporal spikes. CT scan was normal. MRI disclosed a small superficial lesion in the right parietal cortex that was hyperintense on T2 and hypointense on T1 images, without any uptake of gadolinium. The ^{11}C-methionine scan showed a small area of increased activity corresponding to the lesion. This area was hypometabolic on ^{18}FDG scans. Pathological diagnosis was grade I astrocytoma.

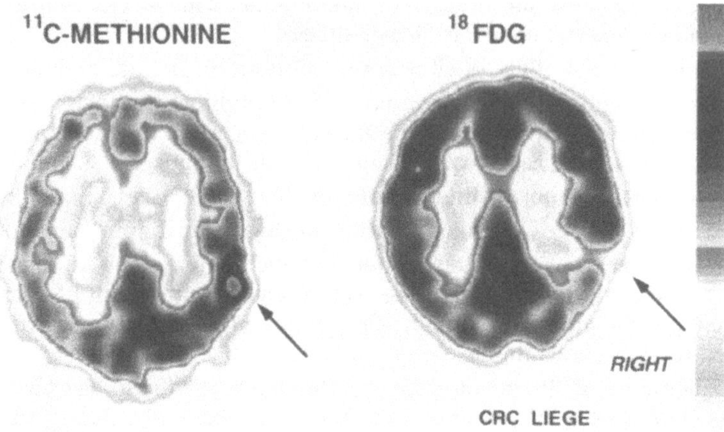

Figure 2. Clinical example 2. Recent and progressive brachiocephalic paresis in a 62 year old man. He had a right fronto-temporal slow wawes focus on EEG. CT scan and MRI revealed a right anterior temporal mass, 4 cm diameter, surrounded by a vast oedema. Captation of contrast agents was homogeneous. At pathological examination of the surgical specimen, this mass corresponded to a glioblastoma. ^{11}C-methionine and ^{18}FDG were increased in the tumor.

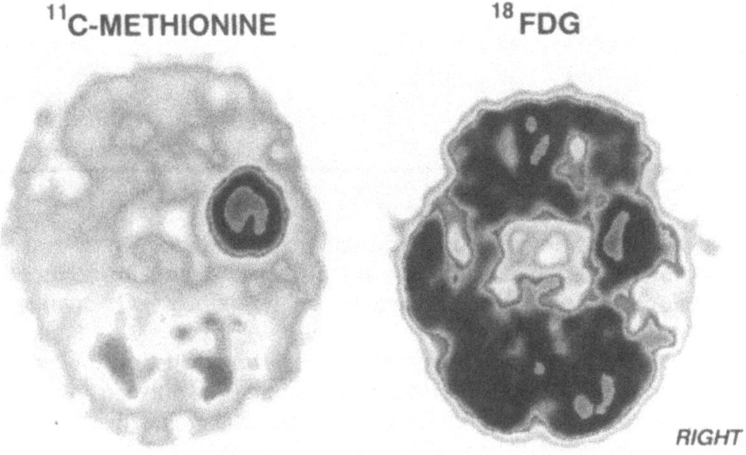

Figure 3. Time-activity curve of ^{11}C-methionine in a patient with a grade III brain astrocytoma. Regions of interest were placed in various parts of the tumor and in the normal cortex. The uptake of ^{11}C-methionine is increased in the tumoral areas as compared to the normal cortex.

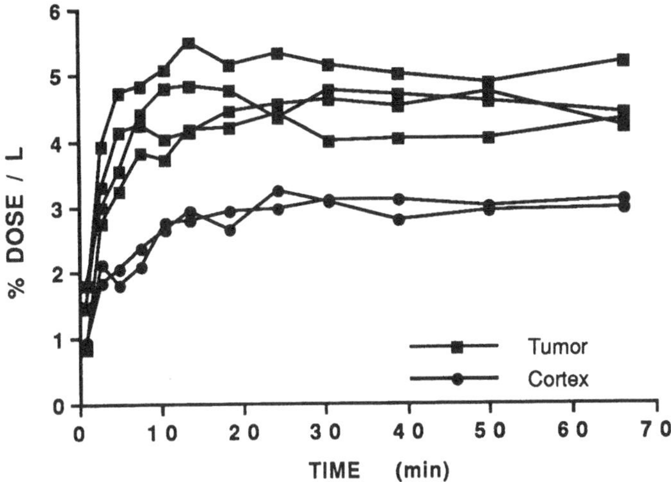

Figure 4. Kinetics of the tumor/cortex and tumor/plasma activity ratio in one patient. Several regions of interest are represented. The shape and pattern of these curves are representative of the curves obtained with the other patients of this series. The data for the calculation of these ratios are those from the same patient as in Figure 3. Plasma counts were not corrected for metabolites.

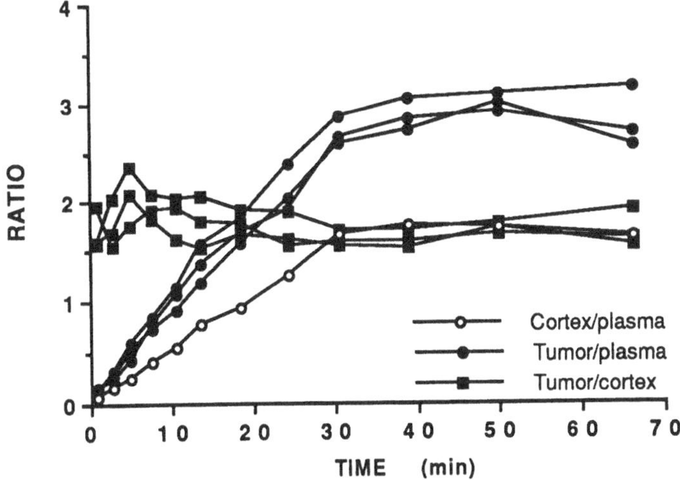

The tumor/cortex ratio was maximum at the beginning of the study, decreased a little up to 30 minutes post-injection and then stabilized. By contrast, the tumor/plasma ratio slowly increased up to 30-40 minutes post-injection and then remained in plateau. Finally, we submitted the brain and the plasma time-activity curves to a Patlak graphical analysis (Fig. 5). The plasma activity was not corrected for metabolites. The curves were curvilinear at all time, particularly at times > 30 minutes, which is an indication of the non applicability of this method to the analysis of our data.

Data collected between 35 and 55 minutes post-injection were averaged for that purpose. The ratio varied between 3.43 in a high grade tumor to 1.66 in a low grade tumor (Table I). The high grade tumors had higher ratios than the low grade tumors. By contrast, the tumor/plasma ratios were highly variable and no relation was found between this ratio and the tumor grade.

Figure 5. Application of the graphical analysis of Patlak to the ^{11}C-methionine uptake data. The brain regions represented here are the same as in Figure 3 and 4. Plasma counts were not corrected for metabolites.

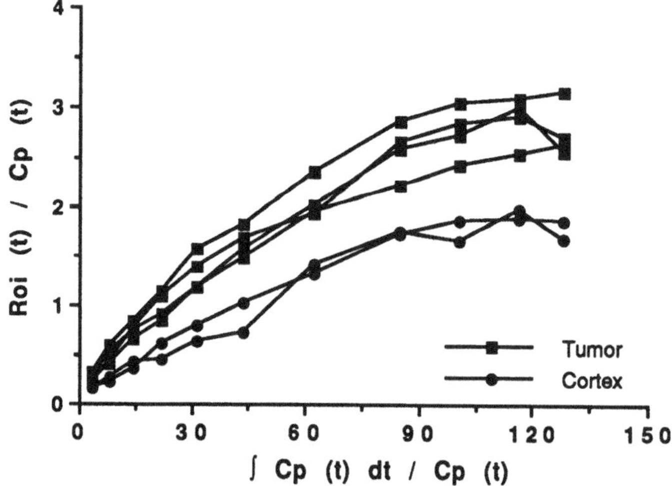

4. Discussion

The major finding of this study is that ^{11}C-methionine uptake was increased in every brain tumor, relative to the normal surrounding grey and white matter. Even a small volume-low grade astrocytoma appeared as a hot spot (Figure 1). As our series is still limited, it is premature to put forward any strong conclusion but these few examples raise some points of discussion.

Little is known about the mechanisms behind the increase of methionine accumulation in these tumors. Disruption of the blood brain barrier is not required for enhanced ^{11}C-methionine uptake since it was also increased in low grade tumors and in parts of high grade tumors with no contrast enhancement on CT scan. The uptake of L-amino acid is supposed to reflect protein synthesis, but the exact metabolic fate of ^{11}C-methionine in the brain in vivo is complex and various phenomena are involved in addition to incorporation into proteins [11]. Therefore it will be difficult to design a compartmental model that will satisfactorily describe the kinetics of ^{11}C-methionine and consider all the complexities of the biology of this tracer.

Because of their increased anabolic activity, tumors have higher needs in esential amino acids and an increased uptake of ^{11}C-methionine is expected. It is however of a surprise that the ^{11}C-methionine tumor/cortex ratio was maximum almost from the beginning of the study and remained stable over the whole period of observation. Indeed, if the rate of ^{11}C-methionine uptake were directly related to the rate of protein synthesis, one might expect this ratio to increase over the time, which is not the case. A better understanding of the fate of methionine in brain tumors is necessary to explain this observation. It could be due to a modification of blood brain barrier, or (and) a rapid metabolism of methionine in tumor tissue. Preliminary results presented in this workshop by Planas et al. [12] may explain this unusual kinetics of ^{11}C-methionine.

Despite these uncertainties, and our limited understanding of what is actually seen on the PET images, this is not a reason for not using a tracer that appears of clinical interest for imaging brain tumors. To quote Di Chiro [13]: "If it were found that shoe polish (to borrow one Lou Sokoloff's favorite analogies) is an effective tracer for diagnosing tumors, it would not matter whether its mode of operation was understood".

The most obvious advantage of ^{11}C-methionine is that it provides a clear delineation of the tumor from surrounding tissues. This is an important advantage over ^{18}FDG. Because normal brain uses glucose as well, tumor limits are not easily seen on the ^{18}FDG images. Furthermore, some tumors, especially low grade tumors, are characterized by a decrease in glucose metabolism, which is a non specific finding, since several pathological processes also result in a local hypometabolism. Thus, ^{18}FDG scans does not always help in the differential diagnosis of brain lesions found on CT or MRI scans. By contrast, ^{11}C-methionine uptake has been found increased only in brain tumors. Because it is more specific and it better shows tumor limits, ^{11}C-methionine scans usefully complement CT or MRI scans, particularly when the tumor is surrounded by oedema.

Some form of quantification of ^{11}C-methionine uptake is necessary to take full advantage of these studies. The graphical analysis of Patlak [14] has been used by some [15, 5]. If the transfert of ^{11}C-methionine from plasma to brain were unidirectional, the relationship between the tumor to plasma ratio and the normalized time should be linear, which is obviously not the case in our study. The most likely explanation for this is that we did not correct blood activity for metabolites or labelled proteins. Thus, without that correction, this method of quantification is not applicable. Furthermore, the tumor/plasma ratios were highly variable and did not correlate at all with tumor grades. Thus, at the present time, blood collection during ^{11}C-methionine studies is not useful and therefore not necessary when studies are carried out for clinical purposes.

The tumor/cortex ratio appeared the most valuable quantitative index. This time period was chosen for it corresponds to the achievement of an equilibrium between tumor and plasma activity; the blood flow influence at that time should be minimized and counting statistics are still satisfactory despite the radioactive decay. High grade tumors were characterized by higher ratios than low grade tumors (Table1). The number of patients in this series is too small to state that there is a direct relationship between tumor activity and this ratio. It is noteworthy that Derlon et al. [6] and Bustany et al. [16] found a correlation between the histological grade and the uptake ratio on larger series of patients. In their pioneering work, Di Chiro et al [17] found that the rate of glucose metabolism in gliomas is correlated with their histological grade. We also found that glucose metabolism was increased relative to the normal cortex in some parts of the high grade tumors while it was decreased in the low grade tumors we studied here. Thus, combining ^{11}C-methionine and ^{18}FDG studies could provide a better assessement of tumor activity than any tracer could do alone.

Newer tomographs have a much improved spatial sampling than the tomograph used for these studies. Because it delineates tumors so well, and if the relationship between the tumor/cortex and the tumor grade holds true, ^{11}C-methionine could be a unique and simple tool to evaluate the biological activity of the tumor (and indirectly its histological grade) and its heterogeneity, as well as a very sensitive test for the detection of post surgical tumor residue or tumor recurrence.

Acknowledgments

This work supported by the National Foundation for Scientific Research (FNRS) and by the Queen Elisabeth Medical Foundation (Belgium). The authors would like to thank all the Cyclotron and Chemistry staff for the productions of the radioisotopes, J. Hodiaumont for excellent technical assistance during the PET studies, C. Degueldre for the maintenance and the calibration of the tomograph.

References

[1] Di Chiro G. Positron emission tomography using [^{18}F] fluorodeoxyglucose in brain tumors. A powerful diagnostic and prognostic tool. Invest. Radio., 1986; 22: 360-371.

[2] Tyler J.L., Diksic M., Villemure J.G., Evans A.C., Meyer E., Yamamoto Y.L., Feindel W. Metabolic and hemodynamic evaluation of gliomas using positron emission tomography. J. Nucl. Med. 1987; 28: 1123-1133.

[3] Glantz M.J., Hoffman J.M., Coleman R.E., Friedman A.H., Hanson M.W., Burger P.C., Herndon II J.E., Meisler W.J., Schold S.C. Jr. Identification of early recurrence of primary central nervous system tumors by [^{18}F]fluorodeoxyglucose positron emission tomography. Ann. Neurol. 1991; 29: 347-355.

[4] Sokoloff M., Smith C. Biochemical principles for the measurement of metabolic rates in vivo In: "Positron Emission Tomography of the brain", edited by W.D. Heiss and M.E. Phelps, 1983, Springer, New York, pp2-18.

[5] Lilja A., Bergström K., Hartvig P., Späbbare B., Halldin C, Lundqvist H., Langstrom B. Dynamic study of supratentorial gliomas with L-methyl-11C-methionine and positron emission tomography. AJNR, 1985; 6: 505-514.

[6] Derlon J.M., Bourdet C., Bustany P., Chatel M., Theron J., Darcel F., Syrota A. [11C]L-methionine uptake in gliomas. Neurosurgery, 1989; 25: 720-728.

[7] Comar D., Cartron S.C., Maziere M., Marazano C. Labelling and metabolism of methionine-methyl 11C. Eur. J. Nucl. Med. 1976; 1: 11-14.

[8] Hoffman E.J., Phelps M.E., Huang S.C. Performance evaluation of a positron tomograph designed for brain imaging. J. Nucl. Med. 1983; 24: 245-257.

[9] Maquet P., Dive D., Salmon E., Von Frenckel R., Franck G. Reproducibility of cerebral glucose utilization measured by positron emission tomography and the [^{18}F]-2-fluoro-2-deoxy-d-glucose method in resting, healthy human subjects. Eur. J. Nucl. Med. 1990; 16: 267-273.

[10] Phelps M.E., Huang S.C., Hoffman E.J., Selin C., Sokoloff L., Kuhl D.E. Tomographic measurements of local cerebral glucose metabolic rate in humans with [F18]-2-fluoro-2-deoxy-D-glucose: validation of method. Ann. Neurol. 1979; 6: 371-388.

[11] Lestage P., Gonon M., Lepetit P., Vitte P.A., Debilly G., Rosatto C., Lecestre D., Bobillier P. An in vivo kinetic model with L-[35S]methionione for the determination of local cerebral rates for methionine incorporation into protein in the rat. J. Neurochem., 1987; 48: 352-363.

[12] Planas A., Kaschten B., Sadzot B., Stevenaert A., DiGiamberardino L., Comar D. In vivo incorporation of labelled mehionin into proteins in brain tumors. 1992. This workshop.

[13] Di Chiro G., Brooks R.A. PET quantitation: blessing and curse. J. Nucl. Med. 1988; 29: 1603-1604.

[14] Patlak C., Blasberg R.G., Fenstermacher J.D. Graphical evaluation of blood-to-brain transfer constants from multiple-time uptake data. J. Cereb. Blood Flow Metabol., 1983; 3: 1-7.

[15] Bergström M., Ericson K., Hagenfeldt L., Mosskin M., von Holst H., Norén G., Eriksson L., Ehrin E., Johnström P. PET study of methionine accumulation in glioma and in normal brain tissue: competition with branched chain amino acids. J. Comp. Assist. Tomogr., 1987; 11: 208-213.
[16] Bustany P., Chatel M., Derlon J.M., et al. Brain tumor protein synthesis and histological grades: a study by positron emission tomography (PET) with C11-L-methionine. J. Neurooncol. 1986; 3: 397-404.
[17] Di Chiro G., De La Paz R.L., Brooks R.A., Sokoloff L., Kornblith P.L., Smith B.H., Patronas N.J., Kufta C.V., Johnson G.G., Mannins R.G., Wolf A.P. Glucose utilization of cerebral gliomas measured by [18F]fluorodeoxyglucose and positron emission tomography. Neurology, 1982; 32: 1323-1329.

^{11}C-LABELED METHIONINE UPTAKE IN GLIOMAS: MODIFICATION AFTER THERAPY AND METABOLIC CORRELATIONS

M.-C. PETIT-TABOUÉ, F. DAUPHIN and J.-M. DERLON

ABSTRACT. ^{11}C-L-methionine can be used for the early assessment of therapy effects in patients suffering of gliomas. In this respect, we developed a specific method of delineation of tumoral region, compatible with intersubject variations, and studied the influence of hemodynamic parameters on ^{11}C-L-methionine uptake. Modifications of tumoral volume seems a better index than changes of tracer uptake for treatment evaluation benefits.

1. Introduction

Positron Emission Tomography (PET) studies provides important research and clinical informations on the evaluation of brain tumor metabolism. Most of the physiological parameters, relevant for a better understanding of the biological behavior of gliomas (hemodynamic parameters, oxidative and glucidic metabolism, amino-acid uptake) can be approached with this technique. Indeed, PET studies of Central Nervous System gliomas are expected to provide an earlier assessment of the biological changes than can do anatomical techniques such as Computerized Tomography (CT scan) or magnetic resonance imaging (MRI) [1]. Moreover, PET studies, using ^{11}C-labeled amino-acids or ^{18}F-fluoro-deoxyglucose, have revealed brain tumoral foci in patients with visually normal MRI examination [2].

Uses of PET in brain tumors were primarily focused on glucose metabolism (^{18}F-FDG; fluoro-deoxyglucose). In this respect, Di Chiro [3] reported a good correlation between increased glucose metabolism and malignancy of the tumor. In parallel, it has been proposed on the basis of in vitro studies, that amino-acid uptake investigations, since reflecting at least partly protein synthesis [4], could provide useful informations in brain tumors. Among the various putative tracers, ^{11}C-L-methionine was the first to be studied [5]. Some authors [6,7] demonstrated a relation between uptake of ^{11}C-L-methionine and histological grade, particularly in differentiating between low (I-II) and high grade (III-IV) gliomas. These authors also underlined the potential interest of such tracer for the follow-up of tumoral lesions after treatments. PET studies give thus important diagnosis and prognostic informations in the management of patients with gliomas and allow the distinction between recurrence of tumor and radiation necrosis. However, in order to obtain a clear and accurate evaluation of the same patient along therapy, a precise, objective and reproducible determination of the tumoral area is needed.

In parallel to in vitro studies with methionine, providing sensitiver information on pathological metabolism than FDG [8], labeled amino-acids PET scans seem noticeably

B. M. Mazoyer et al. (eds.), PET Studies on Amino Acid Metabolism and Protein Synthesis, 255–263.

interesting, giving informations complementary from those obtained with ^{18}F-FDG. However, many problems arised regarding quantification of ^{11}C-L-methionine uptake or mathematical analysis of dynamic data. Among these problems, the influence of hemodynamic parameters might be important, since determination of the tissular radioactivity can be overestimated by the presence of intravascular tracer, and also due to the variable blood flow / blood volume ratio, apparently unrelated to tumor grade.

If amino-acid accumulation is mediated by a facilitated transport at the cell membrane surface and by an increased metabolic demand of the cells themselves [9], the different pathways of tumoral metabolism are not completely known. Part of the tracer can be transformed into various metabolites; for ^{11}C-L-methionine, the importance of the trans-methylation pathway does not seem to be negligeable. In respect to metabolism of the tracer, quantification of ^{11}C-labeled metabolites in plasma is necessary to use a true input fonction in dynamic analysis of data, especially if long acquisition paradigms are used.

However, keeping in mind the limitations of the use of such tracer to obtain absolute value of protein synthesis, ^{11}C-L-methionine can be easily used for the early assessment of therapy effects.

TABLE 1. History of patients before the control study.

Case	Age / Sex	Histological Diagnosis	Duration from first symptoms	CT / MRI Data	Previous Treatments
#1	57 / F	Glioblastoma Premotor Right	4 months	Hypodensity	Surgery Chemotherapy Radiotherapy
#2	54 / M	Glioblastoma Motor Left	4 months	Mild Enhancement	Surgery Chemotherapy
#3	22 / M	Glioblastoma Parietal Left	6 months	Enhancement	Surgery Radiotherapy
#4	20 / M	Oligodendroglioma Temporal Left	2 years	Normal	Surgery
#5	30 / M	Glioblastoma Parietal Right	2 months	Enhancement	Surgery
#6	26 / M	Oligodendroglioma Parietal Right	3 years	Hypodensity Calcification	Surgery
#7	35 / M	Glioblastoma (r) Temporal Left	3 years	Hypersignal Gd-DTPA +	Surgery Radiotherapy

(r): recurrence.

2. Delineation of region of interest in longitudinal studies

Within longitudinal studies, dedicated to the follow-up analysis of tumoral processes in response to treatments (surgery, radiotherapy or chemotherapy) in the same patient, an accurate, objective and reproducible delineation of the tumoral regions of interest (ROI's) represents a crucial issue since the aim is to analyze the treatment effects on both the size of the pathological ROI and the intensity of labeling in this area. Due to the above mentionned differences between metabolic and anatomical visualization of tumoral processes, we have developped a method of tumoral ROI's delineation based on the metabolic parameters (^{11}C-L-methionine uptake) of the surrounding brain healthy tissue for each patient.

Raw data obtained from ^{11}C-L-methionine uptake assessments (high resolution, PET scanner LETI TTV03; 7 planes parallel to the glabella-inion line) are normalized into percentage of injected dose (%ID/ml). The overall normal ^{11}C-L-methionine uptake is evaluated as the pixel per pixel average of tracer normalized concentration from hemispheric ROI's visually excluding high activity areas (vascular structures and supposed tumoral tissue) and cerebellum, over the seven planes. ROI's exhibiting ^{11}C-L-methionine uptake higher than 200% of the tracer uptake in normal tissue are considered as tumoral locations. In the following examples concerning the effects of the various treatments on the ^{11}C-L-methionine uptake in 7 patients suffering from brain gliomas (Table 1), the present method of tumoral ROI's delineation has been used.

One can note that such method, despite less sensitive, improves the specificity of the tumoral area delineation, thus allowing the follow-up of patients along their treatments.

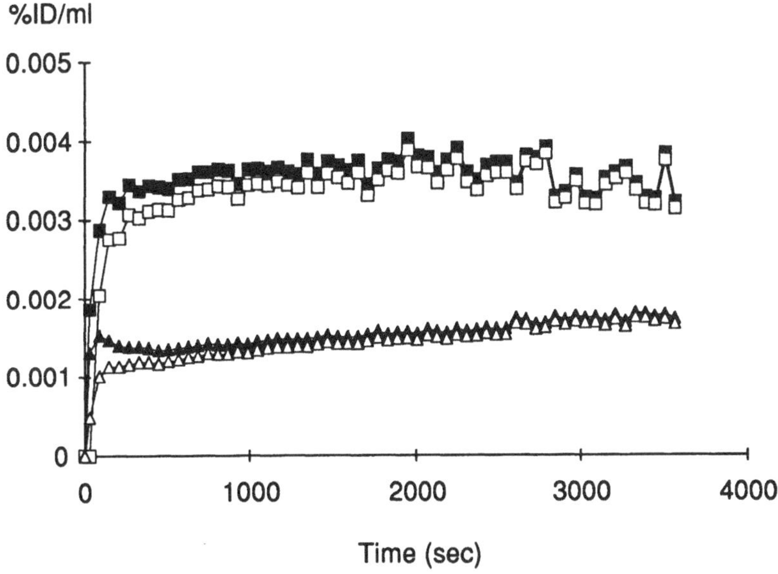

Figure 1. ^{11}C-L-methionine dynamic data in normal (triangles) and tumoral (squares), uncorrected (solid symbols) or corrected (open symbols) for cerebral blood volume. Results illustrate the mean from 3 subjects suffering from glioblastomas (patients 1, 2, 3).

3. Influence of local hemodynamic parameters on brain ^{11}C-L-Methionine uptake

Among the different issues raised by the ^{11}C-L-methionine PET scans data processing (see above), we considered the influence of hemodynamic parameters (cerebral blood flow, CBF; cerebral blood volume, CBV) on ^{11}C-L-methionine brain distribution. CBF and CBV were measured just before ^{11}C-L-methionine injection using the ^{15}O-labeled gas steady state inhalation [10]. CBF and CBV were evaluated in the same ROI's delineated for ^{11}C-L-methionine scanning. Statistical correlation studies were performed between CBF, CBV and ^{11}C-L-methionine data in normal tissue and in tumoral areas (determined as described above) in 3 patients suffering from glioblastomas (patients 1, 2 and 3; Table 1). There was no significant correlation between CBF and methionine uptake. Statistically significant correlation was found between CBV and ^{11}C-L-methionine uptake in normal brain tissue ($n=20$; $p=0.002$) whereas no correlation was found for tumoral tissue ($n=7$; $p=0.746$). Due to this correlation, dynamic ^{11}C-L-methionine PET data were corrected from CBV for further analysis both in normal and pathological tissue. Results of raw dynamic data and CBV-corrected data for ^{11}C-L-methionine are illustrated in Figure 1, showing that the influence of CBV in the determination of tissue radioactivity is especially predominant in the first 15 minutes.

4. Evaluation of glioma treatments

Computerized tomography or MRI are important examinations not only for initial diagnosis but also for assessing the tumor response to therapy. Progress in treatment of malignant gliomas have shown that combined approaches such as surgery and radiotherapy, and adjuvement of chemotherapy improved mildly the evolution or prognosis. However, among all patients within a single histological (and malignancy grade) group, individual response to any treatment regimen is largely unpredictible. Clinical as well radiological evidence of recurrency seem to be often delayed when compared with biological recurrency. Moreover, radiotherapy and chemotherapy can lead to side effects, particularly radiation necrosis of the brain. When neurological symptoms recur or worsen, it is virtually impossible to distinghish radiation necrosis and gliosis from tumor recurrence by conventional imaging techniques or clinical examination, despite their usefulness in the diagnosis and therapy evaluation. PET, using FDG or amino-acids, could help to make such distinction because necrotic areas do not have anymore metabolism.

4.1. SURGERY

^{11}C-L-methionine PET scanning could help in differentiating between tumor proliferation and reactive gliosis whereas anatomical investigations are poorly useful [2,11].

We report a case in which PET study lead to surgery re-intervention in a young boy already operated two years before from a left temporal lobe oligodendroglioma (patient 3, Table 1). Fifteen months after the first operation, complex partial seizures, eventually followed by grand mal seizures appeared several times a week despite an increase in anti-epileptic therapy. CT scan, MRI with gadolinium, ^{99m}Tc-DPTA and thallium SPECT scans were negative. ^{11}C-L-methionine PET scan showed a small "hot spot", located in the area of the previous resection, with a tracer uptake of 44.2 10^{-4} %ID/ml (2.2 fold higher than in the normal brain tissue). A left temporal craniotomy led to the removal of a tumor (1 cm diameter) located at the border of the previous operation site, neuropathologically identified as an oligodendroglioma recurrence. PET scan performed 3 months later did not show anymore abnormal ^{11}C-L-methionine uptake [12].

4.2. RADIOTHERAPY

Follow-up of ^{11}C-L-methionine PET studies allows the evaluation of radiotherapy effects and thus a diagnosis of a possible recurrence. We studied patients suffering from glioblastomas or oligodendrogliomas within 10 days and one year after radiotherapy. For illustration, we report here data obtained in two cases of glioblastomas (patients 2 and 5, Table 1) and one case of oligodendroglioma (patient 6, Table 1).

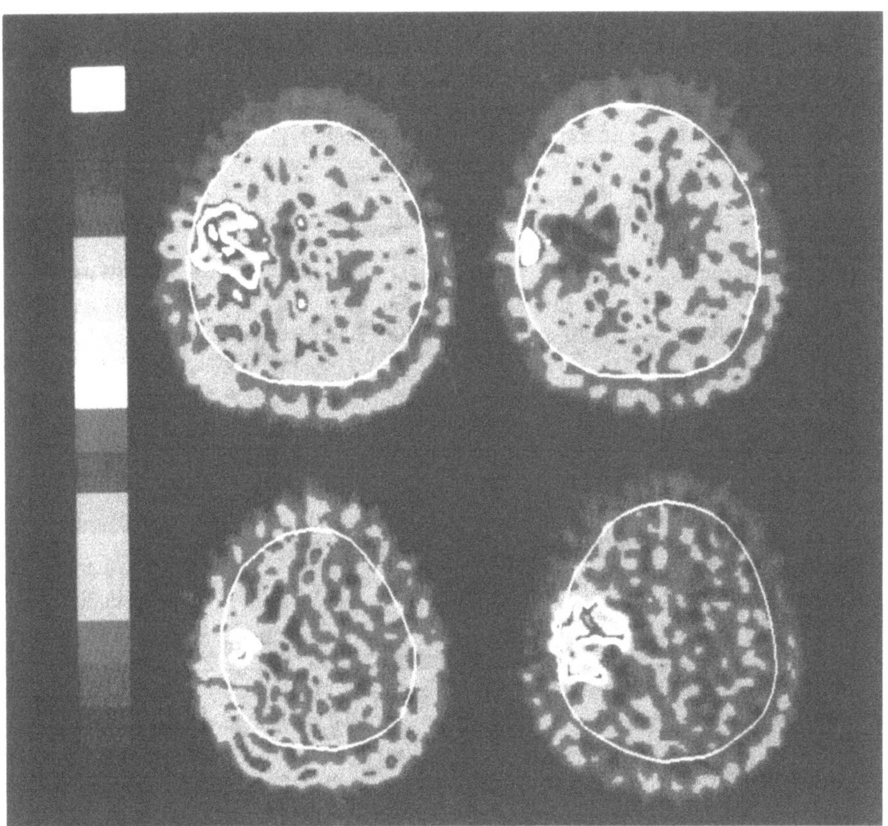

Figure 2. Effect of radiotherapy on ^{11}C-L-methionine uptake in patient 5 (ACPC +47mm) at various time (before radiotherapy, top left; after radiotherapy: 12 days, top right; 5 months, bottom left; 8 months, bottom right).

Table 2 and figure 3 show the changes of ^{11}C-L-methionine uptake and tumor volume before and after radiotherapy (60 Gy, 6 weeks, twice a week) for the patients suffering from glioblastomas. In case 2 (Table 2), if uptake of ^{11}C-L-methionine was slightly decreased between control and post-radiotherapy (10 days) examinations, the tumoral volume increase (+43%) and the evolution was rapid without clinical benefit. On the other hand, the tumoral volume rapidly decreased (-91%; 12 days) in patient 5 (Figure 2; Table 2) and remained stable at 5 months, despite an increase on one plane (+31% versus 12 days) at a time during which the CT scan was normal. From this location started the tumoral reccurency, clearly detectable at 8 months (+676% versus 5 months; Table 2) on PET images, but also on CT scan and SPECT investigations. In parallel, however with less variations, ^{11}C-L-methionine uptake within the tumoral ROI's first decreased at 12 days, remained stable at 5 months and started back to increase at 8 months (Table 2).

TABLE 2. Effects of radiotherapy on ^{11}C-methionine uptake in two cases of glioblastomas (patients 2 and 5) and one oligodendroglioma (patient 6).

	Case 2			Case 5			Case 6
Planes (ACPC, mm)	+23	+35	+47	+23	+35	+47	+23
Control							
Normal Tissue	14.30	15.45	16.00	22.5	23.0	24.5	17.85
Tumoral Tissue	31.15	40.00	44.45	59.0	67.5	60.5	42.85
Tumoral Volume	0.70	9.05	11.30	3.55	3.90	6.50	2.45
12 Days after Radiotherapy							
Normal Tissue	17.25	17.55	17.15	20.0	20.0	22.0	
Tumoral Tissue	39.00	41.75	40.00	-	50.0	47.5	
Tumoral Volume	6.60	9.90	15.35	-	0.90	0.95	
2 Months after Radiotherapy							
Normal Tissue							19.40
Tumoral Tissue							43.30
Tumoral Volume							0.85
5 Months after Radiotherapy							
Normal Tissue				16.5	17.5	18.0	12.45
Tumoral Tissue				-	-	39.0	-
Tumoral Volume				-	-	1.25	-
8 Months after Radiotherapy							
Normal Tissue				13.0	14.5	15.5	
Tumoral Tissue				-	38.5	41.5	
Tumoral Volume				-	1.15	8.55	

^{11}C-methionine uptake is expressed as 10^{-4} %ID/ml both in normal and tumoral tissue. Tumoral volume is measured in ml.

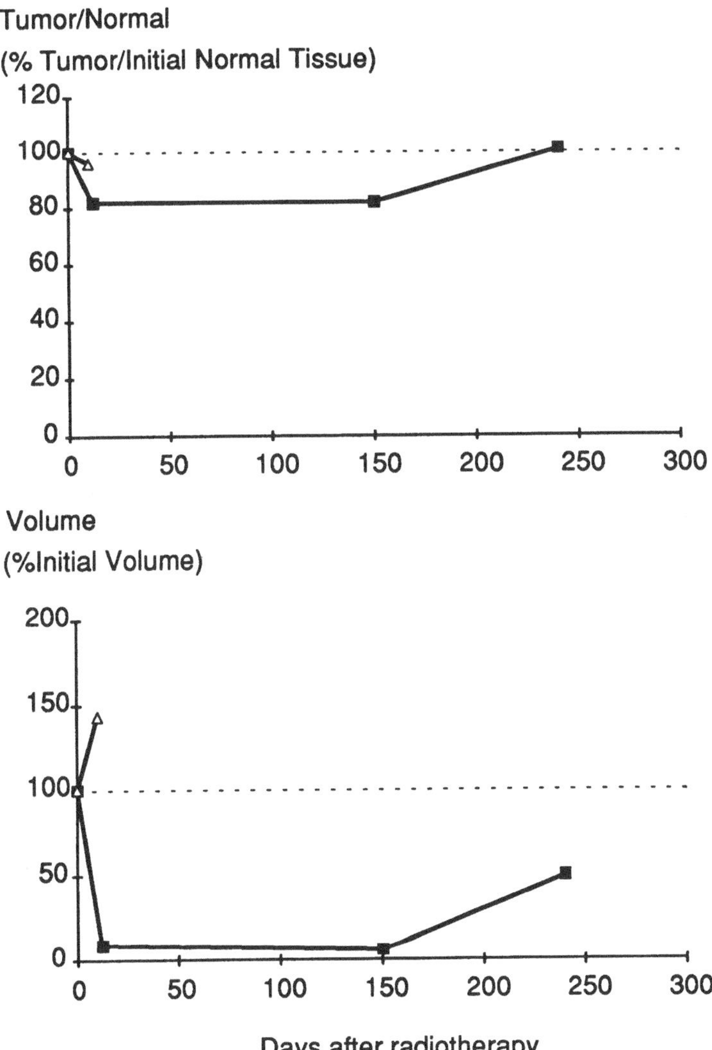

Figure 3. Changes of ^{11}C-L-methionine uptake and tumor volume before and after radiotherapy. Results are expressed in percentage of initial values for individual subjects (triangles: patient 2; squares: patient 5).

Regarding the oligodendroglioma case (patient 6), the ^{11}C-L-methionine uptake was unchanged 2 months after the end of radiotherapy (Table 2). However, the tumoral volume significantly decreased at 2 months as compared to control (-65%, Table 2) and further decreased since no tumoral focus was anymore detectable after 5 months.

4.3. CHEMOTHERAPY

From our limited experience with the follow-up of patients whose recurrency was treated with chemotherapy, we report here the case of one patient having a left temporo-polar astrocytoma (grade II-III), operated and irradiated (Patient 7, Table 1). Three years after this treatment, the patient was operated again, on the basis of temporal seizures reapparition, intracranial hypertension symptoms and MRI conclusion of an expansive temporal processus, histologically diagnosed as a glioblastoma. The patient was then studied three times with ^{11}C-L-methionine: before chemotherapy, 24 hours and 21 days after starting chemotherapy (Fotemustine, i.v.).

One day after the first course of chemotherapy, uptake of ^{11}C-L-methionine in the tumoral area was unchanged as compared to control conditions, but tended to slightly decrease at 3 weeks (-9% as compared to control). In contrast, the tumoral volume rapidly decreased at 24 hours (-10%) and further decreased 21 days after chemotherapy (-33% as compared to control), a profile parallel to the positive clinical evolution.

5. Conclusions

Our new method for the delineation of tumor area seems to be a reliable tool for an accurate assessment of the effects of surgery, radiotherapy and chemotherapy on tumor metabolism. Response to treatment and early determination of tumor recurrency are essentially appreciated by the variations of the tumoral volume rather than the less pronounced changes of ^{11}C-L-methionine uptake.

Further experience is however needed to confirm these preliminary data. In the future, PET methodology will have to be compared with magnetic resonance spectroscopy [13] and SPECT techniques using amino-acid analogs to determine the most accurate and specific way of following gliomas evolution.

Acknowledgements

M.H. Noël and D. Luet are greatly acknowledged for their helpful assistance. The authors thank the technical staff of Cyceron (Cyclotron, Chemistry, Medical and Computer teams). F.D. was supported by a post-doctoral fellowship from the Commissariat à l'Energie Atomique. Supported by Cyceron (MCPT, FD, JMD), Centre François Baclesse (MCPT), CNRS ERS 0019 (FD), CEA DSV/DPTE (FD), Service de Neurochirurgie, CHRU (JMD).

References

1. Dooms GC, Hecht S, Brandt-Zawadski M, Berthiaume Y, Norman D, Newton TH. Brain radiation lesions: MR imaging. Radiology 1986; 158: 149-155.

2. Tovi M, Lilja A, Bergström M, Ericsson A, Bergström K, Hartman M. Delineation of gliomas with magnetic resonance imaging using Gd-DTPA in comparison with computed tomography and positron emission tomography. Acta Radiol 1990; 31: 417-29.
3. Di Chiro G. Positron emission tomography using [^{18}F]fluorodeoxyglucose in brain tumors. A powerful diagnostic and prognostic tool. Invest Radiol 1986; 22: 360-71.
4. Bustany P, Sargent T, Saudubray JH, Henry JF, Comar D. Regional human brain uptake and protein incorporation of ^{11}C-L-methionine studied in vivo by PET. J Cereb Blood Flow Metab 1981; 1: S17-8.
5. Comar D, Cartron JC, Mazière M, Marazano C. Labelling and metabolism of methionine-methyl-^{11}C. Eur J Nucl Med 1976; 1: 11-4.
6. Bustany P, Chatel M, Derlon JM et al. Brain tumor protein synthesis and histological grades: a study by positron emission tomography (PET) with ^{11}C-L-methionine. J Nucl Med 1985; 26: 37-42.
7. Derlon JM, Bourdet C, Bustany P et al. [^{11}C]L-Methionine uptake in gliomas. Neurosurgery 1989; 25: 720-28.
8. Kubota K, Ishiwata K, Kubota R et al. Tracer Feasability for monitoring Tumor radiotherapy: A Quadruple Tracer Study with Fluorine-18-Fluorodeoxyglucose or Fluorine-18-Fluorodeoxyuridine, L-[Methyl-^{14}C]Methionine, [6-^{3}H]Thymidine, and Gallium-67. J Nucl Med 1991; 32: 2118-25.
9. Johnstone RM, Scholefield PG. Amino acid transport in tumor cells. J Neurosur 1990; 72: 110-113.
10. Frackowiak RSJ, Lenzi GL, Jones T, Heather JD. Quantitative measurement of regional cerebral blood flow and oxygen metabolism in man using ^{15}O and positron emission tomography: theory, procedure and normal values. J Comput Assist Tomogr 1980; 4: 727-36.
11. Di Chiro G, Oldfield E, Wright DC et al. Cerebral necrosis after radiotherapy and/or intra-arterial chemotherapy for brain tumors: PET and neuropathological studies. AJNR 1987; 8: 1083-91.
12. Viader F, Derlon JM, Petit-Taboué MC et al. Recurrent oligodendroglioma diagnosed with ^{11}C-L-Methionine and PET. A case report. Eur.Neurology; in press.
13. Heiss WD, Heindel W, Herholz K et al. Positron Emission Tomography of Fluorine-18-Deoxyglucose and Image-guided Phosphorus-31 magnetic Resonance Spectroscopy in Brain Tumors. J Nucl Med 1990; 31: 302-10.

IN VIVO INCORPORATION OF LABELLED METHIONINE INTO PROTEINS IN BRAIN TUMORS

A.M. PLANAS, B. KASCHTEN, B. SADZOT, A. STEVENAERT,
L. DiGIAMBERARDINO and D. COMAR

ABSTRACT. The incorporation of L-(^{14}C-methyl) methionine into proteins was studied in two subjects presenting brain glioma. Comparing to normal brain tissue, an increased methionine metabolic activity was found in the tumors, with most of the tissue radioactivity incorporated into proteins.

1. Introduction, Methods and Results

Labelled methionine is frequently used in PET to visualize brain gliomas. However, it is uncertain whether tumor label uptake reflects incorporation into proteins. We have studied the fate of labelled methionine in brain gliomas in two different subjects, with their consent and that of the ethics committee of the Medical School of Liège. During brain surgery for tumor extraction, L-(^{14}C-methyl) methionine (50 μCi) was i.v. given as a bolus injection. This is an accepted tracer dose of ^{14}C within the range in current use on humans (1). Timed arterial blood samples were withdrawn up to 1 hour after the injection, plasma was separated out of blood and samples were frozen. At different times, one or two samples of tumoral tissue and of apparently normal brain surrounding the glioma (less than 0.08 g of wet weight) were taken out and were immediately frozen into liquid nitrogen for further biochemical studies. Plasma proteins were precipitated out, the total radioactive content was determined and the protein content per volume of plasma was measured. The concentration of free ^{14}C-methionine and the endogenous content of methionine in the plasma was determined after HPLC analysis using pre-column derivatisation, fluorimetric detection and collecting column fractions. Tissues were weighted, sonicated in acid and centrifuged. The supernatant was filtered and further processed for HPLC analysis as the plasma samples. The protein content per weight of tissue and the protein-associated radioactivity were determined from the precipitated fraction.

SUBJECT 1: The endogenous content of methionine in the plasma was 14.4 nmol/ml. Plasma radioactivity expressed as the % of the injected dose/L was 2.3 and 1.4 % at 30 and at 45 min after the injection, respectively. At these times, unmetabolized free ^{14}C-methionine was 35 and 28 % of protein-free plasma label. The tissues studied were: two samples of tumor, T30 and T45 (taken at 30 and at 45 min after the injection) and one sample of cortex, CTX30 (taken at 30 min). Respectively, the endogenous content of free methionine was 18.8, 17.0 and 24.4 nmol/g of tissue and the protein content was 95, 92 and 108 mg/g of tissue. Fig. 1a shows the result of the analysis of tissue radioactivity.

B. M. Mazoyer et al. (eds.), PET Studies on Amino Acid Metabolism and Protein Synthesis, 265–266.

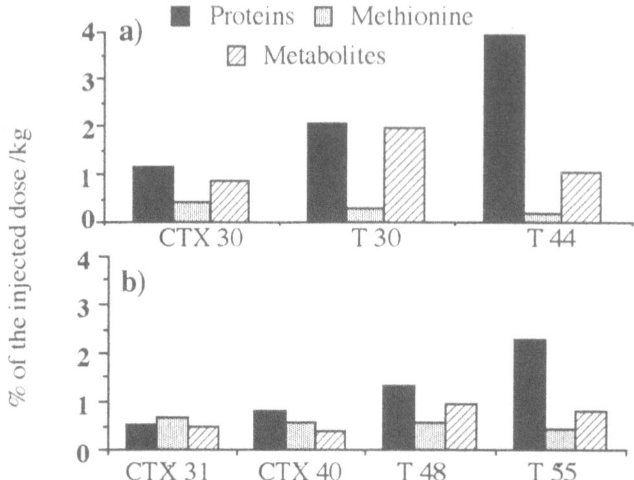

Figure 1. Label distribution in the tissues. Data are from a) subject 1, and b) subject 2.

SUBJECT 2: In plasma: the endogenous methionine was 19.8 nmol/ml and the percent of the injected dose/L was 3.2 and 1.9 % at 15 and 45 min. At these times, unmetabolized ^{14}C-methionine was 54 and 35 % of the protein-free label. The tissues studied were: two samples of tumor, T48 and T55 (at 48 and 55 min) and two samples of cortex, CTX31 and CTX 40, (at 31 and 40 min). The endogenous content of free methionine was 21.6, 24.1, 17.5 and 21.9 nmol/g of tissue and the protein content was 63, 71, 83 and 55 mg/g of tissue, respectively. Fig. 1b shows the results of the radioactivity analysis of the 4 samples of tissue.

2. Discussion

The high accumulation of radioactivity observed in the tumoral tissue is related to the high rates of incorporation of label into proteins. The level of unmetabolized free ^{14}C-methionine was lower in tumor than in cortex. Subject 2 had a lower degree of metabolism, both in cortex and in tumor, with higher contents of unmetabolized ^{14}C-methionine than subject 1. This indicates a non negligible individual variability. In the two studies, labelled metabolites seemed to be washed out of the tissue. However, given the problem of tissue heterogeneity within the tumor, care must be taken comparing tumor samples taken out at different times after injection. In brief, we have shown that, after labelled methionine injection, the higher incorporation of total label in gliomas than in normal brain can be associated to a higher incorporation of radioactivity into tumor proteins. Whether a higher tumor net uptake of label methionine results from blood-brain barrier disturbances, rather than resulting from an increased protein synthesis demand, remains to be studied. In addition, more labelled metabolites seem to be produced in tumor than in normal tissue indicating the presence in the tumor of a high methionine utilization activity.

3. References

1. Cobelli C, Saccomani MP, Tessari P, Biolo G, Luzi L, Matthews DE .Compartmental model of leucine kinetics in human. J Am Physiol Soc 1991; 261:E539-E550.

DISCUSSION

D. Heiss Studies of aminoacid uptake in tumor patients by PET are certainly important to increase our knowledge on functional properties of malignancies in the brain. However, the question arises if such studies really contribute to clinical diagnosis and management. It would be very important to pool all the data presented here and calculate figures of sensitivity and specificity. I suggest that a task force is formed to analyse the existing data and investigate the clinical efficacy of these methods. This task force could be formed among the groups in Hannover, Stockholm, Uppsala and Caen and the other groups are invited to contribute their experience. The problem arises from the still low number of patients studied. The largest experience was collected in Stockholm, Uppsala, Hannover and now in Caen. Comparison between various modalities or tumors suffers from differences of equipment and methods used for the registration. Differences in machine resolutions also play an important role when the extent of a lesion is compared on various slices, and in some instances partial volume effects could be important. The most critical observation mentioned today was that tumor cells are also found outside the region of increased methionine uptake. This is a critical issue for the clinical value of PET of aminoacid accumulation in the diagnosis of tumors and determination of their growth.

A. Lajtha There are a number of markers that could be tried, according to what we think the pathological mechanisms are. Let me mention three possibilities.
One is that there are changes in the blood-brain barrier. If the capillaries are more permeable I would use a marker that normally does not penetrate the brain. If the lesion is large, my choice would be a large molecule such as polyethylene glycol, or a labeled protein. If the brain capillary lesion is small,one would have to choose a small molecule that can penetrate through small lesions - sulfate or phosphate could be tried. One could also try an amino acid if the hypothesis is that the capillaries in pathological tissue are not fully developed. We and others found in the past that metabolic transport (also passive diffusion) is more active in the immature brain, therefore there is increased uptake of essential metabolites during rapid growth. An amino acid that is slowly transported in mature capillaries but rapidly transported in immature ones would be worth trying.
The second possibility is that the change is not at the blood-brain barrier interface, but at cellular membranes where such things as membrane transport are affected. Here one could try substrates that are very actively

B. M. Mazoyer et al. (eds.), PET Studies on Amino Acid Metabolism and Protein Synthesis, 267–268.

taken up D -glutamic acid (because it is not metabolized) or taurine are good markers for active transport , but others could also be tried.
The third possibility is that membranes are not affected, but cellular protein synthesis is increased. My choice for a marker for amino acid incorporation would be valine, which is not metabolized rapidly, or at the other end of the spectrum, arginine, which is not recycled after release from proteins. Since DNA metabolism is not very active in the adult brain, another possibility would be thymidine, or if it is metabolized rapidly, a halogenated analog of thymidine (fluoro -).
Some of these markers may distinguish healthy from pathological tissue better, or may be sensitive enough to measure the spread of the tumor cells beyond the area being considered the only affected one.
I suggest mostly compounds that are present under normal conditions so that toxic side effects are not likely to occur.
What I suggest is that the excised tumor tissue should be studied at some detail. Capillaries can be isolated and their properties compared to those isolated from healthy tissue. Metabolite transport in tissue slices can be examined, and some conclusions on whether capillary tissue transport is altered could be proposed.

Developments in Nuclear Medicine

1. P.H. Cox (ed.): *Cholescintigraphy.* 1981 ISBN 90-247-2524-0
2. P.H. Cox (ed.): *Progress in Radiopharmacology.* Selected Topics. Proceedings of the 3rd European Symposium (Noordwijkerhout, The Netherlands, April 1982). 1982 ISBN 90-247-2768-5
3. M.H. Jonckheer and F. Deconinck (eds.): *X-Ray Fluorescent Scanning of the Thyroid.* 1983 ISBN 0-89838-561-X
4. K. Kristensen and E. Nørbygaard (eds.): *Safety and Efficacy of Radiopharmaceuticals.* 1984 ISBN 0-89838-609-8
5. A. Bossuyt and F. Deconinck: *Amplitude/Phase Patterns in Dynamic Scintigraphic Imaging.* With a Foreword by A. Bertrand Brill. 1984 ISBN 0-89838-641-1
6. M.R. Hardeman and Y. Najean (eds.): *Blood Cells in Nuclear Medicine, Part I.* Cell Kinetics and Bio-distribution. 1984 ISBN 0-89838-653-5
7. G.F. Fueger (ed.): *Blood Cells in Nuclear Medicine, Part II.* Migratory Blood Cells. 1984 ISBN 0-89838-654-3
8. H.J. Biersack and P.H. Cox (eds.): *Radioisotope Studies in Cardiology.* 1985 ISBN 0-89838-733-7
9. P.H. Cox, G. Limouris and M.G. Woldring (eds.): *Progress in Radiopharmacology 1985.* 1985 ISBN 0-89838-745-0
10. P.H. Cox, S.J. Mather, C.B. Sampson and C.R. Lazarus (eds.): *Progress in Radiopharmacy.* 1986 ISBN 0-89838-823-6
11. H. Deckart and P.H. Cox (eds.): *Principles of Radiopharmacology.* 1987 ISBN 0-89838-774-4
12. W.-D. Heiss, G. Pawlik, K. Herholz and K. Wienhard (eds.): *Clinical Efficacy of Positron Emission Tomography.* 1987 ISBN 0-89838-898-8
13. G.B. Gerber, H. Métivier and H. Smith (eds.): *Age-related Factors in Radionuclide Metabolism and Dosimetry.* 1987 ISBN 0-89838-953-4
14. K. Kristensen and E. Nørbygaard (eds.): *Safety and Efficacy of Radiopharmaceuticals 1987.* 1987 ISBN 0-89838-986-0
15. C. Beckers, A. Goffinet and A. Bol (eds.): *Positron Emission Tomography in Clinical Research and Clinical Diagnosis.* Tracer Modelling and Radioreceptors. 1989 ISBN 0-7923-0254-0
16. M. De Schrijver: *Scintigraphy of Inflammation with Nanometer-sized Colloidal Tracers.* 1989 ISBN 0-7923-0272-9
17. Ch. Kessler, M.R. Hardeman, H. Henningsen and J.-N. Petrovici (eds.): *Clinical Application of Radiolabelled Platelets.* 1990 ISBN 0-7923-0729-1
18. H.J. Biersack and P.H. Cox (eds.): *Nuclear Medicine in Gastroenterology.* 1991 ISBN 0-7923-1074-8
19. R.P. Baum, P.H. Cox, G. Hör and G.L. Buraggi (eds.): *Clinical Use of Antibodies.* Tumours, infection, infarction, rejection and in the diagnosis of AIDS. 1991 ISBN 0-7923-1424-7

Developments in Nuclear Medicine

20. J.C. Baron, D. Comar, L. Farde, J.L. Martinot and B. Mazoyer (eds.): *Brain Dopaminergic Systems: Imaging with Positron Tomography.* 1991
ISBN 0-7923-1476-X
21. M.K. Dewanjee: *Radioiodination.* Theory, Practice, and Biomedical Application. 1991 ISBN 0-7923-1491-3
22. P.A. Schubiger and G. Westera (eds.): *Progress in Radiopharmacy.* 1992
ISBN 0-7923-1525-1
23. B.M. Mazoyer, W.D. Heiss and D. Comar (eds.): *PET Studies on Amino Acid Metabolism and Protein Synthesis.* 1993 ISBN 0-7923-2076-X

Kluwer Academic Publishers - Dordrecht / Boston / London

Zeitfracht Medien GmbH
Ferdinand-Jühlke-Straße 7
99095 Erfurt, Deutschland
produktsicherheit@kolibri360.de